MW01482120

Your Sharp & Dohme Representative for:

PRINIVIL, PRINZIDE
ZOCOR, & PEPCID

PRINCIPLES AND MANAGEMENT OF LIPID DISORDERS

A Primary Care Approach

PRINCIPLES AND MANAGEMENT OF LIPID DISORDERS

A PRIMARY CARE APPROACH

Albert Oberman, M.D., M.P.H.
Professor and Director
Division of General and Preventive Medicine
Department of Medicine
University of Alabama at Birmingham

Robert A. Kreisberg, M.D.
Professor and Vice-Chairman
Department of Medicine
University of Alabama at Birmingham

Yaakov Henkin, M.D.
Visiting Assistant Professor
Division of General and Preventive Medicine
Department of Medicine
University of Alabama at Birmingham
Ben-Gurion University
and
Beer-Sheva, Israel

WILLIAMS & WILKINS
BALTIMORE · HONG KONG · LONDON · MUNICH
PHILADELPHIA · SYDNEY · TOKYO

Editor: Michael Fisher
Managing Editor: Carol Eckhart
Copy Editor: Patricia J. Grayson
Designer: JoAnne Janowiak
Illustration Planner: Ray Lowman

Printed in the United States of America

Library of Congress Cataloging-in-Publication Data

Oberman, Albert.
    Principles and management of lipid disorders : a primary care
approach / Albert Oberman, Robert A. Kreisberg, Yaakov Henkin.
        p.   cm.
    Includes bibliographical references and index.
    ISBN 0-683-06623-4
    1. Lipids—Metabolism—Disorders.   I. Kreisberg, Robert A.
II. Henkin, Yaakov.   III. Title.
    [DNLM: 1. Hyperlipidemia—therapy.   2. Lipids—metabolism.   WD
200.5.H8 012p]
RC632.L5024   1992
616.3'997—dc20
DNLM/DLC
for Library of Congress   91-30097
                                                               CIP

92 93 94 95 96
1 2 3 4 5 6 7 8 9 10

ALBERT OBERMAN, MD, MPH

To Marian, Steven, David, and Karen Oberman (all of whom have desirable cholesterol levels) for your encouragement, patience, and unending support in this and other endeavors.

ROBERT KREISBERG, MD

To my children and all of the students, house officers, faculty, and practicing physicians who taught me and have been taught by me.

YAAKOV HENKIN, MD

To Michal, Dor, Ziv, and Nativ Henkin for providing the support and endurance to your husband and father during the numerous hours spent in preparing the book.

PREFACE

Until recently, surveys demonstrated that physicians had not yet accepted the central role of hypercholesterolemia in coronary heart disease (CHD) and were reluctant to treat lipid disorders. Of the major risk factors, managing hypercholesterolemia assumed far less priority than treating high blood pressure or advising against cigarette smoking. This unwillingness to treat anything but the highest levels of cholesterol could be attributed to several factors: A false start some 30 years ago with a hypolipidemic drug, MER 29, which caused numerous adverse effects; the long-term, adverse experience in the World Health Organization (WHO) trial of clofibrate; insufficient direct clinical evidence in humans that lowering cholesterol levels would reduce CHD; the complex nomenclature that prevailed for lipid disorders; the lack of effective, safe, and acceptable drugs for treating lipid abnormalities and the absence of authoritative treatment recommendations on when to treat and how to manage lipids in clinical situations.

Over the past five years a number of key developments have dispelled doubts about the importance and value of reducing cholesterol values. Brown and Goldstein elucidated the genetic defect in familial hypercholesterolemia, establishing a direct cause-and-effect relationship between deficient low density lipoprotein (LDL) receptors and hypercholesterolemia. Results from two primary prevention trials of CHD, the Primary Coronary Prevention Trial and the Helsinki Heart Study, showed that among men who had elevated blood cholesterol levels, treatment with hypolipidemic drugs and diet reduced the risk of CHD.

These trials, along with evidence accumulated from experimental and epidemiologic studies, culminated in a consensus conference sponsored by the National Heart, Lung,

and Blood Institute on lowering blood cholesterol to prevent heart disease. The growing recognition of the need for treating high blood cholesterol plus the development of new hypolipidemic drugs led the National Heart, Lung, and Blood Institute and the American Heart Association (AHA) to believe that new and more useful guidelines for the treatment of high blood cholesterol were in order. The National Cholesterol Education Program (NCEP) created an expert panel on detection, evaluation, and treatment of high blood cholesterol to formulate the needed guidelines that could be adopted for use by practitioners and allied health personnel. The guidelines then were distributed to practitioners and also were publicized widely among the general public, who were quick to respond in seeking medical care for lipid disorders. Concurrently, new drugs became available and microanalytic techniques for the rapid determination of lipids became widespread with expanded population screening efforts. Continuing data on the benefits of treating lipids from various epidemiologic, metabolic, and clinical studies added to the momentum. Furthermore, at least three well-designed arteriographic trials found lack of progression and even regression of atherosclerotic lesions with vigorous lipid-lowering treatment over a period of several years.

In view of these rapid advances in the evaluation and treatment of dyslipidemias, it is our intention to provide the background data, basic pathophysiology, clinical information, and details of treatment to enable the primary care physician to detect and manage effectively patients routinely encountered in practice who have dyslipidemias.

The first chapter deals with the evolution of the lipid hypothesis and a discussion of the basic atherosclerotic process as currently conceptualized. Details are given for the early investigations of lipids and atherosclerosis that led to the clinical trials of efficacy which provided the rationale for treating lipid disorders. **The second chapter** provides a perspective on measuring risk and emphasizes the relative impact of cholesterol, triglyceride, and individual lipoproteins on the development of cardiovascular disease. The contributions of other important risk factors, singularly and in combination with lipids, are delineated as well. The **third chapter** describes the distribution of lipids and lipoproteins in the United States. The factors that influence lipid values are presented within the context of screening procedures. In addition to the current recommendations for screening, approaches to screening techniques and theoretical considerations for population screening and case findings are presented. Finally, information on the trends of lipids over time in this country and the current prevalence of lipid disorders are given.

The next section on pathophysiology considers the transport of lipids in the individual lipoprotein particles (**Chapter 4**). Emphasis also is placed on the pathways for removal of lipoproteins because of the major role this plays in atherogenesis. Other pertinent considerations in this section concern the physiologic changes with age that alter lipid levels and the ability to track high lipid values from youth to adulthood (**Chapter 5**). The influence of sex, including pregnancy and menopause in women, in relation to lipid disorders is examined. The special problems of dealing with lipids in the elderly also are viewed in terms of the anticipated metabolic changes that come with aging. Concepts of metabolism and pathogenesis of lipid disorders, however, continue to evolve. The increasing recognition of the role of lipoprotein subspecies and the apolipoproteins adds to the complexity of managing lipid disorders. Lipoprotein pathophysiology is presented in sufficient detail to provide an understanding of the laboratory evaluation and recognized clinical syndromes.

Clinical concepts dealt with in this book include laboratory measurements (**Chapter 6**), classification of lipid disorders (**Chapter 7**), secondary dyslipidemias (**Chapter 8**), and the clinical evaluation of patients (**Chapter 9**). Because the decisions to evaluate and treat patients rely so heavily on laboratory measurements, special consideration of the sources of variability in lipid values is given. Standardized measurements and details for collecting blood are discussed. The techniques for measurement also are described to gain a better understanding of the laboratory process and to develop criteria for evaluation of individual laboratories. Information on the newer analytic methods and desktop analyzers provides details on procedural technique and the accuracy of these measurements.

A classification of lipid disorders is presented to allow a framework for appropriate treatment. Each of the primary lipid disorders, ranging from hyperchylomicronemia to disorders of HDL metabolism and the newer concepts of hypobetalipoproteinemia, is reviewed. Each of the clinical syndromes is addressed in terms of occurrence, pathogenesis, and diagnosis. Because secondary dyslipidemias are a common source of lipid disorders, diseases associated with abnormal lipid profiles and drug-induced lipid abnormalities are considered in detail. The section on the clinical evaluation bears on the fundamental work-up of the patient who has a lipid problem and the subsequent clinical management. Recommendations from the NCEP and the AHA serve to outline accepted standards for detecting and treating dyslipidemias. The text also deals with clinical presentations and considerations in managing special populations, such as diabetics and the elderly.

In the section on guidelines for treatment, specific approaches to clinical treatment are discussed. **Chapter 10** deals with nondrug therapy, including diet and other health-related behavioral changes. **Chapter 11** discusses drug therapy in terms of general approach, specific hypolipidemic drugs, combination drug therapy, and use of estrogens. This section of the book contains guidelines and information on addressing treatment concerns not necessarily included in the current recommendations from the NCEP. On the basis that elevated cholesterol is a necessary condition for atherogenesis, emphasis has been placed on optimal therapy to effect prevention and regression of atherosclerotic lesions.

The final part of the clinical section (**Chapter 12**) contains a summary of the clinical principles outlined in earlier chapters, with emphasis on a practical approach to the patient who has dyslipidemia. To elucidate therapeutic guidelines, representative case studies are presented and reviewed at progressive stages of treatment. Finally, **Chapter 13** contains available resources and specific educational material that the physician may find useful in the daily management of patients, along with sources for information useful to the patient. Some of the cookbooks applicable to dietary management of lipid disorders are listed as well. Most useful, however, should be the instructional material for specific drugs that may be given to patients. A glossary has been prepared to aid the uninitiated reader in following the commonly used terms in the lipid literature. Special data, such as conversion tables and dietary information that might be needed, are included in the glossary and the appendices.

The text contains sufficient information to treat all but the most complex and obscure lipid disorders. Each chapter contains a recommended reading list for additional information. Because of the rapid strides in detecting and treating dyslipidemias, every effort was made to evaluate and include the most recent data and most current references. An attempt was made throughout the text to adhere to national recommendations for detection, evaluation, and treatment of these lipid disorders, and when recommendations deviated from the NCEP, they were so designated. Because of the current transition from conventional units to the less familiar International System, we have included values according to the *Système International d'Unités* in parentheses in the text as a guide to the reader. The terms *plasma* and *serum* often are used interchangeably, as are the terms *lipids* and *lipoproteins*; however, we have used the appropriate term when specified. Because lipid problems involve elevated *and* reduced values of the lipoproteins, the term *dyslipidemia* has been used to designate lipid disorders in general. *Dyslipoproteinemia* is more appropriate but too cumbersome and, therefore, we elected to use dyslipidemia.

Treatment of common lipid disorders will have a major impact on public health. The majority of future cardiovascular events will not emanate from those few patients who are difficult to treat or who have unusual lipid disorders but from that large segment of the population that have relatively mild lipid abnormalities, in association with other recognized risk factors. Successful treatment of these problems is well within the capabilities of the primary care physician and constitutes the means for preventing and reducing the burden imposed by the high prevalence of atherosclerosis in this country.

The indications and dosages of all drugs in this text have been recommended in the medical literature and conform to the guidelines developed by the Atherosclerosis Detection and Prevention Clinic at the University of Alabama at Birmingham Medical Center. The drug regimens, however, do not necessarily have specific approval by the Food and Drug Administration (FDA) for use in all of the situations described. As standards for usage change, and as new drugs become available, it is advisable to be aware of new recommendations. The current package insert for each drug should be consulted for drug usage as approved by the FDA.

Acknowledgments

Ms. Charlotte S. Bragg and Ms. Heidi C. Hataway reviewed materials for the Educational Resources chapter and advised us on nutritional questions throughout the book.

Dr. James M. Raczynski provided much of the behavioral material used for the section on dietary counseling.

Ms. Paulette W. Guarino and Ms. Sharon Sertell typed the numerous drafts and made many valuable editorial suggestions. Ms. Harriet A. Dean and Ms. Diane Blizard assisted with the editing.

Ms. Mary L. Short contributed greatly with the analyses and development of the graphic materials. Mr. Rod Powers made all the necessary original drawings and photographs.

The authors gratefully acknowledge the contributions of the Atherosclerosis Detection and Prevention group at UAB (Vera Bittner, MD; Bill Bradley, PhD; Frank A. Franklin, MD, PhD; David Garber PhD; Sandra Gianturco, PhD; Alan M. Siegal, MD; and especially Jere Segrest, MD, PhD). Their knowledge and continuous challenges helped us to formulate many of the clinical strategies used in this text.

CONTENTS

SECTION FIVE
EDUCATIONAL RESOURCES

SECTION ONE

A SCIENTIFIC BASIS FOR TREATMENT

CHAPTER 1

Evidence Supporting the Lipid Hypothesis

The Lipid Hypothesis

Over the years high blood cholesterol levels have become inextricably linked with atherosclerosis. Critical levels of blood cholesterol appear to be a necessary condition for the development of atherosclerosis. This intimate association between lipids and atherogenesis has become known as the lipid hypothesis, which proposes that lipid disturbances represent a major contributing cause to atherosclerosis. By inference, the lipid hypothesis states that the risk of coronary heart disease (CHD) may be lowered by modifying the concentration of blood lipids.

Support for the lipid hypothesis rests on an impressive array of findings:

1. Atherosclerosis can be induced in experimental animals by inducing hypercholesterolemia. Whether the hypercholesterolemia is produced by diet, by manipulations such as thyroid ablation, or by inborn errors of metabolism such as deficiencies of low density lipoprotein (LDL) receptors does not seem to matter. The severity and frequency of atheromatous lesions correlate strongly with blood cholesterol levels.

2. Cholesterol accumulation in the arterial wall, a hallmark of experimental and human atherosclerosis alike, originates from circulating lipoproteins.

3. The degree of coronary atheroma at postmortem examination in humans relates to antemortem blood cholesterol values.

4. Substantial associations exist between population values for blood cholesterol and the development of CHD. This risk of CHD is continuously related to blood cholesterol levels, increasing more steeply at the higher cholesterol values. No population yet has been identified with a high CHD rate and low cholesterol levels.

5. Mortality rates of CHD among populations with widely divergent diets corre-

late highly with the intake of saturated fat and cholesterol.

6. Countries that have a high prevalence of risk factors for CHD but a low intake of saturated fat and cholesterol tend to have a relatively low incidence of CHD.

7. Migrants from a low-CHD prevalent country tend to develop CHD at rates more similar to those of their adopted country.

8. Homozygous familial cholesterolemia affords an "experiment of nature" for atherosclerosis. The diminished removal rate of LDL from the bloodstream leads to CHD at an early age in children, despite the absence of other major risk factors or known abnormalities of the vasculature.

9. Sufficient data now exist to show that blood cholesterol reduction can halt progression and even induce regression of atherosclerotic lesions. Large-scale, randomized controlled trials of lipid-lowering agents appear to be effective in the primary and secondary prevention of CHD.

10. The declining cardiovascular mortality in the United States during the past decade can be attributed, at least in part, to secular trends in the population distribution of blood lipids.

11. The biologic mechanisms by which lipoprotein abnormalities contribute to the development of atherosclerosis are beginning to be understood and seem plausible.

The evolutionary changes in our understanding of the atherosclerotic process strongly support the lipid hypothesis.

Early Work on Atherosclerosis

In 1740 Crell, a German physician, contributed perhaps the first work to be recognized as important in understanding the pathogenesis of atherosclerosis. He recognized that the arterial incrustations that commonly were thought to be ossifications were, in fact, not bony. Lobstein, a pathologist working in Straussburg, coined the term arteriosclerosis in 1833 to signify a thickening and hardening of the arteries. Shortly thereafter, in 1845, a German physician, Vogel, determined that cholesterol was a major constituent of atheromatous plaques, while yet another pathologist from Leipzig, Marchand, introduced the term atherosclerosis to indicate the process that affected the intimal coat of the artery.

Virchow, the founder of cellular pathology and "the Pope of medicine in Europe," subsequently observed that the intimal lesions of arteriosclerosis lay under or within, but not on top of, the inner lining of the vessel. He postulated that lipids contributed to these fatty plaques and observed that a gelatinous swelling of the intima preceded the formation of an atherosclerotic plaque. Virchow's view that inflammation was important in the development of atherosclerosis was challenged by the Viennese physician, Rokitansky, who believed that many small blood clots eventually were converted into scar tissue.

The role of cholesterol in atherosclerosis received more attention in the early years of the 20th century. In 1906, a Russian scientist, Anichkov, called attention to the high content of cholesterol in atheromatous aortas. A German physician, Windaus, while working on the structure of cholesterol, found that atheromatous aortas contained up to 20 times as much cholesterol as the normal aorta. Ignatowski, in Russia, while investigating the effects of protein-rich foodstuffs, noted that striking aortic changes were more common in rich, rather than in poor Russians. He attributed these aortic changes to the injurious effect of animal protein (meat, milk, and eggs), because such ingredients were most abundant in the diets of the rich. This was an important hypothesis, despite the misconception that the protein of these foods was responsible. Several years later, in 1913, Anichkov, in collaboration with a fellow Russian, Chalatov, produced typical arterial atherosclerosis in rabbits by adding pure cholesterol dissolved in vegetable oil to rabbit food.

Such observations led to additional feeding experiments in rabbits and clinical studies in humans. The results of clinical and pathologic studies conducted by several investigators finally led Anichkov to conclude that dietary cholesterol was transported into the bloodstream and then deposited into the arteries.

In 1912 in the United States, Herrick had identified the clinical syndrome of myocardial infarction and its association with atherosclerosis and thrombosis and published his observations in a manuscript entitled, "Clinical Features of Sudden Obstruction of the Coronary Arteries," in The Journal of the American Medical Association (JAMA). Around that time a Dutch physician, DeLangen, had noted that the Javanese population had much lower blood cholesterol levels than the Dutch for the most part and related this finding to the striking difference in the frequency of atherosclerosis, phlebothrombosis, and gallstones between the two countries. He advocated a low cholesterol diet to prevent atherosclerosis and published a report in 1916 in an obscure journal that attracted little attention at the time—the first recorded observation on the relationship between cholesterol and atherosclerosis in humans. It was not until the rapid increase in the occurrence of CHD and other clinical manifestations of atherosclerosis, however, that detailed investigations of its causes were pursued. In the 1920s, a German physician, Shoenheimer, reported chemical analyses on aortas that showed that total lipid, especially cholesterol, was increased with atherosclerosis. About this time, CHD became generally recognized as a clinical entity and became listed officially in the 1929 revision of the International List of Diseases and Causes of Death.

Finally, in 1932, the correct structure of cholesterol was established. This came more than a century after Chevreul, in France, had first differentiated between saponifiable and nonsaponifiable lipids and was

credited with the discovery of cholesterol. He derived the original name, cholesterine, from the Greek, chole (bile) and stereos (solid). Later the name was changed to cholesterol after a reactive hydroxyl group was found.

Anichkov, reviewing the topic in 1933, drew the inference that cholesterol was the "materia peccans" of atherogenesis. Several years earlier, Anichkov had studied regression of dietary-induced plaques in rabbits. Mobilization of cholesterol and lipid from plaques had been slow, and several years were required to obtain lipid-free plaques. Interest in these experimental regression studies was revived in the late 1940s and 1950s and have led to the well-controlled angiographic studies in humans described later in this chapter.

Current Concepts of Atherosclerosis

Atherosclerosis refers to the underlying disorder involving the intima of medium and large arteries, which can result in a compromised circulation to the brain, heart, kidneys, and extremities. The term atheroma originally referred only to the fatty mass but now is used to describe the entire atherosclerotic lesion. In addition to producing mechanical narrowing of the lumen, the atheroma also affects function and impairs endothelium-dependent relaxation in isolated human coronary arteries. Such a functional derangement could aggravate the anatomic lesion, further accentuating tissue ischemia and predisposing to thrombosis.

The three basic morphologic stages of atherosclerotic lesions (Fig. 1.1) include

1) The fatty streak—an early stage in which the lesions consist largely of fat-filled macrophages
2) The fibrous plaque—a lesion that consists of foam cells, smooth muscle cells, and varying amounts of collagen, elastins, and proteoglycans
3) The complicated lesion—further progression characterized by necrotic debris,

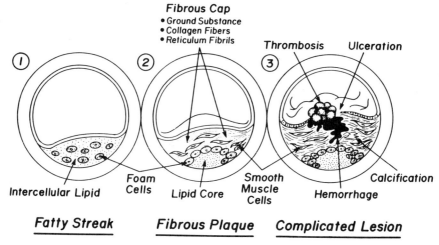

Figure 1.1. Stages of atherosclerosis. Morphologic stages of atherosclerosis: the fatty streak, the fibrous plaque, and the complicated lesion. Lesions generally evolve from a fatty streak characterized by foam cells to a fibrous stage and finally to a lesion involving calcification, ulceration, hemorrhage, and thrombosis. (Adapted from: Grundy SM. Cholesterol and atherosclerosis: diagnosis and treatment. Philadelphia: J.B. Lippincott Company, 1990.)

cholesterol crystals, and calcium, in addition to components of the fibrous plaque. With progression of these lesions ulceration takes place, resulting in hemorrhage and thrombosis.

Pathogenesis

Early theories of the pathogenesis of atherosclerosis emphasized the role of injury, or low grade inflammation, which was conceived as a tissue reaction to filtration of plasma proteins and lipids. This evolved into the lipid infiltration hypothesis. A second theory postulated that small thrombi collected at the foci of endothelial injury and organized into plaques. In the currently accepted view, critical events in atherogenesis center around the focal accumulation of lipids in the vessel wall through a mechanism not necessarily dependent on injury to the vessel wall. Most investigators believe that low density lipoprotein (LDL) is the most atherogenic of the lipids, but there appears to be a growing appreciation for the role of very low density lipoprotein (VLDL) and VLDL remnants.

Although this hypothesis is widely accepted, controversy remains over the precise mechanisms by which the lipid enters the arterial wall. Increased blood levels of LDL or other lipoprotein components may accelerate the rate of lipid influx into the arterial wall by mass action. The LDL particle is too large to penetrate endothelial cell junctions but can permeate through specialized receptors on endothelial surfaces which recognize LDL and modified forms of LDL. The LDL transport also may occur through nonspecific uptake by micropinocytic channels. It also is possible that lipid is ingested by circulating monocytes or macrophages that transport it into the vessel wall. Mechanisms that control the flux of macrophages into and out of atherosclerotic plaques still are poorly understood but obviously important.

Although lipid infiltration may occur in the presence of normal endothelium, the process conceivably is enhanced by endothelial injury, thus removing a natural barrier to the entry of lipoproteins into the arterial wall (Fig. 1.2). Some LDL passes completely through the intimal layer into the vasa vasorum and reenters the circulation while other LDL particles become entrapped in the intima by interacting with local ground substances. Glycosoaminoglycans have a high affinity for apoprotein B-100, preventing some LDL particles from

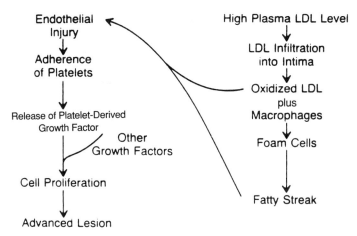

Figure 1.2. Postulated linkage between the lipid infiltration hypothesis and the endothelial injury hypothesis in the development of atherosclerosis. Circulating LDL particles become entrapped in the intima representing a key initial step in atherogenesis. The LDL undergoes oxidation or other modification and is removed by special scavenger receptors on macrophages and mural cells. There is no feedback control on the formation of these receptors and cholesteryl esters accumulate, giving the cell a foamy appearance. The foam cells lead to development of fatty streaks that may cause disruption of the endothelium. Injury also can be primarily from adherence of platelets and an abundance of growth factors, resulting in cell proliferation, fibrous plaques, and, ultimately, advanced atherosclerotic lesions. (From: Steinberg D, Parathasarathy S, Carew TE, Khoo JC, Witztum JL. Beyond cholesterol: modifications of low density lipoprotein that increase its atherogenicity. N Eng J Med 1989;320:915–924.)
"Reprinted, by permission of the New England Journal of Medicine, vol. 320; page 921, year 1989"

leaving the intima. Mechanisms other than endothelial damage and entrapment in the intimal layer, by which cholesterol and cholesteryl ester accumulate in the vessel wall, are unclear but include binding by fibrin in advanced lesions and enzymatic degradation of LDL with leakage into extracellular spaces.

In any event, it is postulated that LDL or VLDL remnants infiltrate the arterial wall and become entrapped in the intima. Here LDL becomes modified, usually through the oxidation of LDL lipids and apoprotein B-100. This modification is induced by macrophages, as well as by endothelial cells and smooth muscle cells, and apparently requires copper or iron. As a result, the LDL particle loses its integrity, can be recognized by specific receptors (the so-called scavenger receptors) that are distinct from the LDL receptors in the liver, and becomes susceptible to engulfment by macrophages. Oxidized LDL stimulates the production of chemotactic proteins that recruit monocytes to the endothelial surface. Oxidized LDL

molecules also inhibit migration factors, thereby increasing the amount of time the macrophages reside in the subendothelial space. Additional modifications of apoprotein B, such as glycosylation or attachment of a malonaldehyde molecule, also may play a role in the above sequence.

Once circulating monocytes adhere to the endothelium, they become transformed into the lipid-laden macrophages present in variable numbers in all stages of human atheromatous plaques. These macrophages (Fig. 1.3) are transformed into foam cells through lipid accumulation that is caused by the presence of at least three receptors on their surface: (1) the beta-VLDL receptor that recognizes VLDL from hypertriglyceridemic patients; (2) the modified LDL receptor (scavenger receptor) that recognizes modified LDL; and (3) the modified VLDL receptor for modified VLDL. Uptake of the lipoprotein by these receptors is followed by its internalization and hydrolysis. The cholesterol is re-esterified and stored as a lipid droplet, producing the foam cell (Fig. 1.4).

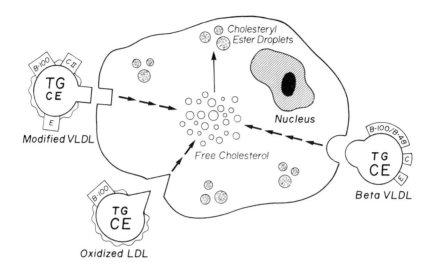

Figure 1.3.   Receptor-mediated lipid accumulation in macrophages. Macrophages possess several distinct receptors for lipoprotein particles. Depicted are the modified VLDL receptor, the scavenger receptor, and the LDL receptor. Putatively, there is a receptor for chylomicron remnants (not shown) that involves LDL receptor-like protein. In addition to the cholesterol deposited as cholesteryl esters, (shown), triglyceride from the lipoprotein particles is metabolized into free fatty acids and stored as triglyceride esters. The lipid accumulation that ensues from the lipoproteins binding to specific receptors transforms the macrophages into foam cells. (Adapted from: Bradley WA, Gianturco SA. Personal communication.)

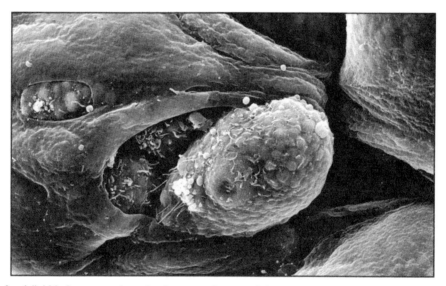

Figure 1.4.   A lipid-laden macrophage is shown coming out of the arterial wall. The accumulated load of lipid and lipoprotein has converted the macrophage to a foam cell. This macrophage (foam cell) is returning to the blood so that it may transport the lipid to the spleen, lung, liver, or lymph nodes, where the macrophage may reside and ultimately digest this material. (From: Ross R. The gallery of the pathogenesis of atherosclerosis. Bristol-Myers Squibb Company, April, 1991.)

Most foam cells are thought to be derived from macrophages, but some may originate from smooth muscle cells. Foam cell formation apparently does not occur through the LDL receptor pathway because the cholesterol released from LDL inhibits the synthesis of new LDL receptors. This feedback mechanism prevents excess cholesterol from accumulating in the cell. When macrophages take up modified LDL through non-LDL (scavenger) receptors, however, there is no downregulation of new receptors. The scavenger receptors, therefore, continue to bind and internalize modified LDL particles. As a result, many large droplets of cholesterol ester accumulate and give the cell a foamy appearance. Increased numbers of these cells deform the underlying endothelium. Over time microscopic separations may expose the underlying foam cells and extracellular matrix, giving rise to sites for platelet aggregation and development of atheroma without overt injury. Oxidized LDL also is cytotoxic and is believed to contribute to foam cell necrosis and endothelial ulceration.

Fatty Streaks

The fatty streaks are the initial lesions associated with atherosclerosis and consist of grossly visible, yellowish, slightly raised longitudinal areas. Microscopically they contain subendothelial aggregates of foam cells made up predominantly of lipid-laden macrophages, as well as modified smooth muscle cells that contain fat (Fig. 1.4). Fatty streaks appear in the aorta of most children after one year of age. The extent of aortic intimal surface covered by fatty streaks increases from about 10% in the first decade to 50% in the third decade. As more mature atherosclerotic plaques appear, fatty streaks decrease in number. The distribution of fatty streaks does not necessarily correspond well to the distribution of later atherosclerotic lesions in adults. It currently is believed that fatty infiltration generates the initial subendothelial lesion, but that other factors control the distribution of future clinically significant lesions. Atherosclerotic lesions are most pronounced in the abdominal aorta and in the initial 6 cm of the coronary arteries, the popliteal arteries, the descending thoracic aorta, the internal carotids, and the vessels of the Circle of Willis in the central nervous system (CNS). The predilection of atherosclerosis for the systemic circulation rather than the low-pressure venous or pulmonary circulation and the greater severity in large arteries emphasizes the importance of mechanical or hemodynamic factors. The distribution of atherosclerotic plaques occurs at arterial sites in which there is high pressure and great turbulence of blood flow. Flow is disturbed at such points, and resulting shear stresses may initiate the process by producing damage to the endothelial cells. Endothelial damage may be facilitated by such diverse factors as hypertension, carbon monoxide, and immune complex disease, as well as by high levels of modified LDL and VLDL remnants.

Fibrous Plaques

Fibrous plaques, characterized by the fibrous cap, begin as slightly elevated lesions in the intima that are composed of lipid-laden macrophages and smooth muscle cells in association with collagen, elastic fiber, and proteoglycan deposition (see Figs. 1.1 and 1.2). Following endothelial damage, the combination of monocytes and endothelial cells may transform the normal anticoagulant vascular surface into a procoagulant surface. Repair of the endothelial damage is characterized by deposition of platelets and infiltration of lipid-laden monocytes into the area of injury. Platelets and macrophages release growth factors that stimulate the migration of smooth muscle cells into the area. This results in synthesis of collagen, which then is deposited in the arterial wall. Fibrous plaques, white in appearance and oval in shape, first appear in the lower aorta.

Complicated Lesions

The fibrous plaque may evolve into a complicated lesion as a result of calcification and thrombus formation on top of an ulcerated surface (see Fig. 1.1). The continued infiltration of lipids, in addition to the release of lipid by degenerating macrophages, leads to the accumulation of extracellular lipid and cytotoxic products. Finally, the plaque surface ulcerates and a thrombus forms on the luminal surface. In addition to monocytes, the characteristic lesion also contains numerous activated T-lymphocytes, the significance of which remains to be determined.

Repetitive damage invokes new cycles of plaque growth. The ingrowth of new blood vessels eventually ruptures and hemorrhages producing cell necrosis, calcification, and mural thrombosis, the so-called "complicated lesion." Ulceration of the luminal surface can cause rupture of the atheromatous plaque, producing microemboli. Computer modeling of these atherosclerotic plaques has demonstrated that the site of ulceration can be influenced by a diminished mechanical strength of the fibrous cap because of focal accumulation of foam cells. Such weak points in the cap explain tears that occur at places other than the site of maximum stress. The complicated lesion gives the disease its ominous clinical significance. A superimposed thrombus in this damaged site may occlude the lumen or become incorporated within the intimal plaque. The more severe the lesions, the

more likely that patients will experience clinical manifestations from the resultant ischemia. As the complicated lesion enlarges it gradually fills the lumen, reducing blood flow to a critical level. Dilation of the artery may occur rather than stenosis if the atheromatous process results in destruction of the media. Rupture of vessels beneath the plaque, resulting in plaque elevation, can lead to rapid occlusion of a vessel by a lesion that was not even physiologically important. It is this rupture of a plaque with adherent platelets and thrombi that obstructs the artery and leads to ischemia and infarction.

In their advanced state (Fig. 1.5), atheromas consist of cholesterol-laden smooth muscle cells and foam cells. The atheroma, or fibrofatty lesion, consists of a raised focal plaque within the intima made up of two components: (1) a thick layer of fibrous connective tissue called the fibrous cap, which is much thicker and less cellular than the normal intima and composed largely of smooth muscle cells that produce collagen, elastin, and proteoglycans; and (2) a core of necrotic calcium, cell debris, cholesterol crystals, and lipid-filled macrophages. As the disease advances, these initially sparsely distributed atheromas become more numerous and sometimes coalesce to cover the entire intimal surface of severely affected

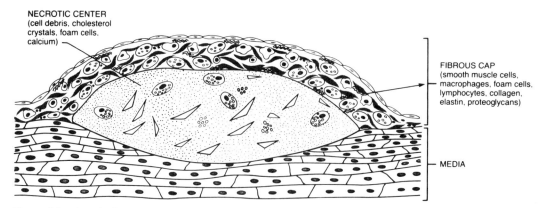

Figure 1.5. A typically advanced atherosclerotic plaque. The lesion consists of a superficial fibrous cap and a deeper necrotic center. The fibrous cap is made up of smooth-muscle cells with a few leukocytes and dense connective tissue. The cellular area beneath and to the side of the cap that consists of a mixture of macrophages, smooth-muscle cells, and T lymphocytes. The necrotic core contains a disorganized mass of cellular debris, lipid material, fibrin, plasma proteins, and foam cells. (From: Cotran RS, Kumar VK, Robbins SL. Pathologic basis of disease. ed. 4 Philadelphia: W.B. Saunders Company, 1989:p.563.)

arteries. As the plaques grow, they progressively encroach on the lumen of the artery, as well as on the subjacent media.

Progression from a simple lesion to the more complicated clinically significant lesions can be found as early as age 20 but becomes highly prevalent by late middle age in susceptible persons. Decades usually are required for the formation of clinically important atheromatous lesions. An accelerated version of this process can occur within months to years in some patients with an unexplained virulent progression, or in those patients who undergo heart transplantation, coronary artery bypass grafting, or coronary angioplasty (Table 1.1). In these situations, denuding endothelial injury appears to be the critical inciting event, followed by platelet activation and thrombus formation, with subsequent organization by smooth muscle cell proliferation.

Why some people survive a lifetime with little evidence of arterial disease, whereas others develop vascular lesions at an early age, is not well understood. Various hypotheses have been proposed, which are by no means mutually exclusive, to explain the development and evolution of these atherosclerotic plaques. In any event, it is conceivable that any number of interventions along this complex and long pathway could retard the progression of lesions or even bring about regression. This is especially important in view of the large proportion of deaths that occur suddenly but in reality can be attributed to severe and extensive atherosclerosis of the coronary arteries.

Population Studies

A German scientist, Rosenthal, was the first to note that the worldwide distribution of atherosclerotic disease corresponds to dietary fat and cholesterol consumption. Anichkov developed these observations in a 1953 paper, in which he ranked countries according to CHD death rates and found that the order paralleled the percentage of total calories provided by fat in the national diet. He then developed experimental evidence to show that a higher ratio of animal fats to plant fats increases the average blood cholesterol concentrations. Subsequent reports confirmed the rarity of CHD in the Far East compared to Western populations, which consumed much larger quantities of meat and dairy products. A rapid decline in cardiovascular disease (CVD) mortality occurred among European countries during World War II, when supplies of butter, eggs, and meat were scarce, re-emphasizing the association between diet and atherosclero-

Table 1.1 Comparison between the Processes of Spontaneous and Accelerated Atherosclerosis[a]

	Spontaneous Atherosclerosis	Accelerated Atherosclerosis
Early platelet involvement	Absent	Present
Early monocyte involvement	Present, if dyslipidemic	Absent
Pathology		
Initial	Lipid deposition, monocyte and platelet adhesion	Thrombosis, intimal SMC proliferation, fibrosis
Late	Intimal SMC[b] proliferation, fibrosis	Lipid deposition
Complication	Plaque rupture, thrombosis	Plaque rupture, thrombosis
Duration of process	Decades	Months to years

[a] Adapted from: Ip JH, Fuster V, Badimon L, Badimon J, Taubman MB, Chesebro JH. Syndromes of accelerated atherosclerosis: role of vascular injury and smooth muscle cell proliferation. JACC 1990;16:1667–1687.
[b] SMC = smooth muscle cells

sis. The finding that CHD levels could drop quickly in such populations lent more credence not only to the concept of the contribution of diet to CHD but also to the potential for slowing the rate of progression and preventing the disease by reducing fat consumption.

Other convincing evidence on the relation of diet to human coronary disease subsequently emerged. The Armed Forces Institute of Pathology reported on the status of the coronary arteries among soldiers who died in the Korean War. Atheromas were grossly visible in 50% of the coronary arteries of 22-year-old American soldiers, whereas Korean and Chinese soldiers, who ate few dairy products and meat, had no lipid in their coronary intimas. The International Atherosclerosis Project systematically studied the severity of atherosclerosis among autopsies of over 20,000 people in twelve countries. Countries were ranked on the basis of severity of atherosclerosis, dietary indices, and serum cholesterol. The severity of atherosclerosis correlated with the percentage of calories from dietary fat and population levels of blood cholesterol. In the Seven Countries' Studies, 12,000 men from eighteen populations exhibited a fourfold difference in the incidence and prevalence of CHD. The incidence rate varied from the lowest (Greece and Japan) to the highest (Finland and the United States). Likewise, saturated fat intake was highest in Finland, the United States, and the Netherlands, representing about 20% of calories, compared with less than 10% of calories in other countries. Figure 1.6 depicts the geographic distribution of CHD deaths according to the median total cholesterol level from sixteen different cohorts within the seven countries.

Migration studies furnish additional evidence on the importance of environment and lifestyle, rather than genetics, as predisposing factors to atherosclerosis. Persons born in a given culture who subsequently move to another country tend to develop CHD rates similar to their adopted country, rather than maintaining those from their country of origin. In the Ni-Hon-San Study,

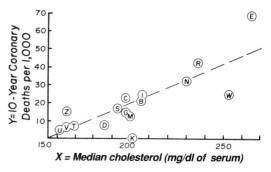

Figure 1.6. Relationship of plasma cholesterol and ten-year coronary heart disease death rate in the Seven Countries' Study. All men were judged free of coronary heart disease at entry. B=Belgrade, Yugoslavia; C=Crevalore, Italy; D=Dalmatia, Yugoslavia; E=East Finland; G=Corfu, Greece; I=Italian railroad workers; K=Crete; M=Montegiorgio, Italy; N=Zutphen, Netherlands; R=American railroad workers; S=Slavonia, Yugoslavia; T=Tanushimaru, Japan; U=Ushibuka, Japan; V=Velika Krsna, Yugoslavia; W=west Finland; Z=Zrenjanin, Yugoslavia. (Adapted from: Keys A. Seven countries: A multivariate analysis of death and coronary heart disease. Cambridge: Harvard University Press, 1980.)

middle-aged Japanese men were compared to similar persons of Japanese ancestry who had migrated to Honolulu or to the San Francisco Bay area. Cholesterol levels were 12% higher in Hawaii and 21% higher in San Francisco than in Japan. Accordingly, CHD mortality was 1.7 times greater in Hawaii and 2.8 times higher in San Francisco than in Japan.

A relationship between the consumption of saturated fat and cholesterol and subsequent risk of death from cardiovascular causes within populations also has become apparent. Seventh Day Adventists who consume lacto-ovo vegetarian diets (milk, egg, and vegetable products) have reported CHD mortality rates that are 72% lower than those found among the general population. Such persons differ, however, in many other ways from those who eat the typical American diet.

Four long-term cohort studies—the Western Electric Study, the Zutphen Study, the Ireland-Boston Dietary Heart Study, and the Honolulu Heart Program—have all found that dietary cholesterol, when assessed reliably at entry and expressed as mg/1000 kcal, relates directly to the risk of CHD after adjustment for other major risk factors. It has been calculated that a dietary cholesterol increment of 200 mg/1000 kcal daily is associated with a 30% greater risk of CHD. In the Western Electric Study, dietary cholesterol was associated with the risk of death from CHD and CVD, independent of its influence on serum cholesterol (Table 1.2). Such results suggest that dietary cholesterol also may affect atherogenesis and CHD by mechanisms other than blood cholesterol concentration. This lends support to the concept that a low cholesterol diet may be worthwhile even for those persons who do not have hypercholesterolemia.

Major findings on the relationships between lipids and CHD were provided by the Framingham Heart Study. This longitudinal study has followed a sample of 2336 men and 2873 women, who in 1948 were residents of the town of Framingham, Massachusetts and were between 30 and 62 years old at their first examination. A routine clinic examination was repeated at two-year intervals, consisting of a medical history, physical examination, blood chemistries, body measurements, vital capacity, chest x-ray, and an electrocardiogram. Mortality and cardiovascular morbidity were documented in detail from these routine exami-

nations as well as hospital records, death certificates, physicians records, and next-of-kin. In this way, a wide variety of information has been collected on the characteristics of the study cohort both before and after the development of cardiovascular disease. This allowed not only documentation for risk factors for CVD but also estimation of CVD incidence and mortality rates. To gain additional data on younger persons and their families, a secondary study, the Framingham Offspring Study, has also begun as an extension of the original study.

The Framingham Study, as well as other longitudinal population studies, have made available extensive information on the risk of CHD by level of total blood cholesterol, apart and in combination with other major modifiable risk factors for CHD—cigarette smoking, diabetes, and high blood pressure. Each of these risk factors was found to contribute independently to the susceptibility for CHD. In addition, the risk factors were multiplicative in the sense that each potentiated the risk from the other factors. Findings among populations, and for persons within populations, consistently exhibit a strong, continuous, and direct association between total cholesterol levels and manifestations of CHD.

The assessment of risk from high blood lipids has been refined even more by reviving

Table 1.2 Age-Adjusted Mortality Rates Per 1000 Person-Years of Observation by Dietary Cholesterol Intake (Western Electric Study, 1958–1983)[a]

Dietary Cholesterol Intake (mg/1000kcal)	Men at Risk	24-Year Mortality Rates/1000 Person-Years, by Cause of Death		
		Ischaemic Heart Disease	Other CV[b] Disease	All Causes
81–186	351	6.9	1.8	16.7
187–214	352	6.2	1.7	14.1
215–245	374	8.3	1.9	16.2
246–288	368	7.1	2.5	16.1
289–590	379	9.6	3.1	19.0
Total	1824	7.7	2.2	16.6
RH[c]		1.38	1.80	1.27
95% Confidence Interval		1.00–1.90	1.00–3.24	1.02–1.58

[a] Adapted from: Shekelle RB, Stamler J. Dietary cholesterol and ischaemic heart disease. Lancet 1989;I(8648):1177–1179.
[b] CV = cardiovascular
[c] RH = relative hazards for fifth versus first quintiles

the concepts of lipoprotein subfractions, first noted in the 1950s by Gofman from the Donner Laboratory. He had used an ultra-centrifuge to measure concentrations of various densities of lipoproteins and had argued that those lipoproteins with a flotation rate of 10 to 20 Sf units were related specifically to CHD. These data also suggested that the high-density lipoprotein (HDL), which carried about 25% of total cholesterol, correlated negatively with risk. At that time, however, HDL-C was dismissed as having little practical importance. Later, Gofman and his associates used data from a carefully designed prospective study with precise lipoprotein quantification to demonstrate again the protective importance of HDL in CHD. Despite this, investigators were still reluctant to accept the concept that a cholesterol of any kind actually could reduce CHD risk. In 1965, Frederickson and Lees had proposed a general classification scheme for blood lipid abnormalities that relied on lipoprotein electrophoresis and was simpler and more reliable than that derived from the quantitative ultracentrifuge. The scheme emphasized abnormalities in LDL and VLDL, but ignored HDL. The initial focus on total cholesterol shifted slowly to the LDL fraction. In the 1970s, however, the importance of HDL reemerged, and the emphasis returned to partitions between various lipoproteins and lipoprotein subspecies.

Finally, the Nobel Prize winning discovery of the LDL receptor by Goldstein and Brown had a profound impact on subsequent clinical investigations into the metabolic effects of various lipoprotein subspecies. More recently, genetic studies in different species have reaffirmed the close relationship between lipoprotein disturbances and atherosclerosis. Genetic forms of dyslipidemia presage early coronary heart disease, even in the absence of major risk factors for coronary heart disease.

Intervention Studies

Despite the large amount of circumstantial evidence in its support, the lipid hypoth-

esis remained controversial, and efficacy data were necessary for the final link between cholesterol and CHD. The task of proving that reduction of the blood cholesterol reduces subsequent cardiovascular risk was not an easy one. Studies in persons who showed no evidence of existing CHD (primary prevention trials) were designed to address this issue from several perspectives. Because the incidence of CHD in such groups is relatively low, however, large study populations and long follow-up periods were required to demonstrate a treatment effect on CHD incidence and mortality. These trials were not designed to test the issue of overall mortality: treatment was not begun until middle age, the reduction in lipids tended to be less than optimal with available therapy, and the period of study, usually about 5 years, was too brief to alter mortality substantially. Consequently, analyses of the effect of treatment on persons who had established atherosclerosis (secondary prevention studies) had more statistical power to impact on total mortality. Two such studies, the Coronary Drug Project (CDP) 15-year follow-up study and the more recent Stockholm Study, have reported a reduction in mortality in the treated groups. Finally, recent trials that assessed angiographic changes in coronary anatomy have now demonstrated that the progression of coronary atherosclerosis and its clinical sequelae can be retarded by lowering the blood cholesterol using dietary, pharmaceutical, and surgical means. These data suggest that the blood cholesterol must be reduced substantially to lower CHD incidence.

Early Dietary Studies

Initial trials to prevent CHD were based on dietary interventions. These studies were hampered by the problems associated with blinding participants and investigators to the mode of therapy (thus creating the potential of applying a different quality of care to the treatment group), by the difficulties in controlling the adherence to the prescribed diets, and by their relatively modest reductions in lipids compared to that achievable with potent drugs. Some of these trials were conducted in special institutions, and their results may not necessarily apply to other free-living populations. The Finnish Mental Hospital Study, initiated in 1959, involved middle-aged patients from two mental hospitals in a cross-over design. A diet low in saturated fat and cholesterol and high in polyunsaturated fat was used in one hospital, while the other retained the usual diet. After 6 years the diets were reversed and the study was continued for an additional 6 years. This intervention diet decreased the mean serum cholesterol levels by 18% for men and 12% for women. Patient turnover in the institutions was fairly rapid, so that the actual follow-up averaged only 3.5 years for each of the periods in the first hospital and less than 2 years in the second hospital. Meals were provided to participants from cafeterias or food service facilities, but no control was exerted over access to foods from snack bars, family members, or other sources. Death rates from all CVD were lower by 39% for men and 14% for women on the restricted diet. Differences in all-cause mortality were small and not significant.

The Minnesota Coronary Survey, which began in 1968, was a 4.5 year doubleblind, randomized clinical trial that was conducted in six Minnesota state mental hospitals and in one nursing home. It compared the effects of a 39% fat control diet (18% saturated fat, 5% polyunsaturated fat, and 446 mg dietary cholesterol per day) with a 38% fat treatment diet (9% saturated fat, 15% polyunsaturated fat, and 166 mg dietary cholesterol per day) and involved 4393 men and 4664 women who were institutionalized and received either diet for a cumulative duration of at least 1 year during the study period. All subjects in the institutions who agreed to participate, irrespective of their entry lipid levels, were enrolled. The average fall in total cholesterol during the low saturated fat diet period (mean duration 384 days) was 14%. For the entire study population, no differences in cardiovascular death or total mortality were observed. A favorable trend for all endpoints occurred in the younger

age groups, who were on the low saturated fat diet for over 2 years, but the numbers were small and the difference did not reach statistical significance.

A randomized controlled trial involving 846 middle-aged and older men in a Veterans Administration Study was first reported in 1969. Many of these men had evidence of atherosclerotic disease at baseline so that the trial actually constituted a primary and secondary prevention study simultaneously. The experimental group received a diet with reduced levels of saturated fat and cholesterol but high levels of polyunsaturated fat, although, by current standards, this diet is not considered to be sufficiently low in fat. No provisions were made to restrict between meal snacks or food outside of regular meals. After 8.5 years of follow-up, the incidence of nonfatal and fatal cardiovascular events was significantly lower among those who had been placed on the special diet; however, differences in overall mortality were small. An excess of cancer deaths was found in the experimental group, but the difference was not related to adherence to the prescribed diet.

The Oslo Study was designed to test whether smoking reduction and dietary lowering of serum lipids among 1232 healthy normotensive men would reduce CHD. Among this population of men at high risk, 80% were cigarette smokers whose initial cholesterol values ranged from 290 mg/dL to 380 mg/dL (7.50 to 9.82 mmol/L). Dietary intervention consisted of advice to substitute polyunsaturated fatty acids for saturated fatty acids, to increase intake of whole grain cereals, and to reduce the total calories in hypertriglyceridemic patients. On the average, the cholesterol concentration decreased 13% and the triglyceride concentrations 20%, whereas tobacco consumption decreased about 45%. After 5 years of observation, the combined incidence of fatal and nonfatal myocardial infarction and sudden death were 47% lower in the treated group than in the controls. Final analyses suggested that the reduction in plasma total cholesterol was the primary reason for the CHD reduction and that changes in cigarette smoking explained, at most, one-fourth of CHD reduction.

Like the Oslo Study, the Multiple Risk Factor Intervention Trial (MRFIT) also began in the early 1970s and intervened on several risk factors—blood cholesterol, blood pressure, and cigarette smoking. This six-year prospective study screened over 350,000 men, aged 35 to 57 years, to identify a high-risk cohort for randomization into a usual or special care group. There was a 5% decrease in blood cholesterol in the special care (diet plus other interventions) group and a 3% decrease in the usual care group. A greater cholesterol reduction took place among hypercholesteremic persons who were nonhypertensive and nonsmokers. This study was complicated by the cholesterol-elevating effect of diuretics in the hypertensive group and the interference of smoking cessation

with weight reduction. The lack of early demonstrable cardiovascular benefit has been attributed to design problems and concurrent risk reduction among the population at large. More recent findings, after 10.5 years of follow-up (3.8 years after the end of intervention), indicate lower mortality rates for men in the special intervention group: 10.6% for CHD and 7.7% for all causes. This was primarily because of a 24% reduction in mortality from acute myocardial infarction (MI).

In addition to the major trials already cited, several other dietary studies were conducted. A plot of findings of commonly cited dietary and drug interventions shows that for each 1% reduction in cholesterol level, the subsequent CHD rate falls by 2% (Fig. 1.7). In those trials that barely lowered cholesterol, the rate of CHD fell minimally,

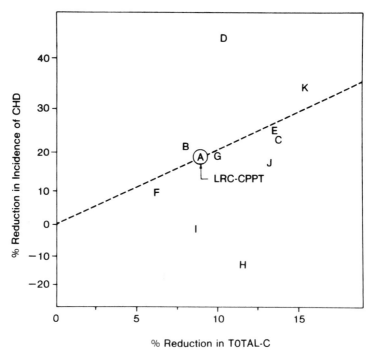

Figure 1.7. Comparison of results of 11 cholesterol-lowering trials with experience of Lipid Research Clinics Coronary Primary Prevention Trial (LRC-CPPT) cholestyramine group. Percent reduction in coronary heart disease (CHD) incidence (logarithmic scale) observed between actively treated and control group in each trial is plotted vs. mean difference in plasma total cholesterol (TOTAL-C) levels (expressed as percent of mean level for control group) resulting from treatment. Reduction in CHD incidence predicted by proportional hazards model for given decrease in TOTAL-C levels within LRC-CPPT cholestyramine group is indicated by dashed line. (From: Lipids Research Clinics Program. The Lipid Research Clinics Coronary Primary Prevention Trial Results: the relationship of reduction in incidence of coronary heart disease to cholesterol lowering. JAMA 1984;251:365–374.)

supporting the concept that the cholesterol level must be lowered sufficiently to induce sizeable changes in the CHD rate. Nevertheless, these findings from the cholesterol-lowering studies certainly presage the major drug trials for lowering cholesterol in which a similar relationship between the blood cholesterol and CHD was observed.

Major Cholesterol-Lowering Clinical Trials

Although the dietary studies suggested potential benefits from reduction of cholesterol values, they were limited in scope and did not achieve the necessary results for substantive conclusions. A second generation of trials (Table 1.3) based on modifying lipids with drugs, generally using larger sample size, longer follow-up, and better methodology, followed about a decade later.

Primary Prevention. One of the earliest trials, and the one that had the most unfavorable impact on the acceptance of lipid therapy, was the World Health Organization (WHO) Cooperative Trial. This double-blind study, which began in the late 1960s, was designed to test whether treatment with the fibric acid derivative clofibrate would alter CHD morbidity and mortality. It was a primary prevention trial and took place in three European centers—Edinburgh, Budapest, and Prague. Average entry serum cholesterol levels of 249 mg/dL (6.43 mmol/L) were reduced by an average of 9% with the medication, resulting in a significant reduction in the number of nonfatal MIs after 5.3 years of follow-up. However, the incidence of fatal MIs was similar in the two groups. Moreover, continued surveillance for almost an additional 8 years after intervention re-

Table 1.3 Major Cholesterol-Lowering Trials With Drugs to Prevent Coronary Heart Disease[a]

Study[b]	Population	Follow-up (Years)	Treatment	Entry Level (mg/dL)	Total Cholesterol (TC) Percent Reduction[c]	CHD Percent Reduction
Primary Prevention						
Coronary Primary Prevention Trial(6)	Men 35–59 yrs, TC $\geq$ 265 mg/dL + LDL-C $\geq$ 190 mg/dL	7.4	Diet + placebo vs. diet + cholestyramine	292	9	17
Helsinki Heart Heart Study(12,17)	Men 45–55 yrs, asymptomatic non-HDL-C $\geq$ 200	5.0	Placebo vs. gemfibrozil	270	9	33
World Health Organization(8)	Persons 30–59 yrs, in the upper ⅓ of TC distribution	5.3	Placebo vs. clofibrate	242	9	20
Secondary Prevention						
Coronary Drug Project(9)	Men 30–64 yrs, survivors of MI[d]	5.0	Placebo vs. nicotinic acid	253	10	20
Stockholm Study(6)	Men below 70 yrs, consecutive MI survivors	5.0	Placebo vs. clofibrate + nicotinic acid	249	13	33
Program on Surgical Control of the Hyperlipidemias(4)	Survivors of first MI: persons 30–64 yrs TC $\geq$ 220 mg/dL or LDL $\geq$ 140 mg/dL (if TC 200–219 mg/dL)	9.7	Partial ileal bypass	252	22	35

[a] Adapted from: Lipid Research Clinics Program. The Lipid Research Clinics Coronary Primary Prevention Trial Results. II. The relationship of coronary heart disease to cholesterol lowering. JAMA 1984;251:365–374.
[b] See Suggested Readings for complete references.
[c] Difference between intervention group and control group events, expressed as percentage of control group.
[d] MI = Myocardial Infarction

vealed more cholecystectomies and deaths from all causes, including a 38% greater incidence of cancer-related deaths in the clofibrate-treated group than in the control group. The committee of principal investigators observed that the excess mortality in the clofibrate-treated group did not continue after the end of treatment but did not provide a reasonable explanation for the excess mortality that occurred during active treatment. Although a drug effect cannot be ruled out, the excess deaths had no apparent association with the extent of reduction in total cholesterol or with the duration of treatment with clofibrate, but were caused by various non-CHD problems, and appeared to be caused by random sampling variations. Despite this, the investigators concluded that clofibrate could not be recommended as a safe and effective agent for lowering cholesterol in the general population.

With the availability of other lipid-lowering drugs, a new enthusiasm for treatment of lipid disorders emerged. Two randomized controlled primary prevention trials clearly have demonstrated a reduction in cardiovascular morbidity and mortality by combined diet and drug therapy for elevated lipids. The first of these was the Coronary Primary Prevention Trial (CPPT) conducted by the Lipid Research Clinics (LRC), which used the bile acid resin cholestyramine to lower cholesterol. In 1984 this study reported the effect of cholesterol reduction on coronary disease events among 3806 otherwise healthy men who had cholesterol levels above 265 mg/dL (6.85 mmol/L) and LDL-C above 190 mg/dL (4.91 mmol/L). The men were placed on a diet calculated to achieve about a 5% reduction in blood cholesterol and then were assigned randomly to receive cholestyramine at a dose of up to 24 g per day or a matched placebo. Total and LDL-C concentrations in the cholestyramine group were reduced by 8.5% and 12.6%, respectively (Fig. 1.8), whereas HDL-C and triglyceride levels were slightly higher in the treated group throughout the follow-up period.

No appreciable difference in fatal and nonfatal events was evident between treatment groups until nearly 5 years after initiation of the study. During the initial 2 years, the endpoint rates were actually higher in the treated group (Fig. 1.9). With little more than 7 years follow-up, the study was halted because of a 19% reduction in fatal and nonfatal MIs in the treatment group. The cholestyramine-treated group also sustained fewer other coronary events, classified as angina pectoris (20% less), new exercise electrocardiographic (ECG) abnormalities (25% less), and coronary artery bypass surgery (21% less). This lesser risk of CHD events with cholestyramine was directly attributed to the reductions achieved in total and LDL-C concentrations. Postrandomization analyses revealed that the reduction in CHD incidence among cholestyramine-treated men was greater in those whose initial HDL-C levels exceeded 50 mg/dL (1.29 mmol/L), relative to those with pretreatment levels below 40 mg/dL (1.03 mmol/L). No similar relationship was noted for LDL-C and CHD. Because of a greater number of accidental deaths in the cholestyramine group, all-cause mortality was only slightly less than in the placebo group. Yet, had follow-up continued, the magnitude of the difference in mortality may have been greater because the curves were diverging at the time the study was discontinued.

The second of these major clinical trials, the Helsinki Heart Study, used a different drug, a fibric acid derivative, gemfibrozil. A total of 4081 middle-aged men with various types of dyslipidemia were followed for 5 years in a randomized, double-blind placebo controlled trial. At entry, these men had a non-HDL-C (i.e., LDL-C plus VLDL-C) concentration of > 200 mg/dL (5.17 mmol/L). Treatment with 1.2 g (600 mg bid) daily of gemfibrozil decreased total LDL-C by about 8%, triglyceride by 35%, and increased HDL-C by 10%. During the initial 2 years, as with the CPPT, no advantage was noted in the drug group; but, thereafter, CHD decreased in the gemfibrozil-treated group compared with the control group. At the end

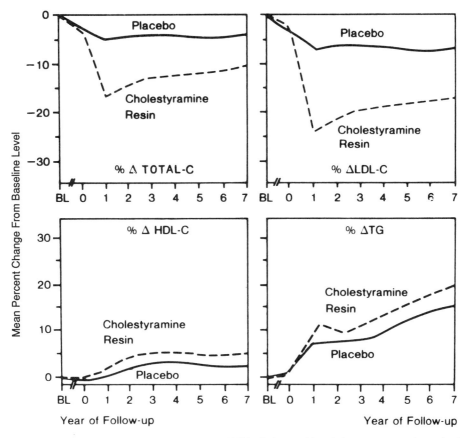

Figure 1.8. Mean percent change in total, LDL, and HDL cholesterol levels by treatment of men in the LRC-CPPT throughout the study period. The treatment differential gradually diminished to 7% in total cholesterol and 10% in LDL-C by the seventh year. The HDL-C levels were consistently approx. 3% higher and triglyceride values were approx. 5% higher in the cholestyramine group. Delta (Δ) indicates change from baseline level; TOTAL-C, plasma total cholesterol levels; LDL-C, low density lipoprotein cholesterol; TG, triglyceride levels; and HDL-C, high density lipoprotein cholesterol. BL represents the prediet period at baseline before initiation of diet and study medication. (Adapted from: The Lipid Research Clinics Coronary Primary Prevention Trial results: The relationship of reduction in incidence of coronary heart disease of cholesterol lowering. JAMA 1984; 251:365–374.)

of the study, nonfatal MI was reduced by 37% and CHD mortality by 26% in the treated group, although all-cause mortality was unchanged. Changes in LDL-C and HDL-C were associated independently and significantly with the reduction in CHD, whereas changes in triglyceride levels were not. Reduction in definite CHD incidence was greatest in those persons who had Frederickson type IIb dyslipidemia (elevated cholesterol and triglycerides) and least in persons who had type IIa dyslipidemia (elevated cholesterol with normal triglycerides). Table 1.4

gives the estimated reduction in CHD related to these lipid changes.

When the major primary prevention clinical trials are analyzed collectively, the results indicate a 2% reduction in CHD risk resulting from every 1% reduction in cholesterol. This relationship is surprisingly similar to that predicted from the results of earlier prospective epidemiologic studies (Fig. 1.10). Based on the lowering of LDL-C alone, the overall reduction in CHD risk in the Helsinki Heart Study should have been approx. 15%. The 34% reduction achieved emphasizes the

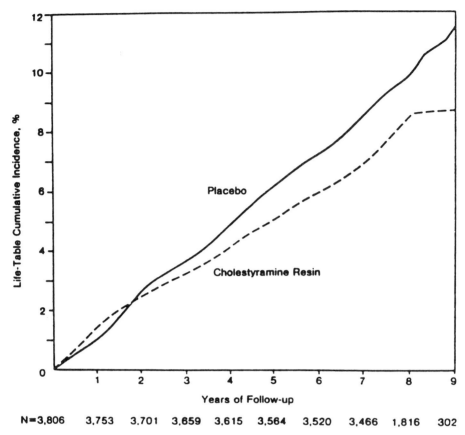

N=3,806   3,753   3,701   3,659   3,615   3,564   3,520   3,466   1,816   302

Figure 1.9. Cumulative incidence of primary endpoint (definitive CHD death or definite nonfatal myocardial infarction) by treatment groups in the Lipid Research Clinics Coronary Primary Prevention Trial (LRC-CPPT). The difference in cumulative incidence between treatment groups widens with time. (From: The Lipid Research Clinics Coronary Primary Prevention Trial results: Reduction in incidence of coronary heart disease. JAMA 1984;251:351–364.)

Table 1.4 Estimated Reduction in Risk of Coronary Heart Disease Attributable to Change in Lipid Level Predicted From the Helsinki Heart Study[a]

Lipid Model	Percent Change in Lipids from Baseline[b]	Percent Estimated Risk Reduction
Single Response		
HDL-C[c]	+8	23
LDL-C[c]	−6	12
Triglycerides (TG)	−24	12
Joint Response		
HDL-C + LDL-C	—	24
HDL-C + LDL-C + TG	—	26

[a] Adapted From: Manninen V, Elo MO, Frick MH, et al. Lipid alterations and decline in the incidence of coronary heart disease in the Helsinki Heart Study. JAMA 1988;260:641–651.
[b] Percent change from second screening to 2-year mean value for the gemfibrozil group minus the placebo group.
[c] HDL-C = high density lipoprotein cholesterol; LDL-C = low density lipoprotein cholesterol.

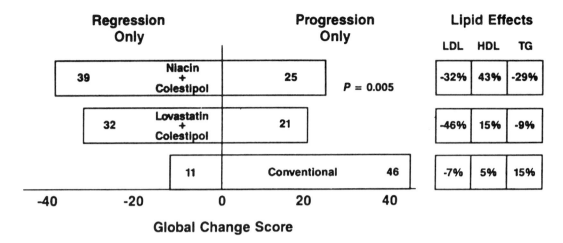

Note: Numbers in horizontal bars on graph represent % of patients.

Figure 1.10. Definite lesion changes in the Familial Atherosclerosis Treatment Study (FATS) according to the three treatment groups. The numbers in the squares on the right represent the lipid and lipoprotein changes in each of the corresponding treatment groups. (From: Coronary heart disease seminar series: lipid regulation and risk prevention. Parke-Davis, Division of Warner-Lambert Company, 1991.)

additional benefit that may have been derived from the increase in HDL-C. As a consequence, the Helsinki Heart Study focused attention on the possible role of HDL in the prevention of CHD. Furthermore, analyses of the gemfibrozil group indicated that increments in HDL-C were the single best predictor of a favorable treatment outcome; the lipid parameters most associated with reduction in CHD risk, in order of significance, were HDL-C, LDL-C, and non-HDL-C (LDL-C and VLDL-C). Reductions in plasma triglyceride levels also related to diminished coronary risk, but this association was not statistically significant.

Secondary Prevention. Until recently the treatment of dyslipidemia in persons who had established CHD was dismissed on the grounds that the process of atherosclerosis was too far advanced and unlikely to be reversible. Several important secondary prevention clinical trials have served to revert this opinion. (see Table 1.3).

The Coronary Drug Project (CDP), conducted between 1966 and 1975, directly evaluated the efficacy of several lipid-lowering drugs in 8341 men aged 30 to 64

years who had a known previous MI. The most productive of these agents proved to be nicotinic acid (niacin); other agents such as estrogens and thyroxine eventually were discarded because of adverse reactions, whereas clofibrate was discontinued because of a lack of effectiveness. Treatment of 3908 men with 3.0 g of nicotinic acid daily diminished nonfatal recurrent MI by 27%. Although the coronary mortality was unchanged after 6 years of treatment, subsequent follow-up of these patients 9 years poststudy revealed additional protection. In the group that originally had received nicotinic acid, total mortality was 11% lower and CHD mortality 12% lower than in the group who had received placebo. These differences in mortality occurred even though no attempt was made to continue drug therapy beyond the initial six-year period. The authors concluded that the 10% reduction in serum cholesterol maintained over the initial 5 to 8 years had slowed progression of coronary atherosclerosis significantly in these patients.

The Stockholm Secondary Prevention Study evaluated the effects of combined therapy with niacin and clofibrate in an open, randomized, placebo-controlled trial of 555 male and female survivors of MI. Fifty percent of the patients had hypertriglyceridemia at baseline, compared to only 13% with hypercholesterolemia. After 5 years of treatment, the serum cholesterol and triglyceride concentrations were reduced 13% and 19%, respectively; this was associated with a 26% reduction in overall mortality in the treatment group and a 36% reduction in CHD mortality. The difference in mortality also was observed for patients above 60 years of age, suggesting that age and preexisting CHD did not preclude successful intervention. In contrast to the results of the Helsinki Study, the benefit was related primarily to a reduction in serum triglyceride concentrations. The reduction in mortality occurred only in patients who had a triglyceride level > 133 mg/dL (1.50 mmol/L), and it was most pronounced in those patients who had a 30% or greater reduction in serum triglyceride concentration. Although it is conceivable that an increase in HDL-C accompanied the reduction in triglyceride, HDL levels were not measured. Reduction of CHD was not related to the degree of cholesterol lowering.

Finally, the fact that the method of lipid-lowering itself is not the major determinant of benefit was demonstrated by the Program On the Surgical Control of Hyperlipidemias (POSCH). This randomized clinical trial determined whether cholesterol lowering by partial ileal bypass surgery would reduce subsequent CHD events among survivors of a first MI. The study population consisted of 838 subjects (women comprised approximately 15% of the group) with an average age of 51 years. Participants were enrolled into the study if they had a total cholesterol of at least 220 mg/dL (5.69 mmol/L) or an LDL-C level of at least 140 mg/dL (3.62 mmol/L), provided that the total plasma cholesterol level was between 200 and 219 mg/dL (5.17 and 5.66 mmol/L) after following a diet for 6 weeks. After a mean follow-up period of 5 years, the surgery group had a total cholesterol level 23% lower, LDL-C 38% lower, and HDL-C 4% higher than the control group. Angiographic evaluation revealed less disease progression in the surgery group at 3, 5, 7, and 10 years. Mortality caused by CHD and confirmed nonfatal MI were reduced collectively by 35%, and the number of patients who required coronary artery bypass grafting or coronary angioplasty was reduced by 62% and 55%, respectively. The number of patients who developed peripheral vascular disease also was reduced by 27%. Although overall mortality in the entire surgical group decreased by 21%, the difference in overall mortality reached statistical significance only in the subgroup of patients who had an ejection fraction $\geq$ 50% (reduced by 36%). These data strongly confirm the benefits of lowering lipid levels among CHD patients who have hypercholesterolemia and raise the possibility of surgery for very high-risk persons who cannot otherwise lower their LDL-C.

Angiographic. The earliest indications that regression of atherosclerotic lesions is possible came from postmortem studies of individuals with catabolic states brought about by wartime malnutrition or chronic wasting diseases. Subsequent animal models provided additional insight into the possibilities of regression.

Hypercholesterolemia is a necessary prerequisite for producing experimental atherosclerosis in various animal species. When this induced hypercholesterolemia is continued over a period of years, some animals develop lesions comparable to those seen in advanced human atherosclerosis. In addition, the relationship of specific blood lipoproteins to animal atherosclerosis has been found to be similar to that of humans. Advanced lesions contain up to 60% of dry weight as cholesterol and phospholipid. Studies have demonstrated that reduction of the elevated blood cholesterol by any means leads to regression of experimental atherosclerosis. A sustained cholesterol lowering to less than 160 mg/dL (4.14 mmol/L) for 12 to 18 months is required before lesions regress in nonhuman primates.

Although cholesterol in human atherosclerotic plaques is in equilibrium with the blood cholesterol, its turnover time is substantially lower (by 1.5 to 3.0 years) than that in most other tissues, including xanthomas of hypercholesterolemic patients. Furthermore, advanced plaques have twice the relative amount of free cholesterol as xanthomas. Because free cholesterol tends to accumulate deeper in the lesion and to crystallize, it is more difficult to mobilize. With the exception of crystalline cholesterol and some cholesterol trapped by altered connective tissue, a constant exchange occurs between tissue and blood cholesterol. Reduction of plaque cholesterol, therefore, should occur when blood cholesterol levels are reduced.

Various nonlipid factors also affect arterial regression. Some of these include flow dynamics, cell membrane receptors, clotting factors, immunologic responses, alteration in proteoglycans, collagen, and arterial permeability. Various factors may influence the local response to atherosclerotic forces, which would manifest itself as reduction of lesions in one coronary segment and unchanged or progressive lesions in another. The growth of atherosclerotic lesions probably is episodic in nature and may be punctuated by periods of relative stability or regression. These factors complicate the assessment of the value of lipid modifications on regression of plaques and necessitate the use of sophisticated techniques. The slow nature of regression, however, should not diminish the value of treating persons who have clinically documented CHD. Even partial regression could improve blood flow because stenosis does not become critical until approx. 75% of the arterial lumen is obstructed. The secondary prevention studies clearly demonstrate the potential for reduction of clinical events and alteration of arterial lesions. Such considerations suggest that even small reductions in growth rate may delay significantly MI and death from CHD.

Furthermore, older persons theoretically could experience greater initial benefit because of the impact on near critical but not yet symptomatic lesions.

Animal regression studies, especially those conducted on nonhuman primates, lend support to the concept that atherosclerosis can be retarded and that lesions conceivably may be reversed with adequate lipid-lowering treatment. Because human lesions cannot be biopsied periodically, arteriography provides the means for studying atherosclerosis within these lesions. Among the earliest human regression studies was a National Heart, Lung, and Blood Institute (NHLBI) intramural study that assessed the rate of angiographic progression and regression in 117 patients who had CHD. The participants were placed on a diet low in cholesterol and high in polyunsaturated to saturated fat ratio ($\geq$ 1.5) and then were randomized in a double-blind protocol to either cholestyramine or placebo. No differences in progression or regression of lesions could be identified between treatment groups at 5 years. A significant inverse correlation was noted, however, between progression and the combined changes in LDL-C and HDL-C in the placebo and cholestyramine-treated groups alike. Patients who had the greatest decrease in LDL-C concentration and the greatest increase in HDL-C concentration after 5 years had about one-third the amount of coronary disease progression as the group with the least change. The HDL-C:LDL-C ratio was the best predictor of lesion change.

An uncontrolled angiographic study conducted in Leiden, Holland from 1978 to 1982 also revealed that a vegetarian diet slowed progression of coronary lesions among 39 patients who had stable angina pectoris. This study also found positive correlations between lesion growth and the HDL-C to total cholesterol or HDL-C to LDL-C ratios. Growth of lesions progressed over a two-year period in patients whose total cholesterol to HDL-C ratio exceeded the median (6.9) but did not progress in those whose ratios were lower than the median. Those whose initial values were reduced significantly by the diet exhibited no progression.

More recently, several important regression trials have been completed (Table 1.5). The Cholesterol Lowering Atherosclerosis Study (CLAS) in Los Angeles was a randomized, placebo controlled trial of drugs and diet designed to impact sufficiently on lipids to bring about regression of atherosclerotic lesions, as assessed by pretreatment and posttreatment angiograms. The drug regimen consisted of combinations of colestipol (a bile acid resin) and niacin in flexible doses. Among nonsmoking men aged 40 to 59 years who had previous coronary bypass surgery, 182 showed a hypocholesterolemic response to a preadmission trial of drug treatment and were assigned randomly to treatment with drugs plus diet or placebo plus diet. The strict treatment regimen produced a 26% reduction in total cholesterol to a mean value of 180 mg/dL (4.65 mmol/L) and a 43% reduction in LDL-C to a mean of 97 mg/dL (2.51 mmol/L); HDL-C was elevated 37% to 61 mg/dL (1.58 mmol/L). Associated with these impressive lipid changes was significantly less progression of lesions in the coronary artery bypass grafts and in the native coronary circulation after 2 years of therapy. Drug treatment also led to an increase in the number of lesions that regressed (16% compared to 2.4%). Extended follow-up to 4 years showed more marked progression of lesions in the placebo group, whereas the drug group remained relatively stable. Perceptible overall improvement occurred in 17.9% of the drug-treated group but in only 6.4% of patients on placebo during the final follow-up period. The treatment benefits persisted even for patients whose entry level cholesterol was below 240 mg/dL (6.21 mmol/L).

A similar drug regimen was used in another trial in San Francisco. Seventy-two patients who had heterozygous familial hypercholesterolemia, characterized by LDL-C > 200 mg/dL (5.17 mmol/L) and a total triglyceride < 275 mg/dL (3.10 mmol/L) received treatment with a low fat, low cholesterol diet and then were randomized

Table 1.5 Regression of Coronary Atherosclerosis—Recent Angiographic Follow-Up Trials

Study[a]	Population	Duration of Followup	Treatment	Findings
Familial Atherosclerosis Treatment Study(3)	Men < 62 yrs apoprotein B ≥ 125 mg/dL documented CHD and family history of vascular disease	30 months	Lovastatin + colestipol or Niacin + colestipol	Nonprogression in 75–79% drug-treated vs. 54% in placebo-treated; regression in 32–39% of drug-treated vs. 11% of placebo-treated.
Cholesterol Lowering Atherosclerosis Study(7)	Nonsmoking men 40–59 yrs had CABG,[b] TC[c] 185–350 mg/dL	48 months	Colestipol + niacin	Nonprogression in 52% drug-treated vs. 15% placebo-treated; regression 18% drug-treated vs. 6% placebo-treated.
Regression of Coronary Artery Disease(14)	Heterozygous familial hypercholesterolemic patients 19–72 yrs, LDL–C ≥ 200 mg/dL and TGY[d] < 275	26 months	Colestipol + niacin	Mean change in percent area stenosis was −1.53 for drug treated and +0.80 for controls.

[a]See Suggested Readings for complete references.
[b]CABG = Coronary Artery Bypass Grafting
[c]TC = Total Cholesterol
[d]TG = Triglyceride

into a control group of 32 patients (treated with diet alone or diet plus low-dose colestipol) and a treatment group of 40 patients treated with niacin and colestipol. When lovastatin became available for clinical use, it was added to the drug regimen of 16 of the patients in the drug-treatment group. After a mean follow-up of 26 months, the lipoprotein changes in the drug-treatment group were −38% for LDL-C, −19% for triglyceride and +28% for HDL-C; the corresponding changes in the control group were −11% for LDL-C and insignificant for triglyceride and HDL-C. This resulted in a significant change in percent stenosis of the coronary arteries. This trial of persons aged 19-72 years corroborated previous studies even more and was the first to include women. Subgroup analyses showed that the women treated actually fared better than the men treated.

Another study conducted in Seattle the Familial Atherosclerosis Treatment Study (FATS), identified 120 men at high risk for CHD and studied them by performing quantitative coronary angiography at entry and 2-1/2 years after treatment for elevated LDL-C levels. All patients were placed on a cholesterol-lowering diet and then assigned randomly to one of three treatment groups: placebo (with the addition of colestipol if the LDL-C remained elevated); lovastatin plus colestipol; or niacin plus colestipol. Both of the groups that received combined drug therapy had a significant reduction in the frequency of lesion progression, as well as an increase in lesion regression (Fig. 1.10). Reduction in apo B (or LDL-C), decreased systolic blood pressure, and increased HDL-C correlated independently with lesion regression. The combined drug group also had significantly fewer clinical CHD events (revascularization procedures, MI or death).

More recently, Blankenhorn and co-investigators demonstrated regression of atherosclerotic lesions by means of coronary angiography, using a single hypolipidemic drug, lovastatin. The 2-year follow-up demonstrated differential improvement in arteriographic lesions in the treated group, as compared with the placebo group. The study was stopped at this point by the Safety and Monitoring Committee so that the placebo group might benefit from treatment.

The studies already outlined suggest that regression of coronary artery lesions is possible by reducing total cholesterol to levels approximating 150 mg/dL (3.88 mmol/L) equivalent to an LDL-C of 100 mg/dL (2.59 mmol/L) or less for at least several years. Indeed, correlative pathologic-chemical studies in animals show that lipids in atherosclerotic plaques are stratified, with recently deposited cholesterol esters present in the luminal part of the intima and older deposits layered in the deeper regions of the vessel wall. The possibility for reversal of lesions most likely is determined by duration and extent of lesions, degree of changes in lipoprotein levels, and other pertinent patient factors, such as blood pressure level, smoking history, glucose tolerance, and physical activity. When the blood cholesterol is lowered sufficiently, cholesterol appears to be mobilized from the lesions, and regression begins. Evidence of plaque regression includes a decrease in intracellular lipids, disappearance of foam cells, and normalization of cellular patterns. The intimal cells decrease in number and no longer accumulate lipids.

An important factor that determines the potential for regression of atherosclerotic plaque is the extent of fibrosis within the lesion. If regression takes place, however, a reshaping of the plaque to a smaller size occurs with time, much in the same way that old dermal scars often diminish after some time. Functional recovery of the endothelium occurs as dilator responses recover and the exaggerated vasoconstrictor responses are lessened. Studies of hypercholesterolemic patients have demonstrated, however, that regression may be markedly different among different vascular beds, perhaps accounting for the variability in angiographic improvement.

Convincing evidence shows that regression can occur as a result of lipoprotein modifications. Retardation of atherosclerosis probably is a more realistic goal than regression for most patients who have atherosclerosis. The degree to which arrest of lesion growth and regression take place in an individual patient depends on the amount of lipid modification and coexistence of other contributory factors. Nevertheless, the concomitant functional improvements in the endothelium, accompanied by a less thrombogenic vascular interface, reduce the likelihood of future clinical sequelae.

Lipid Modification and Mortality. The collective data provided by these trials strongly suggest that lowering total cholesterol and LDL-C with bile-acid resins, nicotinic acid, lovastatin, or gemfibrozil reduces CHD events. No trial has been conducted yet to test directly the benefits of raising HDL-C, and the value of such an intervention currently rests on indirect, although substantial, evidence. Furthermore, additional data are needed to define the relationship of cholesterol and CHD in various unstudied segments of our society, such as women, the elderly, the young, and those at high risk for CHD by virtue of other factors that predispose to CHD progression.

Despite the lack of a significant reduction in total mortality for most studies, these clinical trials provide compelling evidence to show that lowering blood cholesterol reduces a major component of total mortality, CHD. To demonstrate an impact on overall mortality, studies need to be performed on a larger sample size and should be of longer duration, preferably starting in young adulthood. Although only several years are needed to demonstrate stabilization of lesions, and even regression, at least 5 years are needed to show reduction in coronary morbidity and mortality. It should be noted that the CDP required fifteen years of follow-up before the change in total mortality became evident. With the current more rigorous intervention, then, the effect on total mortality may yet become apparent.

Another possibility is that an excess in noncardiovascular mortality compensated for the decrease in CVD mortality that was caused by cholesterol lowering. Some studies noted an increased prevalence of cancer among patients who had blood cholesterol levels below 180 mg/dL (4.65 mmol/L), whereas others have found more accidental or violent deaths in lipid-lowering trials. After analyzing the dropouts from these trials and taking account of known risk factors for these types of deaths (such as a history of alcoholism or psychiatric problems), no evidence was found to support the concept that cholesterol-lowering drugs are associated casually with deaths attributed to homicides, suicides, and trauma.

Some studies suggest that mortality rises again at the lowest cholesterol values, implying a J-shaped relation between cholesterol levels and total mortality. It is highly unlikely, however, that the relatively small decreases in total cholesterol achieved in such trials can account for a noticeable increment in cancer deaths, even if there were such an association. When data from all lipid-lowering trials are combined, no significant differences exist in the incidence of neoplasms or all-cause mortality between groups treated with diet and drugs compared to controls. The most extensive database on these relationships originates from the international collaborative group study on blood cholesterol and cancer—a large population of middle-aged men with over ten years follow-up. It is more likely that patients who had occult cancer also had lower blood cholesterol levels to begin with, rather than the presumption that low blood cholesterol predisposes to the development of neoplasms. A faster catabolism of LDL has been demonstrated in patients who have metastatic prostatic cancer. Although no evidence exists that such a mechanism operates in vivo, it is interesting to note that mitogenic substances such as platelet-derived growth factor stimulate LDL receptor expression in cultured cells. Moreover, in Japan and among groups in whom dietary fat intake is restricted, both the blood cholesterol and the incidence of colon cancer

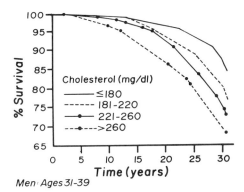

Figure 1.11. Long-term survival by initial cholesterol value in the Framingham Heart Study. A clear trend exists between cholesterol concentration and probability of survival, especially between the highest and lowest cholesterol quartiles. (From: Gotto AM. Diet and cholesterol and coronary heart disease. JACC 1989;13:503–507.)
"Reprinted with permission from the American College of Cardiology (Journal of the American College of Cardiology, 1989;13:503–507)."

are low. Actually, the bulk of evidence points to a positive correlation between colorectal cancer and high dietary fat intake.

Data from Framingham and other epidemiologic studies provide evidence that cholesterol reduction in persons who have "normal" cholesterol concentrations increases longevity (Fig. 1.11). For those persons in the Framingham study below the age of 50 years, there was a direct relationship between blood cholesterol levels and 36-year mortality from all causes and cardiovascular disease. The overwhelming sets of evidence from epidemiologic studies and clinical trials indicate that lowering of cholesterol, when not attributed to disease processes, provides a survival advantage.

Suggested Readings

REVIEWS

Armstrong ML, Heistad DD, Megan MB, Lopez JA, Harrison DG. Reversibility of atherosclerosis. In: Brest AN, ed. Preventive aspects of coronary heart disease/cardiovascular clinics. Philadelphia: F.A. Davis Company, 1990.

Durrington PN. Hyperlipidaemia: diagnosis and management. 1st ed. London: Wright Publishing, 1989.

National Research Council (U.S.). Diet and health - implications for reducing chronic disease risk/ Committee on Diet and Health, Food and Nutrition

Board, Commission on Life Sciences, National Research Council. Washington, DC: National Academy Press, 1989.

Rossouw JE, Rifkind BM. Does lowering serum cholesterol levels lower coronary heart disease risk? Endocrinol Metab Clin North Am 1990;19:279–297.

Smith TH. A chronology of atherosclerosis. Amer J Pharm 1960;132:390–405.

St. Clair RW. Atherosclerosis regression in animal models: current concepts of cellular and biochemical mechanisms. Prog Cardiovasc Dis 1983;XXVI: 109–132.

Stamler J, Shekelle M. Dietary cholesterol and human coronary heart disease—the epidemiologic evidence. Arch Pathol Lab Med 1988;112:1032–1040.

Steinberg D, Parthasarathy S, Carew TE, Khoo JC, Witztum JL. Beyond cholesterol: modifications of low-density lipoprotein that increase its atherogenicity. N Eng J Med 1989;320(14):915–924.

SPECIFIC STUDIES

Blankenhorn DH, Nessim SA, Johnson RL, Sanmarco ME, Azen SP, Cashin-Hemphill L. Beneficial effects of combined colestipol-niacin therapy on coronary atherosclerosis and coronary venous bypass grafts. JAMA 1987;257(23):3233–3240.

Brown G, Albers JJ, Fisher LD, et al. Regression of coronary artery disease as a result of intensive lipid-lowering therapy in men with high levels of apolipoprotein B. N Engl J Med 1990;323: 1289–1298.

Buchwald H, Vargo RL, Matts JP, et al. Effect of partial ileal bypass surgery on mortality and morbidity from coronary heart disease in patients with hypercholesterolemia: Report of the Program on the Surgical Control of the Hyperlipidemias (POSCH). N Engl J Med 1990;323:946–955.

Canner PL, Berge KG, Wenger NK, et al. Fifteen year mortality in Coronary Drug Project patients: long-term benefit with niacin. J Am Coll Cardiol 1986;8:1245–1255.

Carlson LA, Rosenhamer G. Reduction of mortality in the Stockholm Ischaemic Heart Disease Secondary Prevention Study by combined treatment with clofibrate and nicotinic acid. Acta Med Scand 1988;233:405–418.

Cashin-Hemphill L, Mack WJ, Pogoda JM, et al. Beneficial effect of cholesterol-niacin on coronary atherosclerosis: a 4-year follow-up. JAMA 1990;264: 3013–3017.

Committee of Principal Investigators. WHO Cooperative Trial on primary prevention of ischaemic heart disease with clofibrate to lower serum cholesterol: final mortality follow-up. Lancet 1984;2:600–604.

Coronary Drug Project Research Group. Clofibrate and niacin in coronary heart disease. JAMA 1975;231: 360–381.

Dayton S, Pearce ML, Hashimoto S, Dixon WJ, Tomiyasu U. A controlled clinical trial of a diet high in unsaturated fat in preventing complication of atherosclerosis. Circulation 1969;40(Suppl II):1–163.

Frantz ID, Dawson EA, Ashman PL, et al. Test effect of lipid lowering by diet on cardiovascular risk. The Minnesota Coronary Survey. Arteriosclerosis 1989; 9:129–135.

Frick MH, Elo O, Haapa K, et al. Helsinki Heart Study: primary prevention trial with gemfibrozil in middle-aged men with dyslipidemia: safety of treatment, changes in risk factors, and incidence of coronary heart disease. N Engl J Med 1987;317: 1237–1245.

Hjermann I, Velve Byre K, Holme I, Leren P. Effect of diet and smoking intervention on the incidence of coronary heart disease: report from the Oslo Study Group of a randomized trial in healthy men. Lancet 1981;2:1303–1310.

Kane JP, Malloy MJ, Ports TA, et al. Regression of coronary atherosclerosis during treatment of familial hypercholesterolemia with combined drug regimens. JAMA 1990;264:3007–3012.

Keys A. Coronary heart disease: the global picture [Review]. Atherosclerosis 1975;22:149–192.

Lipid Research Clinics Program. The Lipid Research Clinics Coronary Primary Prevention Trial results. I. Reduction in the incidence of coronary heart disease. JAMA 1984;251:351–364.

Manninen V, Elo MO, Frick MH, et al. Lipid alterations and decline in the incidence of coronary heart disease in the Helsinki Heart Study. JAMA 1988;260: 641–651.

Multiple Risk Factor Intervention Trial Research Group. Multiple risk factor intervention trial: risk factor changes and mortality results. JAMA 1982;248:1465–1477.

Multiple Risk Factor Intervention Trial Research Group. Mortality rates after 10.5 years for participants in the Multiple Risk Factor Intervention Trial. JAMA 1990;263:1795–1801.

Report from the Committee of Principal Investigators. A co-operative trial in the primary prevention of ischaemic heart disease using clofibrate. Br Heart J;40:1069–1118.

ATHEROSCLEROSIS

Ip JH, Fuster V, Badimon L, Badimon J, Taubman MB, Chesebro JH. Syndromes of accelerated atherosclerosis: role of vascular injury and smooth muscle cell proliferation. J Am Coll Cardiol 1990;15: 1667–1687.

Ross R, Glomset JA. The pathogenesis of atherosclerosis (second of two parts). N Engl J Med 1986; 314(8):488–500.

Steinberg D, Witztum JL. Lipoproteins and atherogenesis. JAMA 1990;264:3047–3052.

CHAPTER 2

Cardiovascular Risk Assessment

Rationale for Diagnosis and Treatment

Despite dramatic changes in the frequency and distribution of heart disease over the past 20 years, cardiovascular disease (CVD) remains the leading cause of premature death and disability in the United States, being responsible for nearly 50% all deaths in 1987. Among the current United States population, nearly 68 million (approximately one in four) have one form or another of CVD. One out of every three men and one out of every nine women can be expected to develop CVD before the age of 60. This year alone as many as 1.5 million Americans will sustain an MI and more than 500,000 will die, almost one-half before they reach a hospital. The majority (75%) of these CVD deaths occurred in atherosclerotic-related diseases, primarily CHD. Half of the deaths attributed to CHD were caused by acute MI. As a result, almost 20% of hospital days and 10% of physician office visits for diagnosis or treatment of chronic conditions relate to cardiovascular disorders.

The estimated prevalence of CHD for those persons 20 years and older is given in Figure 2.1. The prevalence for men reaches a peak between 55 and 64 years (and a decade later for women). The total cost for CVD in 1990 was an estimated 101.3 billion dollars, of which CHD accounted for 44.5 billion

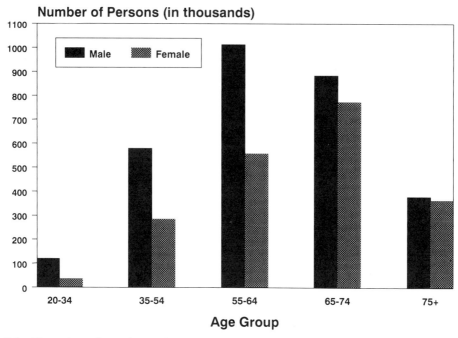

Figure 2.1.   The estimated prevalence of coronary heart disease for persons 20 years of age and older. These data were obtained from multiple sources—National Center for Health Statistics, U.S. Public Health Service, and the American Heart Association—using the 1987 estimates. There is a clear increment in prevalence with age. The prevalence peaks for men in the 55–64 age group and for women in the 65–74 age group. (Adapted from: National Center for Health Statistics, U.S. Public Health Service, DHHS, and American Heart Association.)

dollars (Fig. 2.2). Expenditures for total CVD consisted of 64.7 billion dollars for hospital and nursing home services, 14.9 billion dollars for physician and nursing services, 5.4 billion dollars for medication, and 16.2 billion dollars for lost productivity. Based on the results of epidemiologic and clinical studies (see Chapter 1), even a 10% reduction in serum cholesterol levels through dietary and other nonpharmacologic means ought to reduce CHD rates by about 20%. The resultant annual savings in costs of nearly $9 billion should offset to a great degree the 5 to 8 billion dollar costs associated with the doctor visits and laboratory tests that are required for the dietary management of dyslipidemia

Coronary heart disease mortality rates in the United States cannot be considered uniform, because substantial regional differences exist. As noted in Figure 2.3, the highest rates occurred in the Eastern United States, with the exception of Alabama and

Florida. The Mountain States exhibited the lowest rates overall, although those for the West, Southwest, and Midwest were considerably lower than the Eastern third of the country. Despite its wide prevalence, there has been a continuing decline in CHD mortality—44% from 1968 through 1986 (Fig. 2.4). For the most recent period, 1976 through 1986, the overall age-adjusted death rate for major CVD in the United States declined almost 25%, in contrast to little change for non-CVD. Although the decrease in CHD mortality has involved all races, both sexes, and all ages, the most striking improvements have been observed in the younger and in the higher socioeconomic subgroups. The decline was characterized by fewer out-of-hospital deaths, sudden unexpected deaths, and acute nonfatal MIs. Concurrently, postmortem studies of atherosclerotic lesions have noted less severe atherosclerosis at comparable ages and sex, but there is no satisfactory way to link

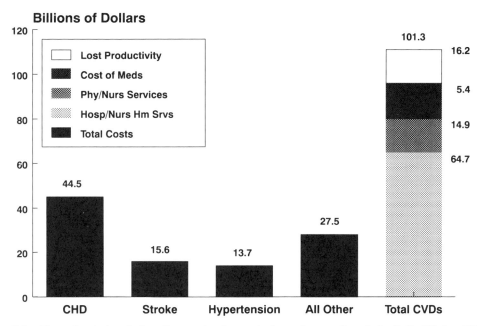

Figure 2.2. The estimated cost of cardiovascular disease by type of expenditure in the United States: 1991 estimate. Overall, cardiovascular diseases account for an estimated 101.3 billion dollars in 1991, with the major cost coming from CHD. With the exception of hypertension, most of the costs of cardiovascular disease are related to hospital and nursing home services. (Adapted from: American Heart Association extrapolation from Hodgson TA, Kopstein AN. Health Care Expenditures for Major Diseases in 1980. Health Care Financing Review 1984:5:4. In: American Heart Association 1991 Heart and Stroke Facts, No. 55–0379[COM].)

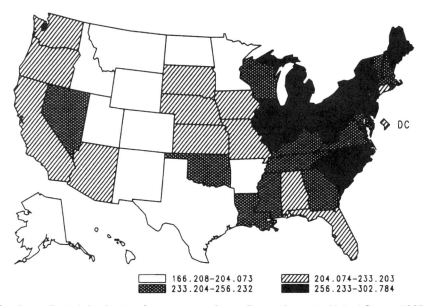

Figure 2.3. Age-adjusted death rates from coronary heart disease by state; United States, 1986. There appears to be a definite geographic trend in mortality rates from coronary heart disease, with the highest rates in the Northeast and the lowest in the Mountain States. (Adapted from: Centers for Disease Control Morbidity and Mortality Weekly Report (MMWR), 1989;38:285.)

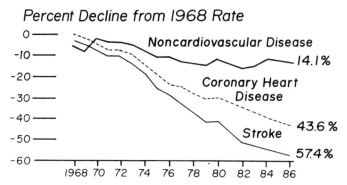

Figure 2.4. The decline in age-adjusted death rates for noncardiovascular disease; coronary heart disease; and stroke in the United States from 1968 to 1986. In contrast to relatively stable rates for noncardiovascular disease, CHD and stroke have declined about 50% since 1968. (Adapted from: The Fifteenth Report of the NHLBI Advisory Council: Progress and Challenge, U.S. Department of Health and Human Services, PHS/NIH Pub. No. 88–2734, September 1988, p. 13.)

these data to the decrease in CHD mortality. Generally, the incidence of a first coronary attack is believed to be declining, and there is improved survival from MI over the short term. The steepest relative decline in CVD mortality has occurred in deaths from cerebrovascular diseases. Yet, because CHD is so much more prevalent, the decrease in CHD has had the greater quantitative impact on the decline. The death rate in the United States continues to fall at the rate of 2% to 3% per year, chiefly as a result of a sustained decline in cardiovascular mortality.

The decline in CHD mortality is not a worldwide phenomenon, and the United States has experienced a more favorable change than most other countries. There are substantial international variations in the prevalence and extent of CHD (Figs. 2.5 and 2.6). In 1985 the highest CHD death rates occurred in Northern Ireland, Scotland, and Finland, whereas the lowest death rates were in Japan and France. A decline in CHD mortality of 10% or more occurred in many countries, and the decrease surpassed 40% in the United States, Australia, Israel, and Canada. Currently, among the English speaking countries, only Canada has a lower CHD mortality rate than the United States. Many industrialized countries, particularly in Eastern Europe, actually experienced an increase in CHD mortality rates between

1970 and 1985. Linking such changes to national cholesterol levels is difficult, but longitudinal data suggest simultaneous changes in cholesterol distribution and CHD mortality. Much of the reduction in mortality in the United States has been attributed to changes in risk factors, particularly dietary changes, increased physical activity, cigarette smoking cessation, and control of hypertension. One review of the possible causes for the decline concluded that almost two-thirds of the "anticipated" deaths were postponed—30% by a reduction in blood cholesterol and an additional 33% from control of hypertension as well as reduction in cigarette smoking.

Most countries that have a low prevalence of CHD also have a low incidence of hypercholesterolemia, even though other risk factors may be common. Conversely, in societies in which the average cholesterol level is below 150 mg/dL (3.88 mmol/L) and levels of low density lipoprotein cholesterol (LDL-C) are low, atherosclerosis and manifestations of CVD remain uncommon. This suggests that an elevated blood cholesterol is necessary for the development of atherosclerosis, even in the presence of other factors that enhance susceptibility to CHD. Although exceptions to this rule are seen occasionally in clinical practice, the occurrence of atherosclerosis in persons who have

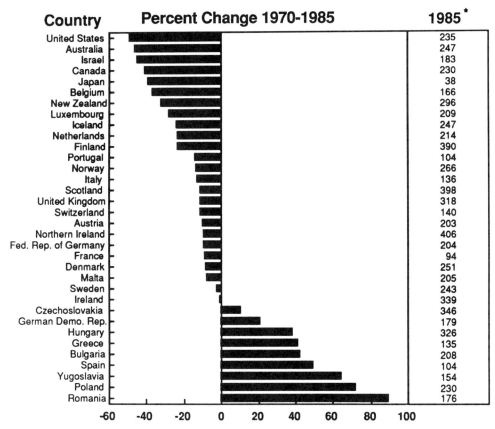

Country	Percent Change 1970-1985	1985 *
United States		235
Australia		247
Israel		183
Canada		230
Japan		38
Belgium		166
New Zealand		296
Luxembourg		209
Iceland		247
Netherlands		214
Finland		390
Portugal		104
Norway		266
Italy		136
Scotland		398
United Kingdom		318
Switzerland		140
Austria		203
Northern Ireland		406
Fed. Rep. of Germany		204
France		94
Denmark		251
Malta		205
Sweden		243
Ireland		339
Czechoslovakia		346
German Demo. Rep.		179
Hungary		326
Greece		135
Bulgaria		208
Spain		104
Yugoslavia		154
Poland		230
Romania		176

-60   -40   -20   0   20   40   60   80   100

Figure 2.5. The percent change in age-adjusted CHD mortality (deaths per 100,000) among 30–69-year-old men from 1970 to 1985.* Age-adjusted mortality from CHD in men for 1985: Although most countries have shown a decline in mortality from CHD, the decrease has not been universal. The largest decline during this time occurred in the United States. (Adapted from: World Health Organization. World Health Statistics Quarterly 1988;41:155–178.)

blood cholesterol levels below 150 mg/dL (3.88 mmol/L) is unusual.

Atherosclerosis is a multifactorial disorder. In the presence of hypercholesterolemia, other risk factors may escalate the likelihood of subsequent CHD. If optimal control of atherosclerosis is ever to be achieved, intervention should be performed on all identified factors rather, than concentration solely on hypercholesterolemia. The majority of CHD incidents occur not in those few individuals who have extremely elevated cholesterol values but, rather in the larger number of persons who have moderately abnormal lipid values in association with other risk factors. This is the most likely reason why some persons remain athero-

sclerosis-free despite an elevated cholesterol level, because they are otherwise free of risk factors.

Theoretical studies indicate that any gain in life expectancy from cholesterol reduction can be measured only in weeks or months. This argument is valid only if increased survival is accepted as the sole measure of health and if no additional benefit from cholesterol reduction occurs after 5 to 7 years of therapy. Life expectancy, of course, is not unlimited. Eradication of all cardiovascular diseases would add only several years to life expectancy at birth, and it is unrealistic to expect that modification of any risk factor over a short time during middle age will have a major effect on life expec-

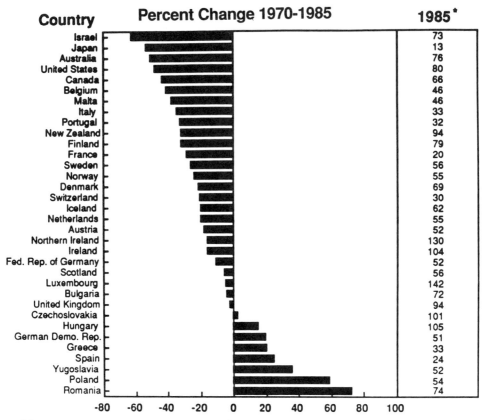

Figure 2.6. The percent change in age-adjusted CHD mortality among 30–69 year old women from 1970 to 1985.* Age-adjusted mortality (deaths per 100,000) from CHD in women for 1985: Even though the CHD mortality rates for 30–69-year-old women are lower than for men, the trends over time are similar to those found in men. The percent change since 1970 is quantitatively the same, and countries appear to be consistent in CHD mortality changes for both men and women. (Adapted from: World Health Organization. World Health Statistics Quarterly, 1988;41:155–178.)

tancy. Added periods of vigorous well-being and prolongation of disease-free life appear to be more rational goals for the health care system, even though they are more difficult to quantify than life expectancy.

The long-term risk for reinfarction and coronary death in survivors of MI also correlates with the blood cholesterol. The Framingham Study demonstrated that in persons who have CHD, an excess risk of about 30% exists for every 50 mg/dL increment in the blood cholesterol level. Among men who have preexisting CVD followed for an average of 10 years in the Lipid Research Clinics (LRC) Prevalence Study, the risk of death from CVD for those with a blood cholesterol >239 mg/dL (6.18 mmol/L) was

3.4 times higher than for those with blood cholesterol levels below 200 mg/dL (5.17 mmol/L). The corresponding risk of death increased exponentially as LDL-C increased or HDL-C decreased (Fig. 2.7). These findings are consistent with those of the Coronary Drug Project and the Stockholm Study. Both of these secondary prevention studies demonstrated that the reduction of lipids reduces the total mortality in persons who have CHD. The probability of a cardiac event is much greater among persons who have known CHD than in those who do not (see Fig. 2.7). Although the contribution of cholesterol to further CHD events (relative risk) is potentially lower in this population, the impact (absolute risk) of even a small re-

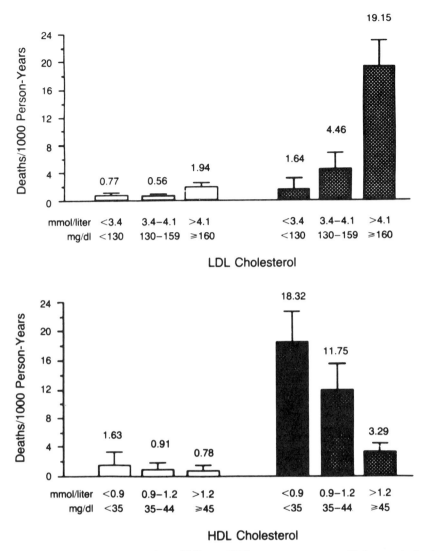

Figure 2.7. Age-adjusted rates of death from CHD per 1000 persons per year of follow-up in the Lipid Research Clinic Prevalence Study. Men who had CVD at baseline (dark bars) had considerably higher rates of death from CHD than those who did not manifest CVD (light bars) at each lipid level. The age-adjusted rates of death from CHD varied directly with LDL cholesterol and inversely with HDL cholesterol. (From: Pekkanen J, Linn S, Heiss G, Suchindran CM, Leon A, Rifkind BM, Tyroler HA. Ten-year mortality from cardiovascular disease in relation to cholsterol level among men with and without preexisting cardiovascular disease. N Eng J Med 1990;322:1700–1704.)
"Reprinted, by permission of the New England Journal of Medicine, vol. 322; page 1703, year 1990"

duction in risk contributed by lipid modification would be substantial because it would affect a large number of persons.

The importance of secondary prevention also extends to patients after coronary artery bypass grafting, because the patency of the graft may depend on the prevailing lipid profile. Data from the Cholesterol-Lowering Atherosclerosis Study (CLAS) clearly demonstrate that it is feasible to retard the progression of coronary artery lesions, and most likely regress preexisting lesions in grafted as well as in the native vessels. Such benefits can be achieved over a relatively brief period by reducing LDL-C and raising HDL-C with appropriate hypolipidemic agents. This is facilitated by the fact that after a CHD event or procedure, patients are more highly motivated to modify coexistent CHD risk factors.

Measurement of Risk

Personal traits and living habits directly related to the probability of developing CHD have been labeled as risk factors (Table 2.1). In addition to hypercholesterolemia, the well-established risk factors include male sex or the postmenopausal state; elevated blood pressure; cigarette smoking; obesity; sedentary life style; diabetes mellitus; a family history of premature CHD occurring in a parent or sibling before 55 years of age; low HDL-C; and the presence of other known atherosclerotic cardiovascular disease. Although not listed as a specific risk factor, the probability of CHD certainly increases with age. Each of these factors is in-

Table 2.1. Major Risk Factors for Coronary Heart Disease Other than Total and LDL Cholesterol Abnormalities

Factor	Criteria for Classification
Sex	Male or female after menopause[a]
Hypertension	SBP ≥ 140 mmHg or DBP ≥ 90 mmHg
Cigarette smoking	Currently smokes cigarettes
Obesity	≥ 30% overweight
Sedentary activity level[a]	No regular strenuous physical activity, an energy expenditure < 2000 kcal/week
Diabetes mellitus	Fasting plasma glucose ≥ 140 mg/dL or clinical diagnosis
Family history of premature CHD	Parent or sibling with manifest CHD or sudden death before age 55
Low HDL level	Below 35 mg/dL in men Below 45 mg/dL in women[a]
Other CVD	History of cerebrovascular or peripheral vascular disease

[a] Not included as risk factor by National Cholesterol Education Program Expert Panel on Detection, Evaluation, and Treatment of High Blood Cholesterol in Adults

dependently related to CHD, and is thought to be directly involved in the pathogenesis of atherosclerosis. Combinations of two or more factors lead to a multiplicative effect on susceptibility toward CHD.

Normal Values

Most risk factors are continuously related to the development of CHD throughout the range of values. Generally, the more abnormal the value, the greater the risk. The normal value for a population may not necessarily be physiologically optimal. Average cholesterol values in Western populations are considerably higher than in Oriental countries that have substantially lower rates of atherosclerosis. Furthermore, the risk curve for total cholesterol starts below 200 mg/dL (5.17 mmol/L), so that the mean serum total cholesterol value for the United States adults, 211 mg/dL (5.46 mmol/L), is certainly higher than optimal from the standpoint of risk. It has been established that the risk for CHD begins to accelerate when total cholesterol levels of 200 mg/dL (5.17 mmol/L) are reached. The same is true for other physiologic traits such as blood pressure—the lower the pressure, the lower the risk for CVD, at least to diastolic levels approximating 80 mmHg, which are well below the standard definition of mild hypertension at 90 mmHg.

The relationship between the levels of most risk factors and CHD incidence is distributed continuously, without evidence of a threshold to demarcate a boundary between those at low risk and those at high risk. It is difficult to claim that the risk of a person who has a cholesterol of 240 mg/dL (6.21 mmol/L), a current clinical decision point, is much different from someone who has a cholesterol of 239 mg/dL (6.18 mmol/L), especially in view of the variability present in cholesterol measurements (see Chapter 6). Nevertheless, clinical practice requires the provision of practical cut-points, and several levels of blood cholesterol have emerged as guideposts for clinical decision making. Metabolic studies estimate that cellular cholesterol homeostasis can be maintained when the blood cholesterol concentration is only 25 mg/dL (0.65 mmol/L). Based on epidemiologic studies, cholesterol concentrations below 200 mg/dL (5.17 mmol/L) clearly indicate a relatively low risk for CHD. In contrast, levels above 240 mg/dL (6.21 mmol/L), which defines the upper 25% of the population, double the risk of CHD and indicate unacceptably high values. Values in the intermediate range between these points must be considered within the context of other risk factors for decision making. The multifactorial nature of CHD also indicates that all modifiable risk factors should receive attention and should be considered in raising or lowering specific cut-points for treatment.

For the individual patient, other factors must be taken into account. Between young adulthood and middle age, cholesterol levels rise progressively. A person who has an elevated cholesterol at a particular age is likely to maintain relatively high values forever. This phenomenon of "tracking", or maintaining a relative rank in the age-sex distribution of cholesterol, would indicate that persons who have high levels of cholesterol at an early age would be more likely to have cholesterol levels requiring treatment in later life. A more detailed discussion of the tracking phenomenon is given in Chapter 5.

Estimate of Risk

The most straightforward estimate of susceptibility for persons who exhibit a trait (exposed to factor) compared to those who do not (nonexposed to factor) is the relative risk, or risk ratio. The relative risk measures the strength of association between a trait and the likelihood of a particular disease and is an important concept for predicting the risk of developing that disease. It has the advantage of not being influenced by different rates of disease among populations. The relative risk usually is determined directly from follow-up studies and is calculated by dividing the occurrence rates of a disease such as CHD for one level of a risk factor (e.g., the highest cholesterol values) by the occurrence rates for lower values (e.g., the lowest cholesterol levels) (Table 2.2). For estimating the usefulness of particular interventions, a more clinically relevant measure is the attributable risk (AR). Attributable risk is calculated as the actual difference (rather than ratio) between the incidence rates in those persons who have high cholesterol (exposed) and those who have low cholesterol (nonexposed). This measure provides an estimate of the disease incidence that is considered attributable to the specific exposure after eliminating the contribution of other factors by subtracting the incidence in persons not exposed to the specific factor. This interpretation is based on the assumption that a cause-effect relationship exists between exposure and occurrence of disease. Carrying this assumption further, the number of cases among the exposed that

Table 2.2 Risk Estimates[a] of Age-Standardized Six-Year CHD Mortality for Groups Exposed to High Total Cholesterol Levels Based on 356,222 Primary Screenees of MRFIT

SERUM CHOLESTEROL	N	CHD DEATHS	SIX-YEAR INCIDENCE
> 181 mg/dL	295541	2062	6.98/1000
≤ 181 mg/dL	60681	196	3.23/1000
Relative risk or risk ratio			$= \dfrac{6.98}{1000} \div \dfrac{3.23}{1000} = 2.16$
Attributable risk			$= \dfrac{6.98}{1000} - \dfrac{3.23}{1000} = \dfrac{3.75}{1000}$
Attributable risk percent			$= \dfrac{6.98/1000 - 3.23/1000}{6.98/1000} \times 100$ $= 53.7\%$

[a] Approximated from available data (Table 1) in Stamler J, Wentworth D, Neaton JD, for the MRFIT Research Group. Is relationship between serum cholesterol and risk of premature death from coronary heart disease continuous and graded? Findings in 356,222 screenees of the Multiple Risk Factor Intervention Trial (MRFIT). JAMA 1986;256(20):2823–2828.

could be eliminated if the exposure were removed also can be calculated.

The AR alone often is misleading, because it does not provide an estimate of the relative magnitude of the problem. For example, a reduction of 2% in a disease that has an incidence rate of 50 per 1000 among those persons exposed to the risk factor is substantial but certainly less important than the same reduction when the incidence rate of the exposed is 500 per 1000. To account for the magnitude of preventable disease, AR often is expressed as attributable risk percent (ARP), that is, AR divided by the incidence rate among the exposed and multiplied by 100. In the example shown in Table 2.2 it is estimated that 54% of the CHD deaths in those persons who have cholesterol values that exceed 181 mg/dL (4.68 mmol/L) are preventable.

When interpreting measures of risk, it is important to realize that relative and attributable risks provide different information. As a measure of the strength of association, the relative risk estimates whether the association is likely to be cause and effect. In contrast, the AR is used to provide an esti-

mate of the quantitative contribution of an exposure to the incidence of a disease and, more importantly, of the effects of removing that exposure on disease prevention. For example, cigarette smoking has a much higher relative risk for lung cancer than for CHD, implying that a higher proportion of lung cancer is caused by smoking than is CHD. Because CHD is more prevalent than lung cancer, however, the actual number of smokers who will develop CHD is greater than the number who will develop lung cancer. The ARP and, consequently, the number of deaths prevented from smoking cessation, therefore, will be far greater for CHD than for lung cancer.

Table 2.3 provides estimates of prevalence, relative risk, population AR, and estimated preventable deaths for the major risk factors. These risk measures emphasize the contrast between the major risk factors from several perspectives. The relative risk or risk ratio (high versus low value) varies from 1.8 for smoking to 3.0 for hypercholesterolemia. Among these risk factors it should be noted that reduction of elevated cholesterol values, by virtue of their high prevalence and

Table 2.3. Prevalence and Measures of Risk for Coronary Heart Disease Mortality for the Major Risk Factors[a]

Risk Factor	Prevalence (%)	Crude Relative Risk	Population Attributable Risk (PAR) (%)	Estimated Preventable Deaths[b]
Cigarette smoking				
Males	31.2	1.9	20.8	64,302
Females	26.5	1.8	16.4	46,570
Hypertension				
>159/>94 mm Hg	17.7	2.8	23.0	136,416
Diabetes mellitus				
Males	5.7	2.1	5.6	17,312
Females	7.4	4.7	21.6	61,336
Cholesterol				
Total ≥240 mg/dL	26.7	3.0	29.8	176,747
HDL[c] <35 mg/dL	11.2	2.4	13.5	80,070
Inactivity	58.8	1.9	34.6	205,216
Overweight RW[c] ≥ 130%	26.6	2.0	18.1	107,353

[a] Adapted from: Deaths from coronary heart disease, United States. Compilation. Chronic Disease Reports from the MMWR, Vols 38 (1989) and 39 (1990). October 12, 1990, p 19. United States Department of Health and Human Services, Public Health Service, Centers for Disease Control.
[b] Estimated preventable deaths calculated as PAR × CHD mortality in the general population, where PAR is the percentage of mortality attributable to the specific risk factor. Because persons may be exposed to more than one risk factor, the numbers are not additive.
[c] HDL = high density lipoprotein; RW = relative weight (actual weight ÷ desirable weight based on 1959 Metropolitan Tables × 100.)

high AR, has a high theoretical potential for reducing the burden of CHD. The relative risk for total cholesterol reportedly declines with age, being a more powerful predictor of CHD before late middle age than later in life. Although the Multiple Risk Factor Intervention Trial (MRFIT) study only spans ages 35 through 57 years, it provides great statistical power to analyze the relationship within that age range between cholesterol and the risk of CHD (Fig. 2.8). The strength of the relationship is clearly evident in the youngest and declines with increasing age. Such data frequently are misinterpreted as meaning that total cholesterol is relatively unimportant in older persons. Yet, the 30-year follow-up data from Framingham indicate that both elevated LDL-C and reduced HDL-C predicted CHD events in men and women through the age of 82 years. More recent data from the Honolulu Heart Program indicate that the relative risk of hypercholesterolemia does not decrease for older men. Several factors could account for the differences between studies. Nevertheless, a single measurement at an older age may underestimate a relationship, and changes in lipids over time may be important to estimate the true risk. Persons who have a given risk factor may have reached older age because they had been more resistant to the

development of atherosclerosis, but that does not necessarily preclude the possibility that they are vulnerable and capable of developing manifestations of CHD at a later age. The causative risk factors maintain their injurious effects throughout life, and part of the apparent decline in predictive power of cholesterol with age may be related to an increased prevalence of other competing risk factors for CHD. Some of these other risk factors require treatment but should not obscure the need for treating dyslipidemia.

The success of secondary prevention trials shows that the presence of advanced atherosclerosis does not, by itself, reduce the benefits of lowering cholesterol values. These studies suggest that retardation of coronary artery lesions can accompany lipid changes within a period of 2 years. Even if the relative risk of hypercholesterolemia declines with age, the absolute risk attributable to the blood cholesterol increases with aging because of the higher prevalence of CHD in the elderly (see Fig. 2.8). In fact, the absolute rates of preventable CHD cases and deaths, expressed as events annually per thousand treated, tends to be greater in older than in younger patients (Table 2.4). A generally aggressive approach to treatment, therefore, may be warranted in the elderly. As little as a 1% fall in CHD mortality rates among

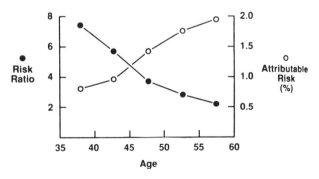

Figure 2.8 Change in risk ratio and attributable risk with age for men in the highest total cholesterol quintile, compared with those in the lowest total cholesterol quintile of the MRFIT. The risk ratio designated by closed circles decreases within the age span depicted whereas attributable risk increases because of the greater absolute risk. The number of CHD events increases greatly with age even though the risk ratio may be less. The greater absolute numbers of persons likely to benefit from cholesterol-lowering, therefore, increases in an older population. "Reproduced, with permission from" (From: Denke MA, Grundy SM. Hypercholesterolemia in elderly persons: Resolving the treatment dilemma. Ann Intern Med 1990;112:780–792.)

Table 2.4 Potential Benefit to Coronary Heart Disease Risk of Reducing Cholesterol Risk from 285 to 200 mg/dL in Different Age Groups

Age Groups (Years)	Decrease in CHD: Risk (%)	Preventable CHD Cases[a]		Deaths[a]	
		M	F	M	F
35–44	77.4	11.8	3.2	1.8	0.7
45–54	61.8	23.4	6.0	6.3	0.9
55–64	46.8	26.0	10.2	8.7	1.8
65–74	33.1	20.4	10.6	9.5	3.8
75–84	22.9	20.4	12.3	12.7	6.5

[a] Per 1000 patients/year (From Gordon DJ, Rifkind BM. Treating high blood cholesterol in the older patient. Am J Cardiol 1989;63:48H–52H.)

those 65 years or older could lead to 4300 fewer CHD deaths per year in the United States alone.

Multifactorial Risk

Although age is the preeminent risk factor, the likelihood of developing CHD varies widely among men and women according to the constellation of other risks present in concert with an inherent vulnerability. Because each risk factor supplies unique information, the best estimates of risk can be made by considering multiple factors simultaneously. Most candidates for CHD do not have marked abnormalities of a single factor but, rather, marginal changes in several factors. The presence of a high blood cholesterol, even at levels above 300 mg/dL (7.76 mmol/L), confers less susceptibility in the absence of other risk factors than a cholesterol of 230 mg/dL (5.95 mmol/L) in the presence of two other risk factors (Fig. 2.9). For this reason, the National Cholesterol Education Program (NCEP) Expert Panel on Detection, Evaluation, and Treatment of High Blood Cholesterol in Adults used nonlipid risk factors in combination with lipid values to set criteria for treatment of lipid abnormalities (see below). Such combinations of moderately elevated risk factors produce high vulnerability levels but frequently escape detection and treatment because no single value is strikingly "abnormal."

National Recommendations for Prevention of Cardiovascular Disease

The Coronary Primary Prevention Trial (CPPT) first reported favorable results on lowering cholesterol in 1984. Interest in diagnosing lipid disorders and actively treating high blood cholesterol levels lagged behind. To address these concerns, The National Heart, Lung, and Blood Institute and The National Institutes of Health Office of Medical Applications of Research convened a consensus development conference in December, 1984 on lowering blood cholesterol to prevent heart disease. A panel of lipoprotein experts, cardiologists, primary care physicians, epidemiologists, biostatisticians, and other biomedical scientists reviewed the evidence for the causal relationship between blood cholesterol and coronary heart disease and the need to reduce blood cholesterol levels. The recommendations from this meeting were as follows:

1. Persons who have high-risk blood cholesterol levels (values above the 90th percentile) should be treated intensively by dietary means under the guidance of a physician, dietitian, or other health professionals. If response to diet is inadequate, appropriate drugs should be added to the treatment regimen.
2. Adults who have moderate-risk blood cholesterol levels (values between the 75th and 90th percentiles) should be treated intensively by dietary means, especially if additional risk factors are present. Only a small proportion should require drug treatment.
3. All Americans (except children younger than 2 years old) should be advised to adopt a diet that reduces total dietary fat intake from the current level of about 40% of total calories to 30% of total calories, to reduce saturated fat intake to less than 10% of total calories, to increase polyunsaturated fat intake but to no more than 10% of total calories, and to reduce daily cholesterol intake to 250 to 300 mg or less.

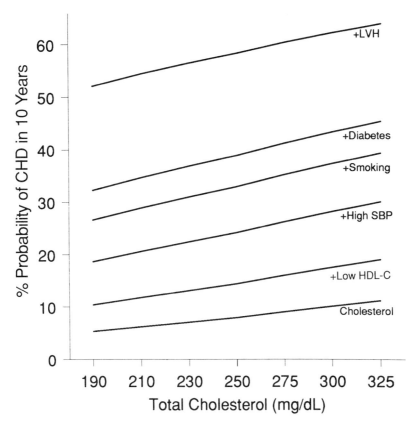

Figure 2.9. The ten year probability (percent) for coronary heart disease for a 50-year old man, based on blood cholesterol alone and with the cumulative addition of other risk factors. The probability varies directly with the level of blood cholesterol. Cumulative addition of other risk factors greatly accelerates the risk of future coronary disease at all levels of blood cholesterol. (Adapted from: Anderson KM, Wilson PWF, Odell PM, Kannel WB. An updated coronary risk profile: a statement for health professionals. Circulation 1991;83:358–359.)
Low HDL-C = 31 mg/dL
High SBP = systolic blood pressure of 180 mmHg
LVH = left ventricular hypertrophy

4. Intake of total calories should be reduced, if necessary, to correct obesity and adjusted to maintain ideal body weight. A program of regular moderate exercise is helpful in this connection.
5. In persons who have elevated blood cholesterol levels, special attention should be given to the management of other risk factors (hypertension, cigarette smoking, diabetes, and physical inactivity).

The panel also made some additional recommendations:

1. New and expanded educational programs should be implemented for physi-

cians and other health professionals on the significance of elevated blood cholesterol levels and the importance of treating lipid disorders.
2. The food industry should be encouraged to continue and intensify efforts to develop and market foods that make it easier for everyone to adhere to the recommended diets. School food services and restaurants should serve meals consistent with these dietary recommendations.
3. Food labeling should include the specific source or sources of fat; content of total fat as well as saturated, and polyunsaturated fat; and cholesterol content plus

other nutritional information. The public should be educated on how to use this information to achieve dietary aims.

4. All physicians should be encouraged to include, whenever possible, a blood cholesterol measurement on every adult patient when that patient is first seen; to ensure reliability of data, standardized methods for cholesterol measurement in clinical laboratories are recommended.

5. More research should be encouraged to compare the effectiveness and safety of currently recommended diets against those of alternative diets; to study human behavior as it relates to food choices and adherence to diets; to develop more effective, better tolerated, safer, and more economical drugs for lowering blood cholesterol levels; to assess the effectiveness of medical and surgical treatment of high blood cholesterol levels in patients who have established coronary artery disease; to develop more precise and sensitive noninvasive artery imaging methods; and to apply basic cell and molecular biology to increase our understanding of lipoprotein metabolism (particularly the role of HDL as a protective factor) and artery wall metabolism as they relate to coronary heart disease.

Despite these recommendations from the consensus panel and mounting evidence from clinical trials, physicians remained reluctant to identify and treat lipid disorders. This unwillingness was attributed to the lack of more specific therapeutic guidelines and the lack of appropriate drugs. Also, the complex terminology that categorized lipid disorders did little to resolve the mystique and associated difficulty of managing dyslipidemias. By way of contrast, The National High Blood Pressure Education Program had been a huge success, and the Heart, Lung, and Blood Institute believed that development of similar therapeutic guidelines for lipids, coupled with an intense educational effort, would promote effectively the treatment of lipid disorders.

The National Cholesterol Education Program (NCEP) was created in 1985 to meet this challenge. Consequently, an expert panel was assembled and met periodically over several years to resolve the issues. Finally, in 1988 the panel issued guidelines for the detection, evaluation, and treatment of high blood cholesterol in adults. The goals of this publication were to simplify the diagnosis and provide a basis for the clinical management of patients who had dyslipidemia. To achieve these objectives, a basic series of proposals were made:

1. Total cholesterol should be measured in all adults at least once every 5 years.

2. Those persons who had established values from 200 to 239 mg/dL (5.17-6.18 mmol/L) were classified as borderline-high blood cholesterol and those who had 240 mg/dL (6.21 mmol/L) and above as high blood cholesterol. These total cholesterol values should only serve, however, as markers for lipoprotein values.

3. Persons who are at high risk because they have borderline-high cholesterol combined with definite CHD or two other CHD risk factors should undergo lipoprotein analysis, as should all those persons who have a high blood cholesterol value.

4. Persons found to have LDL-C levels of 130 to 159 mg/dL (3.36-4.11 mmol/L) with two other risk factors or CHD should begin cholesterol-lowering treatment, as should all persons with levels of 160 mg/dL (4.14 mmol/L) and above.

Criteria from this report served to identify approximately 36% of all adults aged 20 to 74 as candidates for medical advice and intervention based on high blood cholesterol levels and other risk factors. This report was circulated widely by the National Heart, Lung, and Blood Institute to physicians and related health professionals. Concurrently, favorable reports from the Helsinki study and CLAS were reported, and more effective drugs to lower lipids became available. The health care system suddenly was faced with large numbers of patients seeking advice and treatment for lipid disorders, at a time when most physicians were not yet ready to initiate such treatment.

Surveys of Attitudes about High Cholesterol Levels

The National Heart, Lung, and Blood Institute sponsored a survey of practicing physicians and a national probability sample of adults to assess attitudes and knowledge about high blood cholesterol levels. The first survey was conducted in 1983 before release of the CPPT results, and a second survey followed in 1986. A comparison between the two surveys showed a growth in awareness that reducing high blood cholesterol levels could reduce CHD, from 64% in 1983 to 72% in 1986. Also, in a Federal Drug Administration (FDA) survey conducted in 1988, 59% of adults reported that they had their cholesterol levels checked, compared to 46% in 1986 and 35% in 1983. In general, modest gains were made in public awareness and action relating to attitudes about risk from hypercholesterolemia. The awareness percentage, however, lagged behind the approximately 85% who believed that smoking and high blood pressure were important risk factors for CHD. Physicians also had become more convinced of the benefit of lowering high blood cholesterol levels and were treating their patients accordingly. In the 1986 survey, 58% of physicians had indicated a change in their practice habits during the previous two years regarding elevated cholesterol levels. These changes occurred despite the fact that many physicians surveyed were not aware of a consensus conference report, nor had they heard of the results of the CPPT.

A special survey was conducted during January, 1989 in a nationally representative sample of persons 20 years of age or older. Two-thirds of this sample claimed to have had their blood tested for cholesterol at least once. Other studies have shown that those persons 50 years of age or older were more likely to have had their cholesterol checked than younger individuals.

Numerous barriers still exist to implementing adequate therapy for hypercholesterolemia. The public apparently has good knowledge concerning specific food choices but more limited information on distinctions between dietary cholesterol, saturated fats, and unsaturated fats. Many physicians are unsuccessful with dietary counseling, perhaps in part because of the time required and the lack of third-party reimbursement for such efforts. Despite the difficulties in implementing appropriate diets, physicians remain conservative in prescribing available drugs. Nevertheless, it is likely that physicians and other health professionals will be under increasing pressure from a public aware of the relation between lipids and CHD and anxiously seeking advice on modifying lipoprotein patterns.

Risk as a Function of Cholesterol and Other Lipoproteins

Total Cholesterol

Diverse evidence clearly indicates that the total blood cholesterol is a powerful predictor of morbidity and mortality. Some of the strongest evidence for this relation in humans is derived from longitudinal observational studies. The age-adjusted CHD death rates by total cholesterol values among the 350,000 men aged 35 to 57 initially screened in the MRFIT (see Chapter 1) are depicted in Figure 2.10. The six-year mortality rate doubled, from 3.16 to 6.94 per 1,000, as the total cholesterol increased from 153 to 256 mg/dL (3.96 and 6.62 mmol/L); it then doubled again as the total cholesterol increased from 226 to 290 mg/dL (5.84 to 7.50 mmol/L). The resulting risk ratios for plasma cholesterol levels of 250 mg/dL (6.47 mmol/L) and 300 mg/dL (7.76 mmol/L) were 2.0 and 4.0, respectively. A similar association between total cholesterol and risk of CHD death was seen in black men and white men. Data from the Framingham 30-year follow-up and the Framingham Offspring studies revealed that the incidence of CHD varied directly with the total cholesterol levels for men and women alike (Fig. 2.11). This applied to younger persons as well as those between 65 and 94 years of age. The bulk of evidence

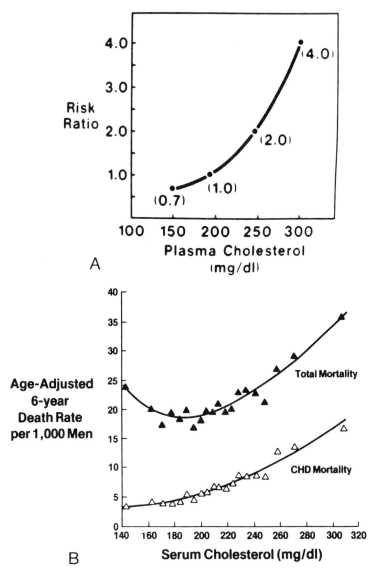

Figure 2.10. Mortality data from six-year follow-up of participants in the MRFIT. *A.* Risk ratios for CHD mortality as a function of the plasma total cholesterol level. (From: Grundy SM. Cholesterol and coronary heart disease: a new era. JAMA 1986;256:2849–2858.) "Copyright 1986, American Medical Association." *B.* Relationship between total mortality and CHD mortality with total serum cholesterol concentration. (From: Martin MJ, Hulley SB, Browner WS, Kuller LH, Wentworth D. Serum cholesterol, blood pressure, and mortality: implications from a cohort of 361662 men. Lancet 1986;2(8513):933–936.) "Copyright The Lancet Ltd."

supports a continuous and direct relation between CHD and the total cholesterol level, beginning at approx. 150 mg/dL (3.88 mmol/L) and upward.

Multiple randomized, controlled clinical trials also have demonstrated a reduction in the incidence of CHD proportional to the differences in total cholesterol levels between treated and control groups. Primary and secondary prevention studies, as well as several angiographic studies, indicate that the magnitude of the difference in CHD risk varies directly with the change in mean total cholesterol (see Chapter 1). The selection of high-risk men for most of these studies involved practical considerations, namely the

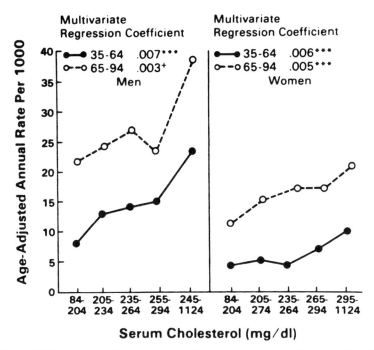

Figure 2.11. Risk of CHD by serum cholesterol according to age in men and women (Framingham Study, 30-year follow-up). The incidence of CHD increased directly with the level of serum total cholesterol for both middle-aged and older men and women. ***p < 0.09; +NS. (From: Kannel WB. Contribution of the Framingham Study to preventive cardiology. JACC 1990;15:206–211.) "Reprinted with permission from the American College of Cardiology (Journal of the American College of Cardiology, 1990;15:206–211)."

excessive cost and problems of conducting such a trial in more general populations with the likelihood of fewer coronary events, and the need to extend the period of observation. It seems reasonable, however, to extend these conclusions to a more general population of men and women who have total cholesterol values of 200 mg/dL (5.17 mmol/L) or greater. Consistent estimates from observational studies and clinical trials project that for persons who have total cholesterol levels between 250 and 300 mg/dL (6.46-7.76 mmol/L) over a wide age range, each 1% reduction in total cholesterol reduces CHD rates approximately 2%. This beneficial effect becomes apparent after several years of treatment, although shorter periods of more intense intervention may suffice to produce a noticeable effect. Systematic data relating blood lipids to the development of stroke remain meager. Most of the population studies indicate a U-shaped relation between total cholesterol and stroke. In the

Framingham Study, the highest risk of stroke occurred at total cholesterol levels below 190 mg/dL (4.91 mmol/L) and above 295 mg/dL (7.63 mmol/L). A similar risk curve was noted in the Chicago Stroke Study and the Honolulu Heart Program. The higher incidence of stroke at low cholesterol values apparently was accounted for by an inverse relation between total cholesterol and hemorrhagic stroke. In the MRFIT, this inverse relation was found only in patients who had total cholesterol levels under 160 mg/dL (4.14 mmol/L) accompanied by diastolic blood pressures greater than 90 mmHg. Studies using computed tomography (CT) scanning, however, have observed a positive linear relation between total cholesterol levels and death ascribed to cerebral thrombosis, explaining the relation between higher cholesterol levels and overall stroke risk. In any event, the attributable risk of the total cholesterol to the incidence of thrombotic strokes and CHD clearly overrides any possible association between low cholesterol levels and hemorrhagic stroke.

In the Framingham Study, men who had higher cholesterol values also were noted to have had a greater incidence of intermittent claudication. Other data have demonstrated a stronger relation between elevated triglyceride and peripheral vascular disease.

Because cholesterol is insoluble in water, it is transported in the blood by different types of lipoprotein particles. However useful the total cholesterol is as a marker for CHD, the more precise definition of risk is derived from the consideration of the individual lipoprotein particles in which it is carried. Both LDL-C and HDL-C levels appear to be better predictors of CHD risk than the total cholesterol level, both among the general population and selected population subgroups.

Low Density Lipoprotein

The lipid hypothesis emphasizes the role of LDL-C because much of the available information has been derived from studies that address this lipoprotein. Findings from patients who have familial hypercholesterolemia (FH), experimental studies, analyses of pathologic specimens, and population studies have dealt primarily with LDL-C. Moreover, LDL carries most (approx. 65%) of the cholesterol circulating in the blood. Current lines of evidence generally indicate LDL to be the most atherogenic of the five major lipoprotein classes.

Variation in LDL-C levels explains much of the difference in individual risk within populations vulnerable to CHD, as seen in numerous observational studies. In the Framingham and the MRFIT studies, as well as among Israeli male civil servants, the change in LDL-C level from baseline has been related strongly to subsequent CHD risk. Similar relations also were also demonstrated in clinical trials. In the hypercholesterolemic men treated with cholestyramine in the LRC-CPPT (see Chapter 1), there was a linear relation between the reduction in risk of CHD and the lowering of LDL-C (Fig. 2.12). A 19% CHD reduction occurred with an 11% reduction in LDL-C, and a 49%

CHD reduction with a 35% LDL-C reduction. Among men who had undergone coronary bypass surgery in the CLAS study (see Chapter 1), a 26% reduction in total blood cholesterol, a 43% reduction in LDL-C, plus a 37% elevation in HDL-C resulted in a significant decrease in the average number of lesions that progressed, as well as in the percentage of patients who developed new atheroma in the native coronaries. Similar findings from subsequent arteriographic studies corroborated these results.

The prevailing concept indicates that excess LDL in the circulation impacts adversely on the arterial wall (see Chapter 1). Because of their relatively small size, LDL particles can infiltrate into the arterial wall, where they release esterified and unesterified cholesterol. The LDL appears to be retained preferentially in the arterial wall at sites destined for atheromatous development, even before there is evidence of cellular infiltration. The LDL particles constitute a rather heterogenous group of particles, ranging in size from 18 to 28 nm, and the size and density of the particles may play a role in determining their degree of atherogenicity. It also has been postulated that the smallest and densest of the LDL subfractions might be the most atherogenic, and persons who have dense (smaller) LDL are likely to be at increased risk of CHD even with normal cholesterol levels (see Chapter 7). Findings from case-control studies indicate that the dense LDL subspecies pattern serves as a marker for a common genetic trait that exerts an array of effects on lipoprotein structure and metabolism and carries a predisposition to CHD. These dense LDL particles are associated with hypertriglyceridemia and reduced HDL-C levels. This may explain in part the presence of premature CHD among some patients who have familial hypertriglyceridemia (FH) or familial combined dyslipidemia, in whom qualitative but not necessarily quantitative changes in LDL-C exist. Also, it is conceivable that FH heterozygotes who do not develop CHD may lack substantial quantities of the dense LDL subfraction.

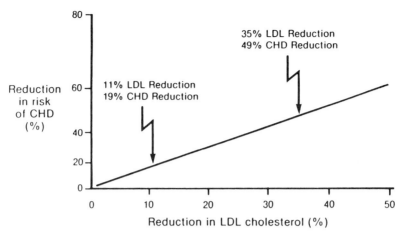

Figure 2.12. Relationship of reduction in low density lipoprotein cholesterol (LDL-C) levels to reduction in CHD risk (logarithmic scale). Risk reduction was estimated by comparing distribution of percent change in LDL-C levels among CHD cases to that among all participants in the same treatment group. The line presents reduction in CHD risk predicted by a porportional hazards model for a given decrease of the LDL-C level among the cholestyramine group. (Adapted from: The Lipid Research Clinics Coronary Primary Prevention Trial (LRC-CPPT) results: II. The relationship of reduction in incidence or coronary heart disease to cholesterol lowering. Lipid Research Clinics Program. JAMA 1984;251:369.)

Although the total cholesterol serves as a surrogate for the LDL-C, the one cannot be extrapolated from the other directly because HDL-C and VLDL-C also contribute to the total cholesterol. In the presence of average HDL-C and VLDL-C levels, a total cholesterol of 200 mg/dL (5.17 mmol/L) corresponds roughly to an LDL of 130 mg/dL (3.36 mmol/L), whereas a total cholesterol of 240 mg/dL (6.21 mmol/L) corresponds to an LDL of 160 mg/dL (4.14 mmol/L). These values serve as the cut-off points in the NCEP guidelines for management of hypercholesterolemia and thus are useful to remember.

High Density Lipoprotein

High density lipoprotein particles consist of several subgroups, differing in size, density, and protein composition (see Chapter 3). In humans, the most abundant of these are termed HDL_2 and HDL_3. The major portion of HDL-C usually is present in HDL_3, although most of the variability in HDL levels generally represents different amounts of HDL_2. The protective effect of HDL-C generally has been attributed to the HDL_2 subgroup; however, some recent prospective studies have shown a stronger relationship between HDL_3 and CHD risk. Technical difficulties in separating out HDL subgroups using current laboratory techniques may account in part for these disparate and confusing results. It also is not yet clear which of the HDL constituents—cholesterol, apoprotein A-I, or apoprotein A-II—is most meaningful for determining CHD risk as well as the beneficial effects of therapeutic interventions. Currently, it appears best to rely on the total HDL-C as a measure of risk.

A low blood level of HDL-C often is the only detectable lipid abnormality in many patients who have premature CHD. Of the various lipid profiles studied at the 7-year Framingham follow-up, a low HDL-C predicted best the CHD risk. An increment in HDL-C of 25 mg/dL (0.65 mmol/L) was linked with nearly a 50% reduction in risk after adjustment for other important risk factors. This consistent inverse correlation between HDL-C and CHD is well recognized, and is stronger among women than men. Furthermore, among older persons, the association with CHD is greater than with LDL-C. Subsets of persons exist in whom low HDL concentrations may be especially atherogenic (Fig. 2.13). One such subgroup includes women who have high triglyceride concentrations in conjunction with low HDL-C.

Although the NCEP considered HDL-C below 35 mg/dL (0.91 mmol/L) to be a major risk factor for CHD, it did not advocate screening for HDL-C or using it to stratify risk; nor did it recommend drug treatment for isolated low HDL-C levels. At the time of the report, the relation between CHD and HDL-C was not as well established as that with LDL-C, nor had there been a therapeutic trial to demonstrate the efficacy of raising HDL-C levels. Furthermore, the existence of populations with low LDL-C levels in whom low HDL-C concentrations do not predispose to CHD is confusing. Certain genetic disorders characterized by low HDL-C, including Tangier disease and apoprotein A-I Milano, are not necessarily accompanied by an increased risk of CHD. Apoprotein A-I Milano is associated with mild hypertriglyceridemia but not with premature CHD. However, homozygotes with Tangier disease may develop premature CHD. Yet, in familial apolipoprotein A-I and apolipoprotein C-III deficiency, a virtual absence of HDL-C is found to be as-

sociated with severe premature atherosclerosis. Also, in certain inbred strains of mice, susceptibility to diet-induced atherosclerosis is related to lower HDL-C levels. Families have been identified who have very high HDL-C levels and increased longevity. Differences in the relative proportions of A-I and A-II apoproteins in the HDL particle may confer different risk and complicate the situation even more.

A protective effect has now been demonstrated conclusively for HDL-C, independent of other lipids and other major risk factors. As the total cholesterol increases from 200 to 250 mg/dL (5.17-6.47 mmol/L), the CHD risk doubles; as HDL-C decreases from 45 to 25 mg/dL (1.16-0.65 mmol/L) this risk also doubles. Results from the Framingham Offspring Study show that low HDL-C, even in the presence of normal total cholesterol and LDL-C, is a potent marker for premature CHD. With isolated low HDL-C < 35 mg/dL (0.91 mmol/L), the odds ratio of subsequent CHD was 2.19 in men and 9.69 in women. In earlier data from the Framingham Study, the risk for developing CHD at any given concentration of LDL-C was modified substantially by the prevailing HDL-C concentration. Figure

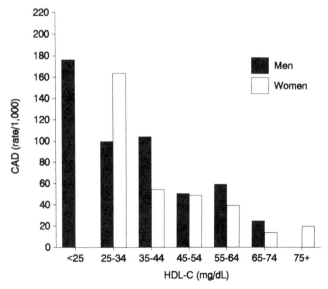

Figure 2.13. Rate of CAD related to baseline high density lipoprotein cholesterol (HDL-C) values for men and women in the Framingham Heart Study. (From: Castelli WP, Wilson PW, Levy D, Anderson K. Cardiovascular risk factors in the elderly. Am J Cardiol 1989;63(Suppl):12–19H.)

2.14 demonstrates the relationships between HDL-C, LDL-C, and CHD. For LDL-C of 100 mg/dL and 160 mg/dL (2.59 and 4.14 mmol/L, respectively), the relative risk exceeds 1.0 (average) only for the subgroups with a low HDL-C (25 mg/dL [0.65 mmol/L]). At a higher LDL-C level, 220 mg/dL (5.69 mmol/L), the relative risk exceeds 1.0 for those persons who have HDL-C of 45 mg/dL (1.16 mmol/L) or below. For those persons who have HDL-C above 65 mg/dL (1.68 mmol/L), the relative risk never reached the average level regardless of LDL-C level. In the 20-year follow-up in the Donol-Tel Aviv Prospective Coronary Artery Disease Study, the HDL-C had the highest predictive value for risk of future CHD. In the Israeli Ischemic Heart Disease Study, the relative risk of HDL was independent of other factors and most marked for men over 50 years old. The Physicians' Health Study clearly demonstrated the importance of HDL-C in predicting the risk of MI even when the total cholesterol level is

not elevated. Such data clearly indicate the protective effects of HDL-C. Of the large prospective studies, only the USSR center of the LRC Mortality Follow-up Study did not show a trend of higher CHD rates with lower HDL-C levels, possibly related to confounding factors such as a high prevalence of alcoholism and cigarette smoking.

Although the effect of raising HDL-C levels in persons who have isolated abnormalities in HDL-C levels has not yet been investigated directly in clinical trials, several primary and secondary intervention studies of lipid-lowering strategies had provided indirect evidence of such effects. In the LRC-CPPT, each increase of 1 mg/dL (0.03 mmol/L) in HDL-C from baseline was calculated to account for a lowering of CHD risk by about 3%. Men whose HDL-C remained under 35 mg/dL (0.91 mmol/L) experienced no significant decrease in heart disease risk even though they had the same degree of LDL-C reduction from cholestyramine. The Helsinki Heart Study demon-

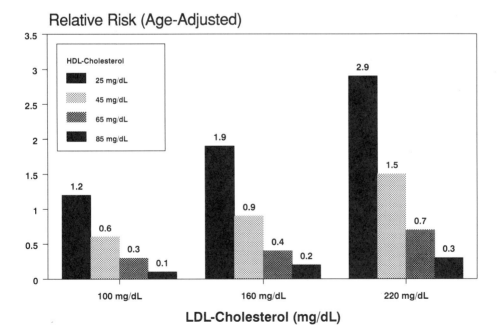

Figure 2.14. Risk of CHD in men aged 50 to 70, according to HDL and LDL cholesterol levels, with a systolic blood pressure of 135 mmHg, over 4 years of follow-up in the Framingham Study. The relative risk of CHD varies directly by level of LDL-C and inversely by level of HDL-C. The relative risk exceeds the average, 1.0, only for men who have HDL-C concentrations lower than 45 mg/dL. (Adapted from: Kannel WB, Doyle JT, Ostfeld AM, et al. Optimal resources for primary prevention of atherosclerosis diseases. Atheroslcerosis study group. Circulation 1984;70:157A–205A.)

strated that gemfibrozil induced a moderate reduction of LDL-C concentrations, decreased triglyceride levels, and moderately increased HDL-C levels. Analyses from this study led the investigators to conclude that both the reduction of LDL-C levels and the increase in HDL-C concentration made independent contributions in reducing CHD rates. In the earlier, long-term results of the Coronary Drug Project, increased HDL-C from use of niacin certainly could have contributed to the observed decreased mortality.

The inverse relation between HDL-C levels and atherosclerosis also was seen in angiographic studies. The Cholesterol Lowering Artherosclerosis Study (CLAS) found that a combination of diet and drugs (colestipol and niacin) that had lowered LDL-C levels raised HDL-C concentrations and reduced triglyceride levels, retarded atherosclerosis, and decreased the appearance of new lesions in bypass grafts. In the Familial Atherosclerosis Treatment Studies (FATS), treatment with a combination of colestipol with either lovastatin or niacin was found to improve the angiographic appearance of coronary vessels as well as the clinical course of the treatment groups. These benefits were correlated independently with the reductions in LDL-C and the elevations in HDL-C.

Because the observations on the protective effects of HDL-C mainly have been by secondary analyses, the benefits that can be derived from raising HDL-C as a preventive strategy still are controversial. New clinical trials and improved laboratory techniques may offer opportunities for more careful evaluation of the relation between HDL-C and CHD and may determine whether elevation of a low HDL-C will be protective beyond any benefit achieved by reducing LDL-C.

The mechanism behind the protective action of HDL-C is attributed mainly to its postulated role in reverse cholesterol transport (see Chapter 4), but this has yet to be demonstrated directly. Other possible explanations exist, such as inhibition of binding of lipids to the collagen matrix, or interfer-

ence with the binding of modified LDL to scavenger receptors. Also, HDL has an effect on prostaglandins, by promoting the release of prostacyclin (PGI_2) from endothelial cells. This is an important action because prostacyclin counteracts the vasoconstrictor and platelet proaggregatory effects of thromboxane A_2. Prostacyclin also exerts more chronic effects on atherosclerosis by inhibiting the release of platelet mitogen (an initial step in atherogenesis); reducing the concentration of cholesteryl esters present in foam cells; and facilitating cholesterol transport to the liver by HDL. It also is possible that low HDL-C levels do not have a direct effect on atherosclerosis but, rather, reflect the coexistence of other atherogenic lipoproteins such as VLDL remnants, intermediate density lipoproteins (IDL), or abnormally small and dense LDL.

Because the mechanism whereby a low HDL-C level places a person in a high-risk category is unclear, the question of whether therapy is indicated or necessary is difficult to answer. The measurement of HDL-C in a man who has a total cholesterol level of 200 mg/dL or more (5.17 mmol/L) is certainly reasonable and can be justified on the basis of known risk estimates, because lowering the LDL-C level in such persons has been considered beneficial. But do persons who have a low HDL-C in the presence of "normal" total cholesterol levels, especially in the presence of other risk factors, also benefit from modulation of their LDL-C and HDL-C levels? If so, then it can be argued that HDL-C should be measured in everyone. Currently, the costs generated by universal screening and the lack of information on therapeutic benefits from raising isolated low HDL-C levels have not convinced the NCEP Adult Treatment Panel to recommend widespread HDL-C screening.

Data from the second National Health and Nutrition Examination Survey (NHANES) indicate that approx. 19% of men and 6% of women have HDL-C concentrations below 35 mg/dL (0.91 mmol/L) in the presence of a total cholesterol level 200 mg/dL (5.17 mmol/L) or less. Using a common cut-point for men and women, however, underestimates the prevalence of low HDL-C values for women, because their distribution of HDL-C is different than that for men. The lower quartile for all women, which corresponds to the 34 mg/dL (0.88 mmol/L) value for men, is 45 mg/dL (1.16 mmol/L). In any case, using an LDL cut-point of 130 mg/dL (3.36 mg/dL) assuredly will miss an appreciable number of person who have low HDL-C if only total cholesterol is tested during screening. One way to reveal those persons who have lower HDL-C values in need of future evaluation is to use selective criteria to identify them. Markers for a potentially low HDL-C include hypertriglyceridemia, obesity, cigarette smoking, physical inactivity, diabetes mellitus, a family history of low HDL-C, and manifestations of premature atherosclerosis.

The potential for raising low HDL-C levels by therapeutic strategies still is unclear, and whether persons who have low HDL-C can achieve levels sufficient to reduce CHD risk has yet to be resolved. It is uncertain whether the HDL-C response to therapy depends on the initial levels of HDL or VLDL and triglyceride, or whether the response depends on the cause of the low HDL-C. We still have much to learn about the therapeutic implications of detecting and treating depressed HDL-C concentrations.

Ratio of Low Density Lipoprotein Cholesterol to High Density Lipoprotein Cholesterol

To balance the opposing effects of LDL-C and HDL-C on CHD, many researchers have used a ratio of LDL-C:HDL-C to better express risk. Even though therapeutic decisions are based on LDL-C, the ratio of total cholesterol to HDL-C is more commonly used because the LDL-C does not have to be calculated. In the Physicians' Health Study, after adjustment for other risk factors, a change in one unit in the ratio of total cholesterol to HDL-C was associated with a 53% increase in the risk for MI. Complete reliance on a ratio may complicate its interpretation, however because the value of the

ratio could be the same at different levels and types of risk. For example, an LDL-C:HDL-C ratio of 4 could be found with the following ratios, each of which requires different therapeutic strategies:

200:50—very elevated LDL-C, normal HDL-C
160:40—elevated LDL-C, low to normal HDL-C
(depending on the person's sex)
100:25—normal LDL-C, low HDL-C

Perhaps even more important is the fact that the risks imposed by LDL-C and HDL-C are independent of each other and should be considered in absolute terms, because each contributes separately to risk estimates. Combining them into a single number conceals information that may be useful to the clinician in much the same way that combining the systolic and diastolic blood pressures would. A rationale for the ratio can be established from the Framingham data, which suggest that a person who has a high LDL-C may not require treatment if the HDL-C is sufficiently high. Also, the simplicity of gauging treatment on a single outcome certainly has merit. Only with the realization of its potential shortcomings can the LDL-C:HDL-C ratio be useful in managing patients.

Triglyceride

In the fasting state, triglyceride is carried chiefly by VLDL and VLDL remnants. Even though hypertriglyceridemia is common in many parts of the world in which mortality from CHD is low, the heterogeneity of triglyceride and VLDL suggests that subsets of hypertriglyceridemia may be important determinants of cardiovascular risk. There is a growing awareness of the atherogenecity of VLDL-C and VLDL remnants, particularly in women. Such cholesterol transported by VLDL may be cleared from the plasma by potentially atherogenic routes, such as macrophages and smooth muscle cells.

Case-control studies consistently show higher triglyceride levels in survivors of MI and in patients with angiographically demonstrable lesions than in corresponding controls. From 20% to 30% of patients who have CHD also have elevated plasma triglyceride concentrations, compared with 5% to 10% of the "normal" population. The association between triglyceride level and CHD remains significant in many of these studies, even after adjustment for LDL-C and HDL-C levels.

Some prospective observational studies, primarily Scandinavian, consistently have shown a strong relation between triglyceride and MI, but a weaker association with angina pectoris. Earlier prospective studies in the United States demonstrated an association of CHD with hypertriglyceridemia; however, this association often diminished when other risk factors were taken into account (multivariate analysis). The least consistent results were obtained in some analyses when adjusted for HDL-C levels, probably because of the strong inverse relation between triglyceride and HDL-C levels. In contrast, the Honolulu Heart Program found serum triglycerides to be the only risk factor independently predictive of the early onset of disease (before age 60). Current data from Framingham indicate that triglyceride is independently predictive of CHD, although more strongly in women than in men. Women also tend to carry more cholesterol in their VLDL fraction than men. Among women, VLDL-C appears to be a stronger risk factor than LDL-C. With longer follow-up in Framingham, the VLDL-C was found to be a significant predictor of CHD in men as well as in women, although not nearly as strong as that for LDL-C or HDL-C.

As with the observational studies, the most consistent clinical trials to show an association between triglyceride and CHD were reported in Scandinavian studies. The Helsinki Heart Study, a controlled clinical trial, showed a 34% reduction in CHD events related to treatment with gemfibrozil, which decreased LDL-C by about 8% and triglycerides by 35% and increased HDL-C by 10%. The reduction in CHD was greatest among those persons who had elevated triglycerides (Frederickson Types IIb and IV dyslipidemia) and least among those who had isolated hypercholesterolemia. However, the association of CHD with triglyceride levels disappeared when HDL was taken into account. In the Stockholm Secondary Prevention Study, treatment of MI survivors with niacin and clofibrate resulted in reductions of the serum cholesterol and triglyceride concentrations by 13% and 19% respectively; this was associated with a 36% reduction in CHD mortality in the treatment group. In this study population, which had a 50% incidence of hypertriglyceridemia at baseline, treatment benefit was related primarily to a reduction in serum triglyceride concentrations. Because HDL levels were not measured in this study, it is unknown whether this relation would have held after adjustment for HDL changes.

Peripheral vascular disease appears to be more closely related to triglyceride than to total cholesterol, LDL-C, or HDL-C. Certainly, among patients who have Type III dyslipidemia, a disproportionate amount of peripheral vascular disease has been reported. A preponderance of hypertriglyceridemia also has been noted in older men and in women among a series of Swedish patients who had peripheral vascular disease.

Whether triglyceride is an important independent risk factor for CHD has not yet been resolved. Part of the problem stems from the great variability in triglyceride measurements, a situation that relates to biologic and laboratory factors both (see Chapter 6). This variability is much greater than that seen for cholesterol measurements (Fig. 2.15) and reduces the statistical power of the analysis. The strong correlations between triglyceride level and other lipoproteins, in particular HDL-C, also hamper the statistical interpretation of the results; even when a univariate association between triglyceride and CHD is found, the association often disappears after performance of multivariate analysis.

Inconsistencies in the association of triglyceride to CHD are not only caused by statistical obstacles but may also reflect differences in the biologic mechanisms that produce hypertriglyceridemia. The VLDL

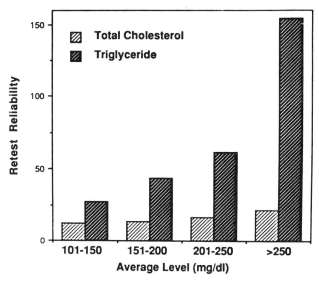

Figure 2.15. Bar graph of intraindividual variation in total cholesterol and triglyceride (mg/dL), based on two measurements from the same persons. Retest reliability (y axis) is equivalent to the SD of paired individual values, and larger values reflect greater intraindividual variation. (From: Austin MA. Plasma triglyceride and coronary heart disease. Arteriosclerosis and Thrombosis 1991;11:2–14.) "By permission of the American Heart Association, Inc."

particles are diverse in size and structure, and it is conceivable that not all VLDL particles have the same atherogenic potential. Premature CHD is absent in familial hypertriglyceridemia (FHT), in which only the triglyceride levels are elevated among family members. In contrast, high triglyceride levels often characterize persons who have other associated lipoprotein abnormalities, such as low HDL (or low apoprotein A) and high apoprotein B levels. This combination is commonly found in the syndrome of familial combined dyslipidemia (see Chapter 7), and is associated with an increased susceptibility to CHD. Type III dyslipidemia (familial dysbetalipoproteinemia), a severe lipid disorder associated with accelerated atherosclerosis and premature CHD, involves both hypertriglyceridemia and hypercholesterolemia caused by an accumulation of chylomicron and VLDL remnants. Finally, high triglyceride levels as well as other lipoprotein abnormalities occur commonly in diseases associated with CHD such as diabetes mellitus, obesity, and chronic renal failure.

Several biologically plausible mechanisms strongly implicate VLDL and triglyceride in the atherogenic process. Hypertriglyceridemia may evolve from various metabolic disturbances and does not represent a single entity. In large part, this is because of the reciprocal transfer of lipids and apoproteins among VLDL, IDL, LDL, and HDL particles throughout the lipoprotein cascade: VLDL is the major triglyceride-carrying lipoprotein, but these particles also contain 20% to 30% cholesterol. Because of its large size, each VLDL particle contains at least five times as much total cholesterol as does one LDL particle; this is important in atherogenesis.

Very low density lipoprotein is catabolized into cholesterol rich IDL remnants as an intermediate metabolic stage in the transition from VLDL to LDL. These remnants contain less triglyceride and proportionately more cholesterol than the VLDL from which they originate and are rich in apoprotein E. The presence of apoprotein E could permit binding of these particles to specific receptors in the endothelial cells where, because of their smaller size, they could filter more readily into the arterial intima. Structurally and functionally abnormal VLDL exists in hypertriglyceridemic patients. In contrast to normal VLDL, these large, abnormal VLDL particles can interact with cell surface receptors on macrophages (see Fig. 1.3). Because one large VLDL particle carries far more cholesterol than one LDL particle, receptor-mediated uptake of abnormal VLDL can be readily injurious to endothelial cells and can facilitate foam cell development.

Defective lipolysis also may be a contributing factor in atherogenesis. The VLDL secreted by the liver releases its triglyceride into the capillaries through the action of the enzyme, lipoprotein lipase (LPL) (see Chapter 4). A mild deficiency in LPL could lead to incomplete catabolism of VLDL, resulting in accumulation of VLDL remnants in the plasma. The VLDL remnants are structurally similar to beta-VLDL, the abnormal lipoprotein subfraction thought to account for the atherogenic lesions noted in familial dysbetalipoproteinemia (Type III). These particles are associated with premature atherosclerosis and play a role in converting macrophages into foam cells. In fact, the only native (unmodified) lipoproteins that have been shown to convert macrophages to foam cells rapidly in vitro are the large VLDLs and chylomicrons from some hypertriglyceridemic patients. This is probably accomplished through specific receptors distinct from the LDL receptor and the acetyl LDL receptor (see Fig. 1.3). Following the uptake of VLDL by macrophages, the triglyceride may be metabolized and mobilized, thus obscuring the source of the remaining cholesterol.

Finally, hypertriglyceridemia is associated with hypercoagulability and deficient fibrinolysis. Triglycerides have been shown to be correlated with factor VII clotting activity and increased levels of tissue plasminogen activator inhibitor (PAI). High levels of circulating triglycerides could initiate thrombosis on a preexisting plaque through a concomitant increase in PAI. The mechanisms underlying factor VII activation by triglyceride-rich lipoproteins remain unclear.

Renewed interest in postprandial dyslipidemia also may provide insight on disorders of triglyceride rich particle catabolism. Lipoprotein particles generally are removed within 6 hours after a meal. A blood sample collected after a 12-hour fast may not reflect whether this rapid clearance of particles has taken place. In certain types of hypertriglyceridemia, chylomicrons persist in the fasting state and are thought to be taken up by macrophages to produce foam cells. Delayed clearance of these intestinal particles may result in a depressed HDL-C concentration or an increase in the concentration of chylomicron and VLDL remnants that are more atherogenic than the precursor chylomicrons. Eventually it may be possible to identify certain types of hypertriglyceridemia that signify disorders of postprandial catabolism and promote atherogenecity of these triglyceride rich particles.

It is important to search for other lipoprotein defects in patients who have high triglycerides (Fig. 2.16). Persons who have so-called borderline elevation of 250 mg/dL to 500 mg/dL (2.82-5.64 mmol/L) may have associated lipid abnormalities such as low HDL-C or high IDL-C levels, low apoprotein A-I and high apoprotein B, and increased VLDL-I remnants, any of which suggests a potentially increased CHD risk. The occurrence of small, dense LDL particles in hypertriglyceridemia already has been discussed. These other defects may only become apparent when the associated hypertriglyceridemia is corrected, for example, with weight reduction or drug therapy.

Patients who have triglyceride values that exceed 500 mg/dL (5.64 mmol/L) must also be considered at risk for developing pancreatitis, because several factors can escalate triglyceride values quickly. Small dietary changes, consumption of alcohol, use of medication (estrogen, retinoic acid, steroids) or some underlying disorders can provoke triglyceridemia to levels over 1000 mg/dL (11.29 mmol/L) in these patients and induce pancreatitis.

A high triglyceride level in any patient should alert the physician to the strong pos-

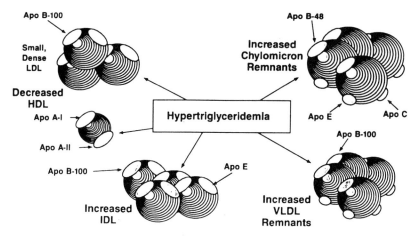

Figure 2.16. Common lipoprotein abnormalities accompanying hypertriglyceridemia. The most common lipoprotein abnormalities produced by hypertriglyceridemia are increased concentrations of chylomicron remnants, VLDL remnants, and intermediate density lipoprotein (IDL), as well as decreased levels of HDL and the presence of small, dense LDL particles. These abnormalities apparently result from metabolic alterations as well as physicochemical changes secondary to the hypertriglyceridemic state. All of them have been implicated in atherogenesis. (From: Coronary heart disease seminar series:Lipid regulation and risk prevention. Parke-Davis, Division of Warner-Lambert Company, 1991.)

sibility of increased CHD risk for various reasons, regardless of whether the association is direct or indirect. Yet, the greater day-to-day variation in plasma triglyceride than in other lipoprotein particles necessitates multiple accurate measurements. Also, blood triglyceride values are required to calculate the LDL-C which is used to direct diet and drug therapy. Hypertriglyceridemia is common in the population and could be substantially normalized with weight loss, diet, and drugs. It therefore is essential to consider such patients for therapy, particularly in the presence of preexisting atherosclerotic disease, diabetes mellitus, or an increased susceptibility to atherosclerosis by virtue of a strong family history of vascular disease or other major risk factors for CVD. Measurement of plasma triglyceride values therefore is important to assess risk for CHD and triglyceride-related pancreatitis. The decision to treat an elevated triglyceride concentration and at what levels to treat it remain somewhat enigmatic.

High Density Lipoprotein and Triglyceride

A strong inverse correlation, ranging from −0.2 to −0.7, exists between HDL-C and tri-

glyceride concentration. This correlation plus the variability of triglyceride levels may account in part for the inability of triglyceride to emerge as a consistent independent risk factor for CHD. Several studies demonstrate that subgroups of men and women who have combined high triglyceride and low HDL-C levels have a substantial risk of CHD. In the Framingham study, the highest CHD rates occurred among the group in the lowest HDL percentile and the highest triglyceride percentile, especially in women. The lowest rates of CHD occurred in those patients who had high HDL combined with low triglyceride. The high triglyceride (> 150 mg/dL [1.69 mmol/L]) and low HDL (< 40 mg/dL [1.03 mmol/L]) levels represent a particularly vulnerable but neglected group.

The combination of increased triglyceride level and low HDL-C is commonly found in association with increased intra-abdominal fat ("central obesity"), high blood pressure, and underlying insulin resistance. This clinical entity has been termed syndrome X and is thought by some researchers to be associated with a diathesis for atherosclerosis (see Chapter 8).

Apolipoproteins (Apoproteins, Apo)

Knowledge about the structure and function of the apoproteins has added greatly to our understanding of the role of lipoproteins in atherogenesis. It is clear that the apoproteins control the activity of enzymes that are responsible for lipolysis, cholesterol esterification, exchange of cholesteryl esters and triglyceride between circulating lipoproteins, and the binding of circulating lipoproteins by cell surface receptors (see Fig. 4.6). As a result, we have come to realize that measurement of cholesterol, triglyceride, and various cholesterol fractions may be a superficial way of assessing cardiovascular risk. Cholesterol and triglyceride are simply transported by the lipoproteins, and the accumulation of these lipids in plasma reflects abnormalities in lipoprotein or apoprotein metabolism and function. Pathologic studies have shown clearly that apoprotein B-containing lipoproteins accumulate in atherosclerotic lesions. Consequently, questions have been raised about whether measurement of apoproteins, especially A-I and B-100, by one of several methods might provide a more accurate assessment of risk.

Apoprotein B-100 (apoprotein B) is the major apoprotein found in VLDL and LDL. Measurement of apoprotein B provides an estimate of the total number of VLDL and LDL particles because there is only one molecule of apoprotein B per lipoprotein particle. Hypercholesterolemic patients almost always have elevated apoprotein B levels and are at high risk for CHD; little is added by measurement of apoprotein B in these patients. Among patients who have high LDL levels, those who have CHD have even higher levels of plasma apoprotein B than those who do not have CHD. Some young to middle-aged adults who have CHD have increased LDL apoprotein B with normal LDL-cholesterol, a condition called hyperapobetalipoproteinemia. In familial combined dyslipidemia, elevated levels of apoprotein B coexist with normal total and LDL-C. Apoprotein B, therefore, might identify persons who have borderline cholesterol levels at risk for premature CHD.

Hypertriglyceridemic patients who have CHD also have higher levels of apoprotein B than other patients. In patients who have moderate hypertriglyceridemia, from 250 to 500 mg/dL (2.82-5.64 mmol/L), a high apoprotein B level indicates an increased number of atherogenic lipoprotein particles. Patients who have small, dense LDL may have apoprotein B measured as abnormal because of methodologic problems in the analysis but do not actually have true hyperbetalipoproteinemia.

It is thought that apoprotein A-I might be the active component in HDL, accounting for the mobilization of cholesterol from the arterial wall to other lipoproteins and the liver. If so, the risk measurement for CHD should be higher for apoprotein A-I. In one review that summarized 12 different angiographic studies, coronary atherosclerosis related to HDL-C and the concentrations of apoprotein A-I and-II. Apoprotein A-I was the best overall discriminator, however, between patients with and without coronary heart disease. Other studies indicated that men who have survived an acute MI or have angina as a result of coronary atherosclerosis have lower apoprotein A-I levels than healthy controls.

Multivariate analysis has indicated that the apoproteins add additional independent information to the more conventional lipoprotein measurements. Because no longitudinal studies have been done with apoprotein A-I or apoprotein B levels comparable to those done with total cholesterol or lipoprotein subfractions, it cannot be established that these abnormalities precede the onset of or predict the development of clinical disease. The lack of standardized techniques for the measurement of apoproteins poses technical problems in interpreting the clinical significance of abnormal values. Furthermore, there are no established cutpoints for elevated apoprotein B levels. It is not yet widely accepted that all apoprotein B-containing lipoproteins are equally atherogenic, so that measurement of total apoprotein B levels can only be a crude index of

CHD risk. Currently, apoproteins can provide only limited information for clinical purposes beyond that provided by blood lipid and lipoprotein cholesterol concentrations.

Lipoprotein(a)

Several epidemiologic studies have found a positive association of plasma lipoprotein a [Lp(a)] concentrations with premature MI. The level of Lp(a) appears to be under strict genetic control at a single locus and is unaffected by sex, age, or diet. Data suggest that the prevalence of elevated Lp(a) levels is higher among blacks and that metabolic determinants for Lp(a) levels may be different among blacks, compared with whites.

As an inherited risk factor, Lp(a) most likely interacts with the trait for familial hypercholesterolemia, possibly explaining why some families are at higher risk of CHD. The relative risk for CHD among persons who have Lp(a) values that exceed 50 mg/dL and concomitantly high LDL-C concentrations may be increased up to sixfold. Whether the Lp(a) concentration is a significant risk factor in normolipidemic patients is unresolved. If Lp(a) is athero-genic in the absence of elevated LDL, it is uncertain whether there is any threshold plasma Lp(a) concentration or whether the risk is continuous.

Lp(a) is assembled from two different components, the apoprotein A and apoprotein B-100 of the LDL particle. The recent discovery of a strong homology of apoprotein (a) to plasminogen allows for two potential mechanisms for the atherogenic properties of Lp(a): (1) by contributing to thrombogenesis and (2) by raising the blood LDL concentration. As a result of its structure, Lp(a) may modulate the clotting process and prolong the time required for fibrinolysis by competitively inhibiting plasminogen activation.

Other Important Risk Factors

Age

The risk of CHD rises steeply with age: CHD prevalence, incidence, and mortality rate approximately double for each five-year period after age 24. Figure 2.17 depicts the number of Americans who experience an MI, according to age and sex. For men the

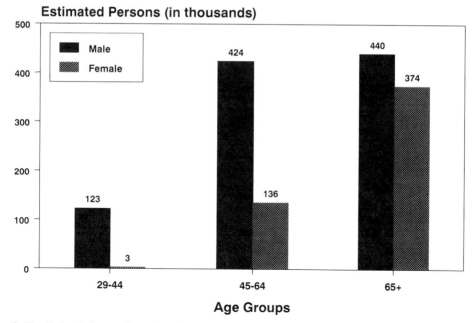

Figure 2.17. Estimated annual number of Americans by age and sex who experience a heart attack. (From: American Heart Association 1991 Heart and Stroke Facts, No. 55-0379[COM].) "Reproduced with permission. Copyright American Heart Association."

occurrence increases abruptly with middle age, but for females there is a modest increment with middle age, with a much larger increase at age 65 and above.

Sex

Generally, women develop clinical manifestations of CHD about 10 years later than men. Unless there are overriding risk factors, women rarely exhibit CHD before menopause. Once menopause has occurred, whether naturally or artificially, women begin to develop CHD at rates similar to men. It is assumed that there is a protective influence of female hormones against atherosclerosis, and such protection has been observed with estrogen replacement therapy. This is possibly because of an estrogen effect on lipids, although a direct effect on vessel wall properties is possible. Nevertheless, nontreated, postmenopausal women should be considered at risk for CHD as much as men.

Although men and women also share the same major risk factors, there are some differences in the contribution of these factors to the relative risk of CHD. Elevated triglyceride levels appear to be a much more potent CHD risk factor among women in the United States, especially in the older age groups. Women who have diabetes are much more likely to develop manifestations of CHD than men with comparable disease. There also appears to be a stronger interaction between elevated triglyceride and low HDL-C levels in women, as demonstrated in the Framingham Study. In the past, the use of oral contraceptives, particularly when associated with cigarette smoking, markedly increased the risk of MI. This may not be true for currently used oral contraceptives, because they contain lower doses of estrogens and androgenic progestins. Women also appear to have a worse prognosis after an MI, which may be related in part to an increased number of associated risk factors that led to the coronary attack.

Elevated Blood Pressure

The most recent National Health and Nutritional Examination Survey (NHANES II),

completed between 1976 and 1980, classified 22% of persons between 25 and 74 years of age as hypertensive on the basis of blood pressures of 160/95 mmHg or greater. Isolated systolic hypertension, defined as a systolic pressure higher than 140 mmHg and a diastolic blood pressure less than 90 mmHg, occurs in more than one-third of persons older than 65. The hemodynamic consequences of hypertension appear to account for the increased predisposition toward atherosclerosis.

At all ages in either sex, the risk of CVD increases in proportion to blood pressure. Although treatment traditionally is based on diastolic elevations, the systolic pressure is more strongly related to the risk of CVD. In the Hypertension Detection and Follow-up Program (HDFP), mortality from all causes increased by 1% with each millimeter increase in systolic pressure in persons who have isolated systolic hypertension. Definite hypertension triples the likelihood of atherothrombotic brain infarction and congestive heart failure, but its impact is relatively less for CHD and peripheral arterial disease. Despite this, CHD is the most common manifestation that results from hypertension because of its higher frequency in the population. Women appear to tolerate elevated blood pressure levels better than men, as evidenced by their lower morbidity and mortality rates at all levels of blood pressure. Blacks have higher blood pressures than whites and are more susceptible to the cardiovascular complications of hypertension.

The risk of CHD increases about twofold with high blood pressure and escalates even more with dyslipidemia. Hypertension is often associated with defects in carbohydrate and lipid metabolism, and untreated patients who have hypertension tend to be insulin-resistant and hyperglycemic. Commonly used drugs for lowering blood pressure, especially diuretics and beta-blockers, actually may exaggerate these problems. Furthermore, a familial predisposition to CHD because of the coexistence of hypertension and dyslipidemia has now been described—familial dyslipidemic hypertension. Studies of families in Utah revealed concordant lipid abnormalities in approximately 12% of those persons who had essential hypertension, a syndrome likely to be more severe than blood pressure elevation alone.

Cigarette Smoking

Reports of the Surgeon General and others have noted consistently that cigarette smoking is the single most preventable cause of premature death in the United States. By 1986, smoking prevalence rates in the United States had declined to the lowest level observed in more than 40 years—only 27% of all adults smoke cigarettes on a regular basis. Smoking rates have declined more rapidly for men than for women, dropping in the past 20 years from 50% to 30% among men but only from 34% to 24% among women.

Evidence that incriminates cigarette smoking in the cause of CVD is substantial and unequivocal. There is a dose-response relation among the number of cigarettes smoked per day, years smoked, and CVD. Regular smoking nearly doubles the probability of cardiovascular disease and total mortality and triples the risk for sudden death, especially among young men. The relative risk of CVD from cigarette smoking decreases with age but persists for men even beyond 65 years of age. Preliminary data indicate that passive smoking also promotes CHD and adversely modifies lipoprotein risk factors.

In contrast to other major risk factors, cigarette smoking is a powerful risk factor in the development of atherosclerosis in the aorta and peripheral arteries. The risk of cerebral ischemia, hemorrhagic stroke, and nonhemorrhagic stroke appears to be greater for smokers than previously reported. Cigarette smoking increases the vulnerability to MI and stroke, particularly among women who use oral contraceptives. Substitution of filter or low-yield cigarettes does not alleviate CHD risk, possibly because smoking habits are altered to obtain the same amount of nicotine. Nonperforated filters actually deliver more carbon monoxide. Cigar and pipe smoking without inhalation have not been demonstrated to have the same relation to CHD as cigarette smoking.

Cigarette smoking relates more closely to acute clinical events than to the primary atherosclerotic process itself. Several possible mechanisms might explain the risk of CHD from cigarette smoking: endothelial injury; changes in HDL and other lipoproteins; an increased propensity for thrombosis; increased vascular resistance; altered oxygen supply and demand; susceptibility to arrhythmias; and related pulmonary dysfunction. Even after adjusting for other differences, those persons who continue to smoke still have a much greater risk of progressive CHD and related complications than those who stop smoking. This is true for persons who do or do not have CHD. Unlike the risk of developing chronic obstructive pulmonary disease or smoking-related cancers, current evidence indicates that the CHD risk declines after cessation, so that, after one year or more, persistent quitters assume a risk similar to nonsmokers. Data from studies of the general population and coronary patients alike suggest a reduction in subsequent CHD of around 50% for men who are able to quit smoking compared to those who continue to smoke.

Family History of Cardiovascular Disease

Several prospective studies have found a positive family history to be an independent predictor of CHD, even after accounting for other known risk factors by multivariate analysis. The magnitude of the attributable risk depends on the definition of a positive family history, that is, 6% to 16% for one or more first degree relatives who had CHD at any age, 35% for one before age 60, and as high as 53% for two or more first-degree relatives who had CHD before age 55.

The susceptibility for development of atherosclerosis is determined mostly by polygenic effects. Although, environmental effects can modify this genetic susceptibility considerably. Factors such as smoking and dietary habits correlate significantly between spouses and blood relatives and can be learned by the offspring. The proportion of variance in occurrence of some risk factors attributable to genetic effects and cultural effects has been estimated in family studies of twins and pedigrees in Utah, as shown by Figure 2.18. It has been estimated that approximately 50% of the variability of these quantitative traits are attributable to genetic traits.

Evidence for a genetic effect through major gene loci has been found for several dyslipidemic syndromes, including familial hypercholesterolemia, low HDL-C, high apolipoprotein B, type III dyslipidemia, Lp(a) levels and, possibly, familial combined dyslipidemia. Other coronary risk factors influenced by genetic predisposition include diabetes mellitus, homocystinuria, and elevated fibrinogen levels.

In a study to determine the frequency of familial dyslipidemia syndromes among families who have premature CHD, it was found that three-fourths of those members who had premature CHD also had lipid abnormalities (cholesterol > 90th percentile or triglyceride levels > 90th percentile or HDL < 10th percentile). The HDL-C and triglyceride abnormalities were twice as common as LDL-C abnormalities, with the most commonly found syndromes being familial combined dyslipidemia, familial dyslipidemic

hypertension, and isolated low HDL-C levels.

Glucose Intolerance

Clinical diabetes mellitus doubles the likelihood of cardiovascular mortality. Although the relative impact is greatest for peripheral arterial disease, CHD is still the most common complication of diabetes. The associated risk is more closely linked to women than to men and is more than additive to other coexisting risk factors. The incidence of CVD among diabetic women is almost triple that of nondiabetic women and nullifies any protective effect of female sex. The elevated insulin levels that commonly accompany noninsulin diabetes mellitus (NIDDM) have been shown to predict future risk from atherosclerotic-related heart disease, independent of other risk factors.

Prediabetic persons usually have an atherogenic pattern of risk factors that precede the development of clinical diabetes by many years. Such persons have higher plasma glucose levels after an oral glucose challenge, higher fasting triglyceride levels, lower HDL-C concentrations and elevated systolic and diastolic blood pressures compared with persons who have normal insulin levels. Furthermore, these persons tend

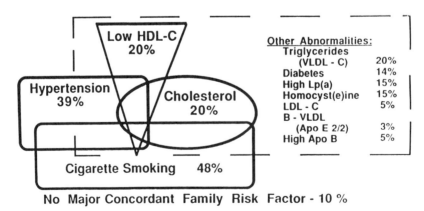

Figure 2.18. Appropriate frequencies of concordant abnormalities in two or more siblings who have CHD under the age of 55, estimated from family studies in Utah. There is considerable overlap between risk factors (i.e., more than one risk factor syndrome often is detected in both affected siblings), and a large percentage of early familial CHD involves syndromes not detected by the simple assessment of total cholesterol, blood pressure, and cigarette smoking. (From: Williams RR, Hopkins PN, Hunt SC. Evaluating family history to prevent early CHD. (Personal Communications).

to be more vulnerable to the consequences of hypertension and cigarette smoking. Insulin resistance and hyperinsulinemia appear to have a central role in the development of the metabolic sequence that leads to increased CHD risk. It may be highly desirable to reduce the cardiovascular risk of those persons who have abnormal glucose tolerance by intervening before the onset of clinical diabetes. Reduction of the risk of cardiovascular complications in diabetics may depend on attenuation of other atherogenic traits more than on the strict control of blood sugar. The present state of knowledge does not demonstrate conclusively that improved control of blood glucose values will lower the risk of CVD.

Physical Activity

The National Health Interview Survey of Health Promotion and Disease Prevention in 1985 revealed that only 43% of all men and 38% of all women exercised or played sports regularly. Longitudinal population studies consistently demonstrate that habitual physical activity protects against CHD. Taking other factors into account, a sedentary man has about twice the relative risk for the occurrence of CHD, in contrast to his more vigorous peers. A correlation between protection from CHD and state of cardiovascular fitness also suggests that the intensity of activity is important. The relation is more apparent with strenuous exercise and appears to be dose dependent. Implicit in this concept of protection from CVD is that the intensity and duration of exercise must be sufficient to stress the cardiorespiratory system and to invoke training adaptations. Yet, several investigators have reported that moderate intensity exercise (e.g., brisk walking, stair climbing, or gardening), if done on a regular basis, will suffice.

Possible mechanisms for preventing or alleviating coronary artery disease by regular vigorous exercise include decreased myocardial work and oxygen demand, retardation or regression of atherosclerosis, improved myocardial oxygen delivery or use, enhanced myocardial function, reduced vulnerability to arrhythmias, and favorable metabolic and lipid adaptations. Exercise reduces the overall risk of CVD by helping to maintain desirable body weight, producing greater glucose tolerance, improving lipid patterns, lowering blood pressure, and, possibly, enhancing fibrinolysis.

A growing number of controlled clinical trials have demonstrated reduced mortality in coronary patients randomized to a physical training program. Pooled results from these major trials indicate a 19% reduction in total mortality over a period of three years for the exercise group (compared to the nonexercise group). Sufficient evidence exists to justify the recommendation of regular, frequent isotonic exercise as part of a healthy lifestyle. But exercise is not without hazard, including sudden death during and immediately after a work-out. This emphasizes the need for a thorough cardiovascular evaluation before initiating exercise programs among those persons at special risk. Nevertheless, sudden cardiac death is substantially less common among active people. Several reports suggest that those persons who exercise regularly and sustain a MI are more likely to survive the event.

Obesity

The primary criterion for overweight is a body mass index (weight/height^2) of 25 to 30 Kg/m$^{.2}$ or a relative weight (distribution of weight according to height) of 20% to 40% above the median weight. Higher values of either body mass index or relative weight define obesity as opposed to overweight. About 26% of adult men and 30% of adult women in this country were found to be overweight in the NHANES II.

Obesity has long been known to predispose to CHD, but the mechanism was thought to be related primarily to its influence on other risk factors. Some of the major risk factors for CHD, including hypercholesterolemia, decreased HDL-C, hypertension, and diabetes mellitus are strongly associated with increased body weight. But even after adjusting for other major risk factors, overweight significantly increases the predicted

CHD incidence in men and women alike (Table 2.5). Risk of coronary heart disease increases progressively after body mass exceeds the average by 10% to 19%. This association of weight with CVD, most pronounced before the age of 50 years, increases with duration of overweight, so that the death rate from CHD is about 50% greater for those who remain overweight throughout adult life. Large fluctuations in body weight, especially during young adulthood, may be more detrimental toward CHD than weight gain without losses. The relation between total mortality and obesity is characterized by a U-shaped relation. The increased mortality noted for those markedly underweight most likely represents persons who have chronic diseases, unfavorable health traits such as smoking, or older persons who have disease-related nutritional problems.

Reduction in weight, even to levels still above the desirable weight, generally will decrease associated risk factors. This may not be true for all lipid changes. During weight loss, serum triglyceride values fall to more desirable levels within a period of days or weeks, but it may be necessary to achieve near ideal weight to raise HDL-C concentrations. HDL-C levels may actually decrease transiently during the hypocaloric phase of active weight loss. Linear regression analyses indicate HDL-C increases by 1 mg/dL for each decrement in body mass of about 0.1 kg/m². Data obtained in Framingham suggest that a 10% reduction in relative weight for men is associated with a fall in

Table 2.5 Relation of Body Mass Index to Relative Mortality Risk

Body Mass Index[a]	Risk Category	Mortality Risk Ratio
20–25	Desirable	1.0
25–30	Mild Risk	1.0–1.25
30–40	Moderate	1.25–2.5
> 40	High Risk	> 2.5

[a] Body Mass Index = wt/ht² in kg/m². (From Grundy SM, Greenland P, Herd A, et al. Cardiovascular and risk factor evaluation of healthy American adults: A statement for physicians by an ad hoc committee appointed by the steering committee, American Heart Association. Circulation 1987;75: 1340A–1362A.)

serum glucose of 2.5 mg/dL, a fall in systolic blood pressure of 6.6 mmHg, and a fall in serum uric acid of 0.33 mg/dL. For men, these data predict an anticipated 20% decrease in the CHD for each 10% reduction in body weight.

Recent data indicate that body fat distribution relates better to CHD than either body mass index or relative weight. Among men, the increased adipose tissue usually involves principally the abdominal region (central obesity), whereas, in women, the excessive fat usually accumulates in the buttocks and thighs (peripheral obesity). The ratio of the circumference of the waist to that of the hips (waist-to-hip ratio) may be used to assess whether body fat distribution is of the male or female type. The male pattern of excessive abdominal fat, presumably because of regional differences in circulation or metabolism, predisposes more to elevated blood pressure, hypertriglyceridemia, and glucose intolerance. This android pattern is detrimental for both men and women. In men the risk of CHD increases at waist-to-hip ratios above 1.0, and in women, at ratios above 0.8.

Psychosocial Factors

Extensive data exist on reported associations between CHD and anxiety, stress, coping, life events, social mobility, and poor social support. Attention has focused on the behavior pattern known as Type A. The Type A behavior pattern is characterized by extreme competitiveness, impatience, time urgency, aggressiveness, striving for achievement, and abrupt, tense speech and movement. Claims about the independence of the Type A behavior pattern as a risk factor for CHD have yet to be documented convincingly. Other analyses of Type A behavior have isolated specific components that relate more strongly to CHD than the pattern as a whole; these components—anger and hostility—may be the key elements associated with CHD risk.

Other Risk Factors

A number of other factors have been associated with an increased probability of

CHD. Gout and hyperuricemia have long been thought to be independently related to CHD. Among women, both menopause and the use of oral contraceptives appear to be related to an increased risk of developing CHD. Newer preparations of oral contraceptives may not confer an additional risk for CHD (see Chapter 8). The question of whether postmenopausal hormones, primarily estrogens, can diminish the risk for CHD has not been settled, but most evidence suggests that they do.

Although alcohol intake has been established to be asssociated positively with blood pressure levels, the relation of alcohol intake to cardiovascular mortality is more complex. Moderate alcohol consumers (no

more than two drinks per day) have a lower risk of CHD than either nonusers or heavy consumers (often attributed to higher HDL-C levels). Using alcohol even in moderation as protection against CHD cannot be advocated because of the lack of substantiation of cause-and-effect and the larger potential problem of inducing alcoholism in a segment of the population. Data on coffee intake remain inconclusive.

Inadequate attention has been given to thrombotic tendencies, but new evidence implicates abnormal levels of plasma fibrinogen, platelet aggregation, and even blood viscosity as CHD risk factors. Effects of trace metals and immunologic factors on atherosclerosis have been suggested but remain in-

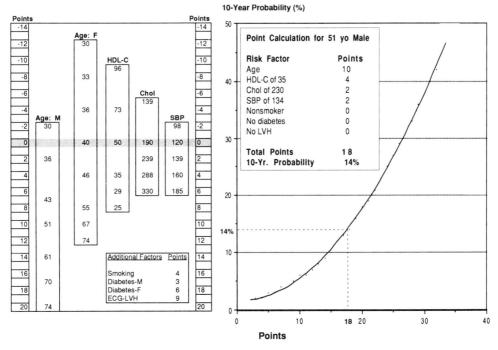

Figure 2.19.   Estimated 10-year probability of coronary heart disease based on the calculation of risk factor points. Also, this presentation allows comparison of the quantitative differences among risk factors. To calculate the 10-year probability of CHD, determine the number of points associated with each risk factor value and add them. For example, being male (M) aged 51 equals 10 points; an HDL-C of 35 mg/dL, 4 points; a total cholesterol (Chol) of 230, 2 points; a systolic blood pressure (SBP) of 134, 2 points; being a nonsmoker without diabetes or ECG-LVH contributes no additional points. After summing these seven factors, the 10-year probability of CHD can be estimated from the graph. For the 18-point total in the example, the 10-year probability of CHD is 14%. This nomogram provides a rapid approximation of points; however, it is unlikely that this would alter the estimated 10-year probability. For the most precise estimate, the reader should refer to the original source from which this figure was derived. (F=Female) (From: Anderson KM, Wilson PWF, Odell PM, Kannel WB. An updated coronary risk profile: a statement for health professionals. Circulation 1991;83:356–362.)

conclusive. Various subclinical manifestations of CHD presage future clinical disease: electrocardiographic abnormalities at rest, especially left ventricular hypertrophy; carotid bruits; and ischemic signs and symptoms on exercise testing, especially in the presence of other conventional risk factors. Also a striking inverse relation between vital capacity and cardiovascular mortality has been reported from the Framingham Study.

Multivariate Risk Function

Atherosclerosis generally is considered to be a multifactorial disease. The risk associated with any single lipid abnormality varies with the number and magnitude of other risk factors present. To truly estimate risk, multivariate risk assessment is essential. First, no single risk factor inevitably leads to the development of CHD. Reasonable estimates of the likelihood that persons will develop an atherosclerotic cardiovascular event can best be made by applying risk formulations that involve multiple components. All probabilities are conditional on the particular set of risk factors chosen and the population to whom they are applied. Second, by using clinical cut-points set at high levels, those persons who are at high risk because of multiple marginal abnormalities often are missed. Multiple borderline elevations lead to greater risk than any isolated, even markedly elevated risk factor.

Cholesterol certainly interacts in a synergistic fashion with other risk factors to potentiate the risk for CHD (see Fig. 2.9). In particular, cigarette smoking and hypertension produce higher mortality from CHD when combined with hypercholesterolemia than when predicted on the basis of each risk factor separately. There is no question that the major risk factors are interrelated and enhance each another. To facilitate the assessment of risk with major risk factors, handbooks have been compiled and appropriate tables derived from multivariate risk equations used by the American Heart Association and other sources (see Chapter 13). Figure 2.19 shows the calculation of the probability of developing CHD in 10 years for men and women according to coexistent risk factors derived from the 30-year follow-up of the Framingham Study.

Suggested Readings

REVIEW ARTICLES

Anderson KM, Wilson PWF, Odell PM, Kannel WB. An updated coronary risk profile: a statement for health professionals. Circulation 1991;83:356–362.

Austin MA. Plasma triglyceride and coronary heart disease. Arteriosclerosis and Thrombosis 1991;11:2–14.

Breslow JL. Apoplipoprotein genetic variation and human disease. Physiol Rev 1988;68(1):85–132.

Castelli WP, Wilson PWF, Levy D, Anderson K. Cardiovascular risk factors in the elderly. Am J Cardiol 1989;63:12H–19H.

Garber AM, Sox HC, Litenberg B. Screening asymptomatic adults for cardiac risk factors: the serum cholesterol level. Ann Intern Med 1989;110:622–639.

Grundy SM, Vega GL. Role of apolipoprotein levels in clinical practice [Editorial]. Arch Intern Med 1990;150:1579–1581.

Hopkins PN, Williams RR. Human genetics and coronary heart disease. A public health perspective. Ann Rev Nutr 1989;9:303–345.

Lefer AM. Prostacyclin, high density lipoproteins, and myocardial ischemia. Circulation 1990;81:2013–2015.

Stampfer MJ, Colditz GA. Estrogen replacement therapy and coronary heart disease: a quantitative assessment of the epidemiologic evidence. Prev Med 1991;20:47–63.

POSITION PAPERS

Atherosclerosis Study Group. Optimal resources for primary prevention of atherosclerotic diseases. Circulation 1984;70:155A–205A.

Exercise standards: A statement for health professionals from the American Heart Association. Circulation 1990;82:2286–2322.

Gotto AM. AHA conference report on cholesterol. Circulation 1989;80:715–744.

Grundy SM, Greenland P, Herd A, et al. Cardiovascular risk factor evaluation of healthy American adults—a statement for physicians by an ad hoc committee appointed by the Steering Committee, American Heart Association. Circulation 1987;75:1340A–1362A.

National Research Council (U.S.). Diet and health: implications for reducing chronic disease risk. Committee on Diet and Health, Food and Nutrition Board, Commission on Life Sciences, National Research Council. Washington, DC: National Academy Press, 1989.

Report of the National Cholesterol Education Program Expert Panel on Detection, Evaluation and Treatment of High Blood Cholesterol in Adults. Arch Intern Med 1988;148:36.

U.S. Department of Health and Human Services, Public Health Services, Centers for Disease Control, Center for Chronic Disease Prevention and Health Promotion, Office on Smoking and Health. Reducing the health consequences of smoking: 25 years of progress. A report of the Surgeon General. DHHS Publication No. (CDC) 89–8411, 1989.

OTHERS

Badimon JJ, Badimon L, Fuster V. Regression of atherosclerotic lesions by high density lipoprotein plasma fraction in the cholesterol-fed rabbit. J Clin Invest 1990;85:1234–1241.

Goldman L, Weinstein MC, Williams LW. Relative impact of targeted versus populationwide cholesterol interventions on the incidence of coronary heart disease—projections of the coronary heart disease policy model. JAMA 1989;80:254–260.

Reaven GM. Banting lecture 1988: Role of insulin resistance in human disease. Diabetes 1988;37: 1595–1607.

The Pooling Project Research Group. Relationship of blood pressure, serum cholesterol, smoking habit, relative weight, and ECG abnormalities to incidence of major coronary events: final report of the Pooling Project. J Chron Dis 1978;31:201.

Williams RR, Hopkins PN, Hunt SC, Wu LL, Hasstedt SJ, Lalouel JM, Ash KO, Stults BM, Kuida H. Population-based frequency of dyslipidemia syndromes in coronary-prone families in Utah. Arch Intern Med 1990;150:582–588.

CHAPTER 3

Epidemiologic Considerations

Distribution of Lipids in the United States Population

Because of the central role of blood cholesterol and lipoproteins in the development of atherosclerotic disease, their distribution in our population becomes extremely important. It provides useful reference values for comparing patients with their age- and sex-matched counterparts and facilitates detection of changes in their levels over time relative to anticipated changes. Finally, knowledge of blood cholesterol and lipoprotein distribution makes it possible to estimate the number of persons who will be over or below a specific cut-point value when screening various populations.

Reference laboratory values traditionally are considered normal if they fall within two standard deviations of the mean, or between the 5th and 95th percentiles of the normal distribution curve. For the prevention of CHD it is important to distinguish between "normal" and "desirable" or "optimal" values. A normal value generally refers to the average, not necessarily the desirable level. An average total cholesterol value in the United States would be considered abnormal in Japan, where the average value is below 200 mg/dL (5.17 mmol/L). This situation is analogous to that found in various other traits, such as body weight, blood pressure, and maximal oxygen capacity, in which the average American population value differs from the optimal value.

Unfortunately, many physicians tend to be guided by their laboratory's normal reference range, even when this range does not make sense in the clinical and epidemiologic sense. By one common definition, which considers the 95th percentile to represent the upper normal range, only 5% of the population is hypercholesterolemic. Yet the majority of CHD occurs among that half of the population who have borderline to high cholesterol values, which lie well within these reference values. Even though the relative risk may not be as high at these levels, the greater number of persons who have cholesterol levels in this range leads to a higher attributable risk and more CHD morbidity and mortality. The average total cholesterol found in patients who sustain an MI ranges from 220 to 230 mg/dL (5.69-5.95 mmol/L), values that would not be considered markedly abnormal. Small, but not trivial reductions in cholesterol values for the population as a whole can result in sizable reductions in CHD for that large number of men who have cholesterol levels initially in the borderline range. It is important to note that about one-third of middle-aged Americans have serum total cholesterol values of 240 mg/dL (6.21 mmol/L) or above, the recently established "cut-point" for an abnormal value (Fig. 3.1).

Several studies have found that merely by lowering the reference values for "abnormal" as designated on laboratory reports, more persons were provided appropriate evaluation and treatment of lipid disorders. Clinical decision making requires availability of lipid reference values for specific population groups classified by age, sex, and race. The optimal lipid value for any given patient is best assessed in conjunction with the history, physical examination, and the presence or absence of other risk factors for cardiovascular disease (CVD). With this additional information, the physician can better judge the prognostic significance of a lipid value and the intensity with which treatment should be undertaken.

Two major studies have defined the lipid distributions of adults in this country: the Prevalence Study of the Lipid Research Clinics (LRC) Program, which was conducted from 1972 to 1976 by the National Heart, Lung, and Blood Institute (NHLBI), and the second National Health and Nutrition Examinations Survey (NHANES II), which was conducted from 1976 to 1980 by the National Center for Health Statistics. Because of dissimilarities in selection of the population, data collection and technical procedures, some differences in lipid values exist between these two studies. Methodologic differences between the studies include the exclusion of persons who had triglyceride levels greater than 200 mg/dL

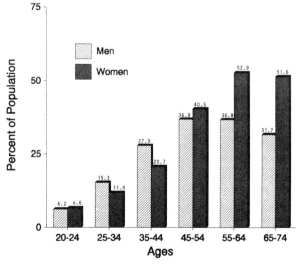

Figure 3.1. The estimated number of Americans who have serum total cholesterol of 240 mg/dL or more. The prevalence of hypercholesterolemia based on a level of 240 mg/dL or more ranges from slightly over 6% in the 20 to 24 age group to over 50% of the women and about one-third of the men 55 years and older. It is apparent that there is an increase in the prevalence with age for women and for men alike, although there is a slight decrease for men in the oldest age group, 65 to 74 years. (From: American Heart Association 1991 Heart and Stroke Facts, No. 55-0376 [COM]. Reproduced with permission. Copyright American Heart Association.)

(2.26 mmol/L) in the LRC Prevalence Study and recruitment of a representative population in NHANES II. In general, slightly higher average lipid values were observed in the NHANES II Study. The data selected for reference values were taken from the LRC Prevalence Study. Although this study is not as representative of the American population as was NHANES II, rigorous attention was given to blood collection techniques and fastidious standardization of methods for lipid determinations. It is one of the few large data sets in which information is available on fasting lipids and lipoproteins. Furthermore, it provided information on women who were and were not receiving hormones. These data provide excellent reference values for age and sex but not for race because of an inadequate sample size. The curves have been smoothed to simplify estimation of these reference values. Figures 3.2 through 3.6 show the distribution of total cholesterol, triglyceride, and lipoprotein-cholesterol by age groups for both men and women by selected percentiles.

Blood levels of total cholesterol increase with age in both men and women, but the effect of aging is much more marked among women (see Fig. 3.2). For women 50 years of age and older, the 50th percentile (median value) is approximately 230 mg/dL (5.95 mmol/L), whereas for men it is approximately 210 mg/dL (5.43 mmol/L). Various lipoprotein cholesterol values by age are shown in Figures 3.3 through 3.5. The median value for LDL-C does not generally exceed 150 mg/dL (3.88 mmol/L), even for older persons.

The 50th, 25th and 10th percentile values for HDL-C are provided, rather than the high percentiles, because of its inverse relation with CHD. As expected, adult women have consistently higher HDL-C values than men, even after menopause. Whereas women show a gradual modest increase of HDL-C with age, the increase in men occurs primarily after age 50. Because these are prevalence data, the higher values among older men may represent an artifact of differential survival, because men who have

lower HDL-C values tend to die at earlier ages.

The VLDL-C increases consistently with age in women, but in men it peaks in middle age and then decreases. The trends and variability in plasma triglyceride levels are similar to those in VLDL-C. This is not surprising, because VLDL consists predominantly of triglyceride. Trends in plasma triglyceride levels are shown in Figure 3.6. The values are higher in men than in women at younger ages but become similar after 50 years of age. The median value (50th percentile) remains well below 150 mg/dL (1.69 mmol/L) for both men and women at all ages. The variability in triglyceride level is much greater than that for plasma cholesterol and is affected to a greater extent by an incomplete fasting period. (See Fig. 2.15).

Factors That Influence Lipid Values
Smoking

Cigarette smoking may alter lipoprotein levels. The effect is more prominent in women than in men, especially below 50 years of age. There is a dose-response relation between the number of cigarettes smoked and the lipoprotein changes. Cigarette smoking has its greatest impact on HDL-C levels, which can decrease by 2 to 9 mg/dL (0.05-0.23 mmol/L). In contrast, smoking cessation is associated with increases in HDL-C. Smoking also may raise the total cholesterol levels, presumably by increasing the level of LDL-C. Consequently, smoking is one of the few habits whose alteration can lead to modified lipid profiles. Recent studies also incriminate passive smoking in altering lipoprotein profiles; it has been shown that preadolescent children exposed to long-term passive cigarette smoke, primarily from smoking mothers, have lower average HDL-C than children from nonsmoking mothers.

Obesity

Obesity is the only factor that consistently shows a positive relation with LDL-C and VLDL-C and a negative association with

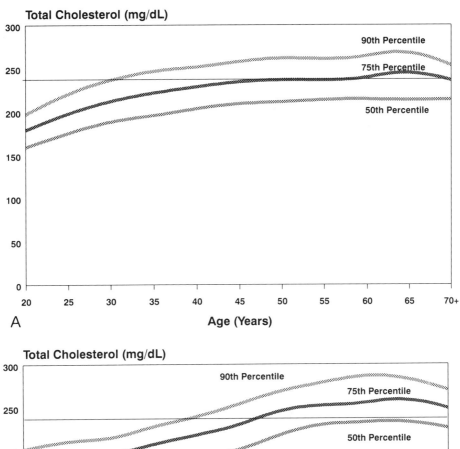

A

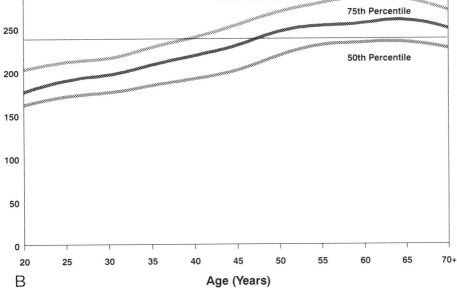

B

Figure 3.2. Selected percentiles of plasma total cholesterol by age for white males (3.2A) and females (3.2B) (nonhormone users). Horizontal line represents the commonly used cut-point. The curves are smoothed. (Adapted from: United States Department of Health and Human Services. The Lipid Research Clinics Population Studies Data Book, vol. 1, The Prevalence Study. NIH Publication No. 80-1527, 1980.)

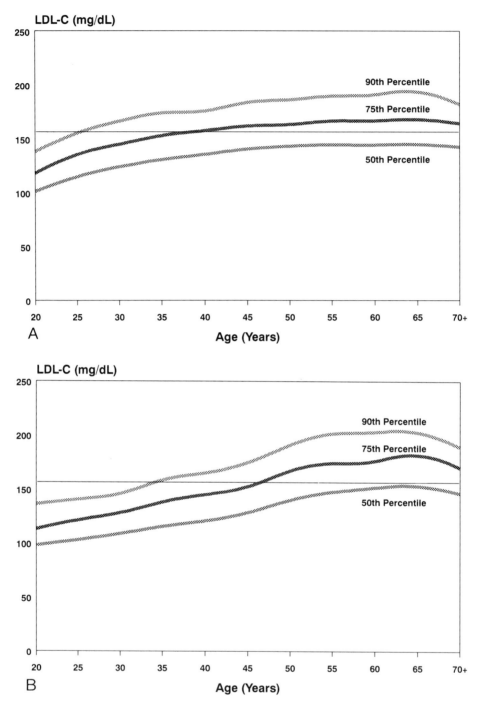

Figure 3.3. Selected percentiles of plasma LDL cholesterol by age for white males (3.3A) and females (3.3B) (nonhormone users). Horizontal line represents the commonly used cut-point. The curves are smoothed. (Adapted from: United States Department of Health and Human Services. The Lipid Research Clinics Population Studies Data Book, vol. 1, The Prevalence Study. NIH Publication No. 80-1527, 1980.)

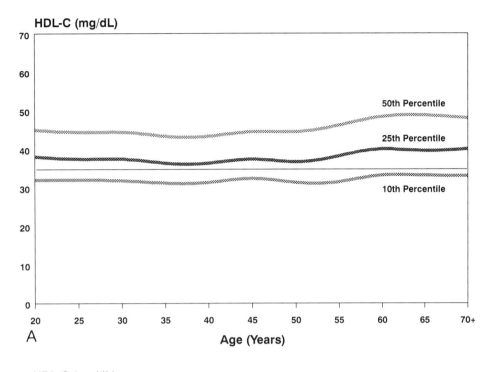

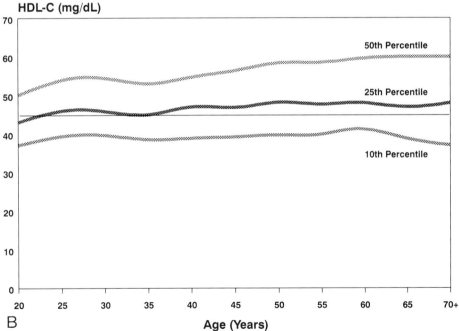

Figure 3.4. Selected percentiles of plasma HDL cholesterol by age for white males (3.4A) and females (3.4B) (nonhormone users). Horizontal line represents the commonly used cut-point. The curves are smoothed. (Adapted from: United States Department of Health and Human Services. The Lipid Research Clinics Population Studies Data Book, vol. 1, The Prevalence Study. NIH Publication No. 80-1527, 1980.)

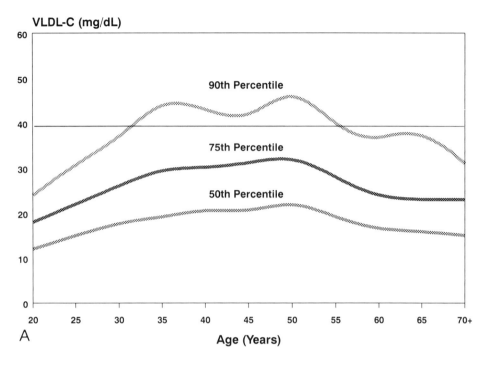

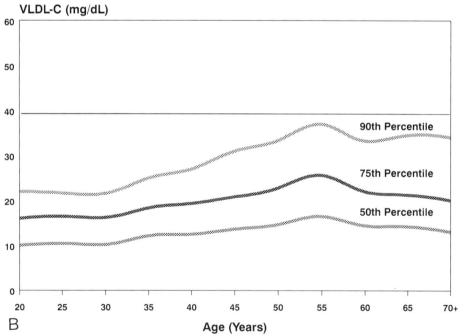

Figure 3.5. Selected percentiles of plasma VLDL cholesterol by age for white males (3.5A) and females (3.5B) (nonhormone users). Horizontal line represents the commonly used cut-point. The curves are smoothed. (Adapted from: United States Department of Health and Human Services. The Lipid Research Clinics Population Studies Data Book, vol. 1, The Prevalence Study. NIH Publication No. 80-1527, 1980.)

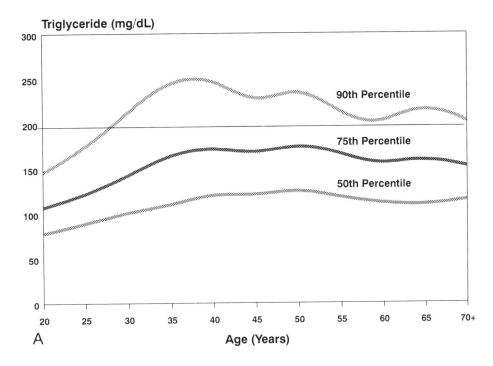

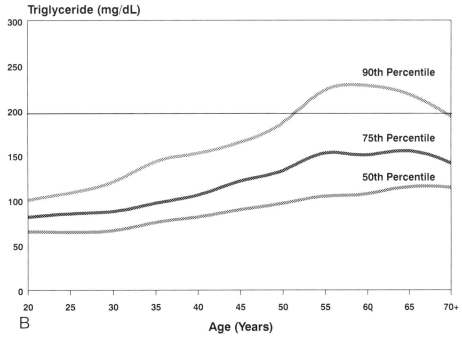

Figure 3.6. Selected percentiles of plasma triglyceride by age for white males (3.6A) and females (3.6B) (non-hormone users). Horizontal line represents the commonly used cut-point. The curves are smoothed. (Adapted from: United States Department of Health and Human Services. The Lipid Research Clinics Population Studies Data Book, vol. 1, The Prevalence Study. NIH Publication No. 80-1527, 1980.)

HDL-C. Not all obese persons have lipoprotein abnormalities. Usually, compensatory mechanisms override the increased metabolic load and enhanced VLDL synthesis. But dyslipidemias can be produced with obesity when these regulatory processes are impaired on a genetic basis or otherwise. An increase in body fat is largely responsible for the increment in total cholesterol and the fall in HDL-C that occur from the age of 19 to 29 years. The high VLDL levels associated with obesity promote the transfer of apo A-I from HDL to VLDL, lowering the number of available HDL particles. Genetic regulation of lipoproteins, body habitus, and increasing age all influence the association of obesity with VLDL, LDL, and HDL.

Regional obesity, particularly intra-abdominal fat deposition, influences blood lipoprotein distribution more than obesity alone. With abdominal obesity, free fatty acids are released directly into the portal circulation, providing the liver with an abundant substrate for triglyceride synthesis. Apparently, adipocytes in this area are especially sensitive to stimuli for mobilization of free fatty acids. Consequently, excess portal free fatty acids perfuse the liver and result in enhanced triglyceride synthesis and VLDL secretion, leading to an increased concentration of VLDL, LDL, and apo B-100. This excess of free fatty acids also leads to increased hepatic gluconeogenesis and interferes with hepatic clearance of insulin. Consequently, there will be elevated blood glucose and insulin levels.

The waist-to-hip ratio generally is used as an index of abdominal fat. A ratio greater than 1.0 for men and 0.8 for women is considered to be high. Studies indicate that a decrease in this ratio is associated with an increase in HDL-C. Much of the HDL-C variance among obese persons can be explained by the combined effect of three interrelated variables: (1) the waist-to-hip ratio; (2) the plasma insulin level; and (3) the degree of glucose intolerance. The sex differences in HDL-C concentration also have been attributed in large part to the differences in fat distribution. It is estimated that obesity results in HDL-C levels that are 5 to 10 mg/dL (0.13-0.26 mmol/L) lower than average, and that the average overweight male would need to reduce body weight and maintain this lower weight to raise his HDL-C noticeably. In contrast to LDL-C, weight may have to be reduced to near ideal weight to affect HDL-C. Only when caloric restriction is sufficient to reduce excess adipose tissue does the HDL-C begin to rise. The fact that body fat is largely responsible for the suppression of HDL can be inferred from the equivalent increases in HDL with fat loss, whether from diet or exercise. Lipoprotein lipase increases with loss of body fat, a possible explanation for the increased HDL. It appears that HDL-C can be restored in overweight persons if lower body weight can be maintained for a prolonged period. The magnitude of the increment realized would depend on many factors, including the original HDL-C value, the degree of overweight, and the extent of weight reduction.

Obesity also is associated with an overproduction of apoprotein B-100, which results in increased levels of VLDL and LDL. In obesity, levels of LDL apo B-100 tend to be higher than the measured LDL-C concentration. The LDL particles are small and dense, so that for a given amount of cholesterol there must be more particles and, consequently, more apo B-100. Apparently, some obese individuals can compensate for this by raising LDL receptor activity. This occurs despite the suppressing effects of excessive saturated fat and cholesterol on LDL receptors. Among those unable to compensate appropriately, the combination of increased fat intake and obesity can bring about a marked elevation of LDL-C concentrations. For the average man who gains 5 to 10 pounds, it is anticipated that the total cholesterol value would increase up to 10 mg/dL (0.26 mmol/L).

Physical Activity

Habitual exercise without dietary changes has little if any effect on the total and

LDL cholesterol but can reduce the trigly-ceride concentration and increase HDL-C (see Chapter 10). These exercise-induced changes in lipids and lipoproteins are associated with heightened levels of apoprotein A-I and lipoprotein lipase (LPL) activity. Habitual endurance training in young men also diminishes postprandial lipemia. The ability of exercise to reduce triglyceride and raise HDL-C depends on the initial lipid level and the amount and intensity of activity performed. Whether there is a threshold for these effects is unclear. In general, a minimal energy expenditure of 800 to 1000 kcal/week, equivalent to 8 to 10 miles of jogging per week, appears necessary for these metabolic changes; HDL-C appears to increase linearly with training, up to 4500 kcal or more per week. Often, plasma volume is increased with training; generally, there are 20 to 50% more circulating HDL particles because of both a higher HDL concentration and plasma volume. Induction of lipoprotein changes requires a period of months, often a year or more for older or overweight persons.

Lower levels of exercise may benefit sedentary persons when associated with a decrease in body fat. To maintain these exercise-induced lipid changes, exercise must be continued on a regular basis. When physical activity stops, lipid levels revert to former values within several weeks.

It has been demonstrated that exercise combined with diet results in better weight loss and maintenance, as well as more favorable lipid changes, than with diet alone. It is well established that physical activity increases insulin sensitivity and improves glucose tolerance. Weight loss by diet combined with physical activity appears to be the best nonpharmacologic approach to improve lipoprotein levels, glucose tolerance, and insulin sensitivity.

Stress

Several decades ago it was noted that severe occupational stress increased the blood cholesterol level. This relation between stress and cholesterol has been well documented, with increments of blood cholesterol ranging from 8% to 65% having been noted during stressful periods. Triglyceride levels show even greater variability during periods of stress, ranging from reductions of 33% to increments exceeding 100%. These lipid alterations apparently relate to changes in neurohormonal activity, such as the release of epinephrine, norepinephrine, growth hormone, and ACTH and their influence on fatty acid release and lipoprotein production. No well-controlled data are available on modification of stress and resultant changes in lipid values.

Blood Pressure

The relation between blood pressure and lipids has been established only recently. Preliminary findings from several studies indicate that dyslipidemia occurs with a greater frequency than expected among hypertensives, especially in the younger age groups. The proportion of hypertensive women who also have lipid abnormalities exceeds that of hypertensive men. This relation was noted in untreated as well as treated hypertensives. In the Framingham Offspring Study, hypertensives aged 20 to 44 years had higher levels of total cholesterol and LDL-C and lower levels of HDL-C. A new syndrome, familial dyslipidemic hypertension, has been defined as hypertension in two or more siblings before age 60 associated with abnormal lipids (LDL-C or triglycerides above the 90th percentile or HDL-C below the 10th percentile). In the usual sequence of events, dyslipidemia develops 10 to 20 years before the onset of hypertension, with subsequent predisposition to CHD. This diathesis has been related putatively to an underlying disturbance in insulin sensitivity (see Chapter 2).

Alcohol

The lipoprotein influenced most by alcohol is VLDL, and, consequently, plasma triglyceride and VLDL-C levels are affected. Heavy alcohol use is a major cause of hypertriglyceridemia, but even small amounts can bring about a pronounced increment in susceptible individuals. Generally, triglyceride elevations occur in persons who drink more than three to four drinks per day, especially in obese persons. Susceptibility to alcohol-induced triglyceride elevation relates to the fat content of the diet, fasting triglyceride level, integrity of liver function, and a genetic predisposition. The mechanism of triglyceride elevation is unclear but most likely involves multiple pathways. Excess production of VLDL particles appears to be the predominant mechanism. Differences in the capacity to clear triglyceride-rich lipoproteins probably account for the variations of triglyceride levels among heavy drinkers.

The HDL-C is increased by alcohol by a magnitude of approximately 1.0 mg/dL (0.03 mmol/L) for each ounce of alcohol (or about 2 drinks) consumed. Women tend to raise their HDL-C levels more readily than men in response to alcohol consumption. Metabolic studies show that both HDL-C and apoprotein A-I increase across the range of low to moderate alcohol intake; it has been suggested that this increment is greatest in those persons who have the highest initial values. The HDL-C response to alcohol involves both HDL_2-C and HDL_3-C subfractions, although the most marked effect appears to be on HDL_3. This alcohol-related increase in HDL results in absence of the usual inverse relation between triglyceride and HDL-C levels. It is one of the few causes (in addition to estrogen use) for a combined elevation of triglyceride and HDL-C.

Coffee

A strong relationship between consumption of coffee and blood cholesterol concentration has been reported in some cross-sectional studies. Coffee consumption of six or more cups per day is reportedly associated with higher blood cholesterol and apoprotein B levels. Evidence that relates decaffeinated coffee to lipids has been inconsistent, although some studies suggested that it increases blood cholesterol to a slightly greater degree than regular coffee. In most studies, tea or caffeine-containing beverages did not affect blood cholesterol levels.

The dose and method of brewing coffee appear to be primary determinants of this effect. Populations from Scandinavian countries, who drink mainly boiled coffee, have demonstrated the strongest correlation between cholesterol levels and coffee intake. Apparently, the length of time the coffee grounds remain in hot water determines the extent of this effect. Similar methods for preparing coffee to obtain percolated or Turkish coffee are likely to increase cholesterol as well. This suggests that coffee contains one or more substances that affect the metabolism of cholesterol; however, the precise mechanism is unknown. Although no correlation has been noted between coffee and triglyceride or HDL-C, abstention from coffee in a small trial of healthy male coffee drinkers resulted in small decreases in HDL-C and apolipoprotein-I (A-I) but did not affect LDL-C.

Seasonal Variation

Cholesterol levels are higher in winter than in summer. The placebo group in the Coronary Primary Prevention Trial (CPPT) offered a unique opportunity to study this phenomenon because carefully standardized lipid profiles were obtained at bimonthly intervals over a 6.5-year period at twelve United States centers. These analyses demonstrated a clear seasonal variation of cholesterol levels, higher in winter, lower in summer, with the increase amounting to $\pm 3\%$ of mean plasma cholesterol levels, or 7.4 mg/dL (0.19 mmol/L). These findings could not be explained by seasonal variation in diet or weight. Similar patterns emerged for LDL-C and HDL-C levels but were of lesser magnitude. Triglyceride levels tended to be highest in the autumn. This seasonal variation could have epidemiologic implications if lipid comparisons are made between studies conducted in different seasons. Such seasonal variation most likely will not create any difficulty in assessing anyone who has repeated measurements over time; however there is the potential to misclassify individual trends when only two data points, winter versus summer, are available.

Strategies for Screening

Two basic screening strategies can be used to prevent CHD:

1. The patient-based or "case-finding" strategy—this strategy is aimed at identifying persons at particular risk of CHD through "case-finding" techniques directed primarily at those persons who seek medical care or have a family history of premature CVD.
2. The public health or "population-based" strategy—this strategy has the more general goal of lowering the blood cholesterol levels of the community at large but also serves to detect persons who would benefit from a more intensive evaluation and treatment program. Early detection provides an awareness of a condition before symptoms are apparent, so that prevention and intervention are possible.

Both types of screening have advantages and disadvantages. Because they are complementary and interact to bring about the desired changes in blood lipids, both should be used by the practitioner.

Case-Finding Strategy

In the high-risk, or case-finding strategy, candidates who have high levels of cholesterol are identified and given appropriate advice and treatment. Such persons can be detected through risk-factor assessment as part of their full clinical examination, during employment physicals, or during consultation for concurrent medical problems. Case finding can be made more efficient by selectively screening persons who have a family history of CHD in first-degree relatives younger than 55 years old and a positive family history of lipid disorders or physical signs such as xanthomas, xanthelasmas, or corneal arcus. In addition, young hypertensives, smokers, diabetics, and those who are

obese would also be more likely to have abnormal lipids. Case finding is best conducted at the primary health care level as well as in hospitals, because this affords an opportunity to reach as large a proportion of susceptible persons as possible. By extending the process over several years, the impact on clinical workload and laboratory cost is minimized.

Over the course of several years, the goal is to identify most patients who have levels of risk that require treatment. By doing so, the treatment can be tailored to the individual patient. By evaluating and treating within the context of the usual practice situation, compliance with advice and treatment is much more likely because both patient and physician are more motivated toward treatment. The importance of this approach has been demonstrated by the success of programs aimed at smoking cessation. The availability of primary health care services for the greater part of the American population makes this strategy feasible and desirable. Another advantage of the high-risk approach is the more cost-effective use of resources. It is more effective to concentrate limited medical services at a time when the need and anticipated benefits are greatest. Also, in this situation, the benefits are likely to be greater for patients than for the more healthy population at large.

One disadvantage of the high-risk approach is that the power to predict future disease is not optimal. Most of those persons identified as being at risk, in all probability, will not develop clinical manifestations of disease in the near future. Despite this they will require prolonged surveillance and management and often will be required to adhere to a diet and take medication that may be difficult to tolerate for a prolonged period. An additional problem is the labeling of persons as being abnormal. This can lead to undue anxiety concerning lipid levels and has been noted to be especially stressful to women because of the wide publicity of "the cholesterol issue" in the media. Even women whose elevated total cholesterol levels are the result of high HDL levels can become unduly concerned. This anxiety over lipid values can reach severe proportions and might even be termed "dyslipidemia nervosa." Also of concern is the "rating" of health insurance policies for such healthy, asymptomatic persons.

Assessment of other major risk factors is an integral part of this approach. This will help identify the large number of persons who are at risk of CHD by virtue of moderate abnormalities in several risk factors rather than isolated extreme hypercholesterolemia. Treatment of this much larger number of persons at moderate risk would prevent more cases of CHD than a major intervention in the small number of persons at high risk. The case-finding strategy is best understood as the incorporation of preventive medicine into clinical practice.

Population-Based Strategy

None of the growing body of knowledge concerning the mechanisms of atherogenesis supports the view that threshold levels of risk factors are biologically plausible. In other words, the risk from cholesterol is continuously distributed—the more abnormal, the greater the risk.

The population-based strategy attempts to shift the entire population distribution curve for cholesterol to lower values. This is done by dietary and lifestyle recommendations for the population as a whole. For example, shifting the distribution of cholesterol toward a lower level that more closely resembles the average values found in Japan (154 mg/dL; 3.99 mmol/L) would result in a major reduction in CHD in this country, because the distribution of cholesterol values would now fall on the low-risk portion of the risk function curve (Fig. 3.7). However, the benefit offered by this approach to the individual in a population may be relatively small. This situation has been called the prevention paradox—namely, that a preventive measure may bring public health benefits to the population as a whole but may yield little for individual efforts. Immunization against infectious disease and wearing of seat belts represent other examples of widespread efforts for the benefit of only a small segment of the population at risk. Because of the high prevalence of moderately elevated levels of cholesterol in this country, however, a population strategy is the only practical means of reducing risk in the large number of people affected.

Population strategies have been compared with high-risk strategies in computer simulations. Findings reveal that targeted pro-

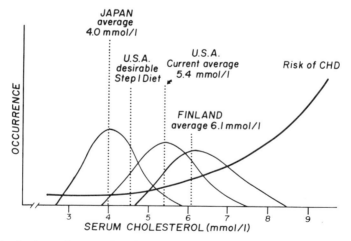

Figure 3.7. The distribution and mean values for serum cholesterol of the Japanese, the American white population, and the Finnish population in relation to the curve that shows the relationship between serum cholesterol and the risk of CHD. With a shift in the natural diet there should be a shift in the cholesterol distribution for the United States to the low-risk segment of the CHD risk curve. (Adapted from: Finnish Heart Association. For the healthy heart, 1988 (a pamphlet).)

grams are likely to have a greater effect in reducing projected CHD incidence among younger than older men, because a higher proportion of new cases of CHD would be predicted to occur in younger men who have cholesterol levels of more than 250 mg/dL (6.47 mmol/L). It is interesting that the percentage change in incidence secondary to cholesterol intervention for women who have cholesterol levels over 250 mg/dL (6.47 mmol/L) is projected to be 2.5-fold greater than for men. This is because CHD in women is more concentrated among those who have high cholesterol levels, and more cases of CHD occur in this high cholesterol setting, whereas men usually have additional risk factors. (Table 3.1).

According to computer simulations, a population-wide reduction in cholesterol levels of 5% (approximately 10 mg/dL, or 0.26 mmol/L) in men would result in the same reduction in CHD as that achieved by a targeted program to reduce cholesterol values from above 250 mg/dL (6.47 mmol/L) to 250 mg/dL (6.47 mmol/L) for high-risk individuals. In women, a 23-mg/dL (0.59 mmol/L) reduction in population cholesterol levels would be necessary to achieve the same benefits likely to accrue from such a targeted case-finding approach. It is expected that a 10-mg/dL (0.26 mmol/L) reduction in average blood cholesterol levels may be approachable, but that

the 23-mg/dL (0.59 mmol/L) population-wide reduction needed in women would be almost double of what could be expected by current national trends and the reported effects from population studies on mass educational intervention. The advisability of advocating mass population screening has been criticized by many authorities. Dietary advice does not require a preliminary blood test, and it is possible to lower lipid levels by educational efforts, without conducting mass screening efforts. Mass screening may lead to false-negative as well as false-positive results. A false-negative finding is more likely to occur in younger persons because of the use of fixed cut-points for all ages, even though lipid levels tend to increase with age. It is conceivable, therefore, that younger persons might gain a false sense of security and continue current lifestyle habits that may lead to elevated cholesterol levels in later years. It also is possible that the individuals screened as negative would do little to reduce their lipid levels, bringing about a situation in which only part of the population is motivated to improve lipid values. The cost for being screened falsely as positive includes monetary expenses related to additional lipid assessment and intangible costs associated with the anxiety or discomfort of being screened as positive. The recent commercialization of massive screening efforts also could have serious drawbacks because the screening is separated from the follow-up care. For screening efforts to be successful, coordinated follow-up care is essential. Finally, the economic burden to the community of conducting large numbers of tests is not an insignificant consideration.

The NCEP is attempting to address these strategies by recommendations from expert panels. A population-based strategy for alteration of lipid patterns in the population has been issued. A separate panel has issued recommendations about screening evaluation and treatment of lipid disorders among high-risk children. All of these reports will impact on the screening process and population strategies. Current NCEP recommenda-

Table 3.1. Percent of All Cases of New Onset Coronary Heart Disease That Are Predicted to Occur in Persons Who Have Cholesterol Levels of More than 250 mg/dL[a]

Age (yr)	Men (%)	Women (%)
35–54	36	56
55–64	30	57
65–74	27	50
75–84	24	39
Overall	30	50

[a] Based on projected population distribution in the year 2015. (From: Goldman L, Weinstein MC, Williams LW. Relative impact of targeted versus populationwide cholesterol interventions on the incidence of coronary heart disease—projections of the coronary heart disease policy model. Circulation 1989;80:254–260. By permission of the American Heart Association, Inc.)

tions for detection and evaluation of lipid disorders can be applied either in population screening programs or in medical examinations. These guidelines will be described in more detail in Chapter 9.

Time Trends

The NHANES, conducted from 1960 to 1980, provided the means to monitor trends over time for blood cholesterol values. Participation rates among surveys ranged from 68% in the 1976–1980 survey (NHANES II) to 86% in the 1960–1962 survey (NHANES I). Although different analytic methods were used for measurement of cholesterol during the surveys, each was standardized to serum with known values determined by the Abell-Kendall method (see Chapter 6).

During the period from 1960 to 1980, notable changes in nutrition and lifestyle had the potential to influence blood cholesterol levels in the population. These included concurrent social changes and the increased availability of fast foods; the increased popularity of eating out; decreased consumption of milk, cream, butter, and lard and the in-

creased consumption of margarine, poultry and fish; the initiation of health promotion and disease prevention activities, including altered smoking habits; changes in physical activity and alcohol consumption; and trends in the use of sex hormones and altered composition of oral contraceptives. All these possible influences complicate the interpretation of changes in lipids during this period. At the same time, new lipid-lowering agents were introduced by the pharmaceutical companies (Fig. 3.8). Taken altogether, these social, dietary, and drug changes were likely to have had an influence on lowering blood cholesterol levels in the United States.

Mean serum cholesterol levels adjusted by sex and age for these three national surveys are shown in Table 3.2. From the early 1960s to the mid-1970s, the mean cholesterol levels for men declined by 1 to 6 mg/dL (0.03-0.16 mmol/L), with some variation, depending on the age group. The decreases in mean cholesterol levels for women were slightly greater than those for men, ranging from 4 to 7 mg/dL (0.10-0.18

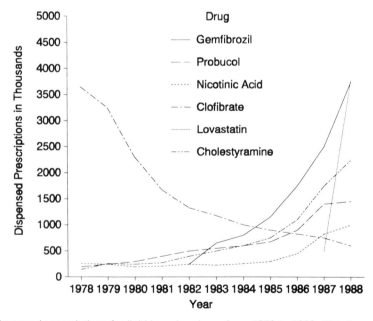

Figure 3.8. Dispensed prescriptions for lipid-lowering drugs from 1978 to 1988. With the exception of clofibrate, there has been a marked increase during the period 1986 through 1988. (From: Wysowski DK, Kennedy DL, Gross TP. Prescribed use of cholesterol-lowering drugs in the United States, 1978 through 1988. JAMA 1990;263:2185–2188.)

Table 3.2. Age-Adjusted[a] Mean Serum Cholesterol Levels of Adults Aged 20 to 74 Years by Sex and Race

	Mean Serum Cholesterol, mg/dL		
	1960-1962	1971-1974	1976-1980
Sex			
Males	217	214	211
Females	223	216	215
Race and Sex			
White Men	218	214	211
Black Men	211	213	209
White Women	224	216	215
Black Women	216	218	214

[a] Age-adjusted by the direct method to the United States population at the midpoint of 1976 to 1980 study. (From: National Center for Health Statistics, National Heart, Lung, and Blood Institute Collaborative Lipid Group. Trends in serum cholesterol levels among U.S. adults aged 20 to 74 years: data from the National Health and Nutrition Examination Surveys, 1960-1980. JAMA 1987;257:937-942.)

mmol/L). Except for the youngest age group, the decrease for women was uniform across all ages during this time. Over the entire 20-year period, the decreases were even greater: the decline in cholesterol values for men was 3 to 10 mg/dL (0.08-0.26 mmol/L) and for women, 2 to 20 mg/dL (0.05-0.52 mmol/L). The age-adjusted mean cholesterol reductions for whites were greater than those for blacks.

These trends observed among the three health and nutrition examination surveys are consistent with other studies. A longitudinal comparison of cholesterol levels between parents and off-spring in the Framingham Study showed a difference of about 9 mg/dL (0.23 mmol/L) between 1950 to 1970. Because of the complex nature of the potential lipid and lipoprotein changes, it is important to consider not only total cholesterol but also HDL-C and other lipoprotein levels. Such data may help explain trends in CHD in this country.

Prevalence of Lipid Disorders

The NHANES II Study estimated the serum total cholesterol and HDL-C levels for the United States population, based on a representative sample of over 11,000 adults aged 20 to 74. Serum triglyceride levels also were measured among 2,804 persons who complied with the recommended fast for at least 12 hours. These data have been applied to the NCEP guidelines to gauge the percent of adults with desirable, borderline high, and high blood cholesterol values. Calculations also were made for the numbers with LDL-C levels and associated risk factors that determine whether persons are candidates for medical advice and intervention after initial screening.

The NCEP has recommended a two-stage process for detection of persons who have lipid abnormalities (see Recommendations for Screening, Chapter 9). In the first stage, serum total cholesterol values > 200 mg/dL (5.17 mmol/L) are used as a marker for screening to detect abnormal lipoprotein values. Nonfasting total cholesterol is a relatively good predictor of those persons who ultimately will require medical advice and intervention for lipoprotein disorders. As noted in Table 3.3, it can be estimated from NHANES II that 57% of all persons will have undesirable total choles-

Table 3.3. Percent Distribution of Population by Risk Groups According to Total Cholesterol Levels and Presence of Two or More Risk Factors

TC (mg/dL) CHD No. Risk Factors	<200	200–239 Absent <2	200–239 Present ≥ 2	≥ 240	Percent of Population Needing Lipoprotein Analysis
All Persons	43	16	14 +	27 =	41
Race					
Black	48	13	15 +	24 =	39
White	42	17	14 +	27 =	41
Sex					
Male	44	11	20 +	25 =	45
Female	42	21	8 +	28 =	36
Age Group (years)					
20-39	61	17	9 +	13 =	22
40-59	28	18	17 +	37 =	54
60-74	22	13	20 +	44 =	64

TC = Total cholesterol
(Adapted from: Sempos C. Fulwood R, Haines C, et al. The prevalence of high blood cholesterol levels among adults in the United States. JAMA 1989;262:45-52.)

terol values $\geq$ 200 mg/dL (5.17 mmol/L). More than 50% of those persons (30% of the total population) will have borderline-high values defined as 200 to 239 mg/dL (5.17 to 6.18 mmol/L), and slightly less than 50% (27% of the total population) will have high values >240 mg/dL (6.21 mmol/L). The percentage of the population with desirable cholesterol values is only slightly different by race and sex but varies considerably by age.

The number of persons at high risk for CHD (14% with borderline-high cholesterol values coupled with CHD or two risk factors plus 27% with high values of cholesterol) determines the total percent who need lipoprotein analyses. In the total population of all adults aged 20 to 74, approximately 41% require lipoprotein analysis. As anticipated, there was a substantial increase with age; for those patients aged 20 to 39, the rate was 22%; for those aged 40 to 59, it was 54%; and for those aged 60 to 74, it was 64%. The percent of persons who needed lipoprotein analyses increased with age because of the higher levels of lipids, the greater prevalence of CHD, and the greater number of CHD risk factors among older persons.

The higher rates of men who require lipoprotein analysis (45% for men compared to 36% for women) can be attributed in part to the fact that male sex is considered to be an independent risk factor for CHD. Men require only one other CHD risk factor in addition to a borderline elevated cholesterol to be considered at high risk of CHD by criteria of the NCEP.

Additional calculations define the percent of candidates for whom medical advice and treatment are needed. It is important to stress that therapy is a function of the LDL-C level rather than total cholesterol and is modified by the presence of CHD or two other risk factors for CHD. For those persons who apparently do not need lipoprotein analysis, approximately 8% who would not be picked up by the total cholesterol values advocated for screening actually require medical advice and treatment on the basis of increased LDL-C levels. This implies that a low HDL-C coexists and is partially responsible for the acceptable total cholesterol with a high LDL-C. Approximately 88% of the 41% who need lipoprotein analyses, or about one-third of the total population, turn out to be candidates for medical advice and treatment for high levels of blood lipids. It is readily apparent that most people who need lipoprotein analyses will end up being candidates for medical advice and treatment.

The NHANES II sample provided a unique opportunity to estimate the numbers of people who would be potential candidates for medical advice and treatment in the United States. Because of the survey methodology used, however, there are several recognized limitations. The results tend to underestimate those persons who have clinical disease (CHD, diabetes mellitus, or cerebral vascular disease). Also, no data were available on peripheral vascular disease or family history of premature CHD. Finally, assessing cholesterol levels and blood pressure on only one occasion often overestimates the frequency of these conditions because of the likelihood of higher values at an initial screening. Persons who have high values usually have lower values on repeat evaluations, a phenomenon known as regression to the mean.

The NHANES II also provided data on the percentage of men and women who have low HDL-C <35 mg/dL (0.91 mmol/L) for selected serum cholesterol cut-points for various age groups (Fig. 3.9). For blacks and whites alike age 20 to 74 (age-adjusted), 18.6% of the men and 6.1% of the women who had desirable total cholesterol values of <200 mg/dL (5.17 mmol/L) also had low HDL-C concentrations. It must be taken into consideration that this value for women represents a greater departure from normal, because women generally have higher HDL-C values (approximately 10 mg/dL; 0.26 mmol/L) than men. It is also noteworthy that the percent for men was especially low in the borderline cholesterol

% of Population

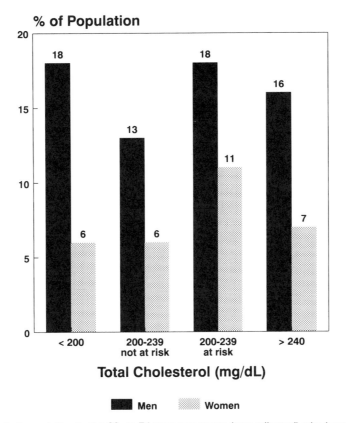

Figure 3.9. Percent of population in the 20- to 74-year age group (age-adjusted) who have low HDL-C (< 35 mg/dL) in the various risk groups for CHD defined by the NCEP. (From: Second Nutritional Health and Nutrition Examination Survey (1976–1980, National Center for Health Statistics. Sempos, C: Personal communication.)

group not otherwise at risk of low HDL-C, and the prevalence for women was especially high in the borderline cholesterol group at high risk.

In another analysis from a different population, patients who had low HDL-C and cholesterol lower than 200 mg/dL (5.17 mmol/L) frequently were diabetics, smokers, or were being treated with beta-blockers. If these other factors were used for low HDL-C markers, only about 2% of those persons who had a desirable cholesterol concentration but a low HDL-C value would have been overlooked. Elevated serum triglyceride should also raise the suspicion of a low HDL-C. Screening only for high cholesterol can leave a false impression that low total cholesterol is uniformly associated with a good health risk profile. It appears

appropriate to recommend routine HDL-C measurements among persons who have CHD, diabetes mellitus, high serum triglyceride, or who take beta-blockers or who are at high risk for CHD otherwise.

Suggested Readings

American Academy of Pediatric, Committee on Nutrition. Indications for cholesterol testing in children. Pediatrics 1989;83:141–142.

Carleton RA, Dwyer J, Finberg L, et al. Report of the expert panel on population strategies for blood cholesterol reduction: a statement from the National Cholesterol Education Program, National Heart, Lung, and Blood Institute, National Institutes of Health. Circulation 1991;83:2154–2232.

Franklin FA Jr, Brown RF, Franklin CC. Screening, diagnosis, and management of dyslipoproteinemia in children. Endocrinol Metab Clin North Am. 1990; 19: 399–449.

Garcia RE, Moodie DS. Routine cholesterol surveillance in childhood. Pediatrics 1989;84(5):751–755.

Goldman L, Weinstein MC, Williams LW. Relative impact of targeted versus populationwide cholesterol interventions on the incidence of coronary heart disease policy: projections of the Coronary Heart Disease Policy Model. Circulation 1989;80:254–260.

National Center for Health Statistics—National Heart, Lung, and Blood Institute Collaborative Lipid Group. Trends in serum cholesterol levels among US adults aged 20–74 years. JAMA 1987;257:937–942.

National Heart, Lung, and Blood Institute, National Institutes of Health. National Cholesterol Education Program: report of the expert panel on population strategies for blood cholesterol reduction. Arch Intern Med 1988;148:36–39.

Steinberg D, Pearson TA, Kuller LH. Alcohol and atherosclerosis. Ann Intern Med 1991;114:967–976.

United States Department of Health and Human Services. The Lipid Research Clinics Population Studies Data Book, Volume I: The Prevalence Study. Washington, DC: NIH Publication No. 80-1527, 1980.

United States Department of Health and Human Services, Public Health Service. Chronic disease reports: coronary heart disease mortality, United States. MMWR 1989;38:269–288.

Wilson PWF, Christiansen JC, Anderson KM, Kannel WB. Impact of national guidelines for cholesterol risk factor screening, The Framingham Offspring Study. JAMA 1989;262:41–44.

Wynder EL. American Health Foundation Monograph. Coronary heart disease prevention: cholesterol, a pediatric perspective. Prev Med 1989;18:323–409.

SECTION TWO

PATHOPHYSIOLOGY

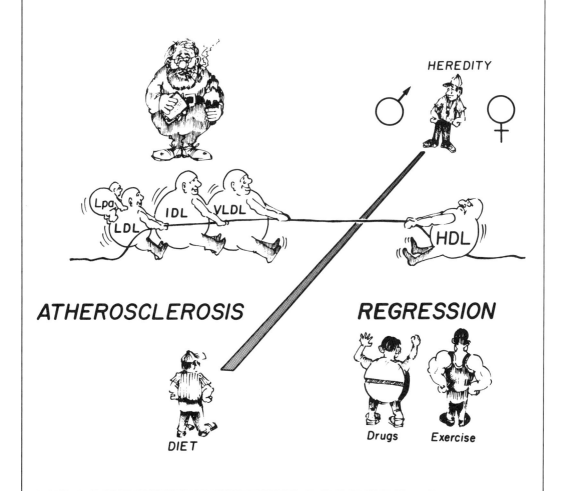

HEREDITY

Lpa LDL IDL VLDL HDL

ATHEROSCLEROSIS

REGRESSION

DIET

Drugs Exercise

CHAPTER 4

Lipoprotein Transport

Physiology and Pathophysiology of Lipid Transport

Lipoproteins are a heterogeneous collection of particles that range in size from 80 to 1500 angstroms (15- to 20-fold) (Fig. 4.1). They transport cholesterol, triglyceride, and other lipid constituents through the circulation to peripheral tissues. The basic lipoprotein structure is illustrated in Figure 4.2. It consists of a central core of nonpolar lipid-containing hydrophobic cholesteryl esters and triglyceride, surrounded by a hydrophilic, or polar monolayer of phospholipids, free cholesterol, and proteins. The hydrophilic coat or "membrane" allows the transport of otherwise insoluble cholesteryl esters and triglyceride in the aqueous phase of plasma.

There are six major lipoprotein classes (Table 4.1). Each class is heterogenous in size, composition, and electrophoretic mobility. The subgroups have physiologic as well as anatomic significance. The lipoproteins often are secreted in one form and then are transformed rapidly into other subgroups or into other classes of lipoproteins (Fig. 4.3). This dynamic process occurs as the lipoproteins interact with circulating enzymes, lecithin-cholesterol acyltransferase (LCAT), endothelial-bound enzymes, lipoprotein lipase (LPL), hepatic triglyceride lipase (HTGL), and other lipoproteins (Table 4.2).

Two pathways predominate in lipoprotein metabolism: (1) The exogenous pathway—from dietary lipid through the gastrointestinal (GI) tract and (2) the endogenous pathway—from hepatic synthesis of VLDL.

The exogenous pathway involves absorption of dietary cholesterol and triglyceride-derived free fatty acids from the intestinal tract (see Fig. 4.3) and their transport to peripheral tissues and the liver. Approximately 100 grams of fat and 500 mg of cholesterol are absorbed from the intestines each day. Dietary cholesterol and triglyceride fatty acids are synthesized into chylomicrons in the intestinal cell. The rate of secretion of chylomicrons by the intestine into the plasma reflects the rate of fat absorption. In

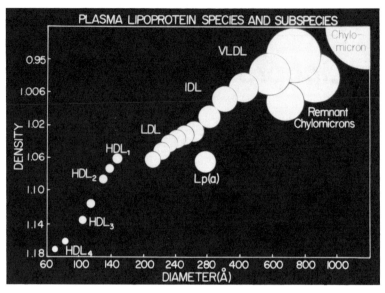

Figure 4.1. The lipoprotein cascade that depicts the heterogeneity of the lipoprotein classes. Note that each lipoprotein class consists of multiple subclasses that vary in size and composition. (From: Segrest JP. Personal communication. [Copyrighted]. Atherosclerosis Research Unit. University of Alabama at Birmingham. "Reprinted with permission of Jere P. Segrest. Copyright UAB 1976."

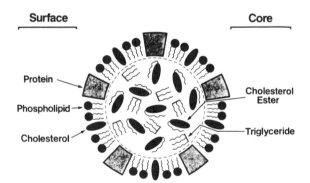

Figure 4.2. Basic lipoprotein particle. Generalized cross-sectional structure of human lipoproteins is spherical and varies in diameter from 10 nm to as much as 1000 nm, depending on the particular proteins and lipids. Each of the lipoprotein classes contains a neutral lipid core composed of triglyceride and cholesteryl ester. Around the core is a layer of protein, phospholipid, and cholesterol that is oriented with the polar portions exposed to the surface of the lipoprotein. (From: American Heart Association, Cardiovascular Research Report (CRR): Reporting on heart and blood vessel diseases and stroke; No. 32, 1989. "Reproduced with permission. Copyright American Heart Association."

the endogenous pathway, lipoprotein particles are synthesized in the liver and transport triglyceride and cholesterol to extrahepatic tissues (see Fig. 4.3). Both exogenous and endogenous pathways provide energy to peripheral tissues by the hydrolysis of triglyceride and the release of free fatty acids (FFA) and also provide cholesterol to satisfy the metabolic and structural needs of the cells.

When cholesterol and triglyceride are measured in blood, they are stripped from the lipoproteins by organic solvents. Consequently, one cannot refer to a specific lipoprotein as being abnormal solely as a result of an abnormality in the measured amount

Table 4.1. Lipid Composition And Electrophoretic Mobility for Major Lipoproteins

Sf[b] (d=1.063)	Density (g/ml)	Lipoprotein Fractions (Particle Size)	Major Lipids (%) CE[a] C[a] TG[a] PL[a] P[a]					Major Apoproteins	Electrophoretic Mobility
400+	<0.98	Chylomicron (75-1200 nm)	3	1	90	4	2	B-48, C, E	Origin
20-400	0.98–1.006	VLDL (30-80 nm)	12	6	60	14	8	B-100, C, E	Prebeta
12-20	1.006–1.019	IDL (25-35 nm)	26	10	30	20	14	B-100, C, E	Broad Beta
0-12	1.019–1.063	LDL (15-25 nm)	40	11	5	22	22	B-100	Beta
—	1.063–1.21	HDL (5-12 nm)	18	5	7	25	45	A, C, D, E	Alpha
—	1.050–1.120	LP(a) (25 nm)	33	9	3	22	33	(a)	Prebeta

[a]CE = Cholesteryl ester; C = Cholesterol; TG = Triglyceride; PL = Phospholipids; P = Protein.
[b]Sf = Svedberg units

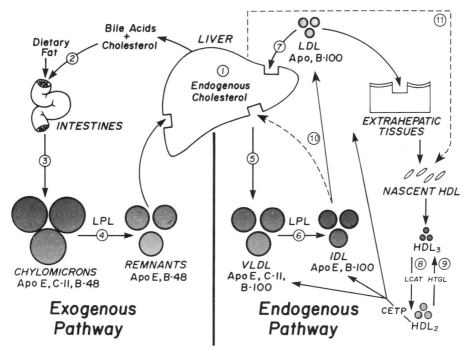

Figure 4.3. Diagrammatic representation of lipoprotein metabolism. Lipoproteins are depicted as circles, with their major apoproteins (apo) noted below. In the exogenous pathway, chylomicrons are produced in the intestines from dietary fats (3), whereas the exogenous pathway VLDL is produced in the liver (5). Both particles are hydrolysed by LPL to chylomicron remnants (extrinsic pathway [4]) or IDL and LDL lipoproteins (intrinsic pathway [5]). IDL (10) and LDL (7) particles bind to hepatic receptors. Nascent HDL is produced in hepatic and extrahepatic tissues (11). The HDL₃ is converted to HDL₂ by the action of LCAT (8), whereas the reverse reaction is enhanced by HTGL (9). Transfer of cholesteryl esters between HDL and other lipoproteins occurs through CETP. HTGL = hepatic triglyceride lipase; LCAT = lecithin:cholesterol acyltransferase; CETP = cholesteryl ester transfer protein.

Table 4.2. Major Enzymes Involved in Lipoprotein Transport

Enzyme	Location	Function
Lipoprotein lipase (LPL)	Endothelium of adipose tissue and skeletal muscle	Hydrolyzes triglyceride in chylomicrons and VLDL, releasing fatty acids and reducing triglyceride content
Hepatic-triglyceride lipase (HTGL)	Endothelium of the liver	Hydrolyzes triglyceride in IDL, LDL and HDL, releasing fatty acids and reducing triglyceride contents
Lecithin-cholesterol acyltransferase (LCAT)	Synthesized in the liver and released into the circulation	Esterifies cholesterol in the HDL particle
Cholesteryl ester transfer protein (CETP)	Plasma	Transfers cholesteryl esters from HDL to chylomicron remnants and IDL and triglyceride from these lipoproteins to HDL

of cholesterol or triglyceride. Lipoproteins vary in the amount and kind of lipids and proteins that they transport (Fig. 4.4). The composition of the various lipoprotein classes is given in Table 4.1.

Because chylomicrons contain little cholesterol and mostly triglyceride, marked increases in the blood chylomicron fraction usually are associated with normal or only mildly elevated cholesterol levels. Although VLDL is also rich in triglyceride, it contains considerably more cholesterol than chylomicrons (20% of the lipid compared to 1% of the lipid). Consequently, an increase in the VLDL fraction can be associated with marked hypercholesterolemia as a consequence of the cholesterol that is cotransported with triglyceride. The LDL and HDL contain such little triglyceride that marked elevation in either of these lipoproteins is never associated with hypertriglyceridemia, only with hypercholesterolemia.

Lipid Constituents

Apoproteins

The protein components of the lipoproteins are called apolipoproteins or apoproteins; they have important structural and functional properties (Table 4.3). The apoproteins have polar and nonpolar regions that confer amphipathic properties and allow solubilization of lipid in aqueous solutions. The polar face of the apoprotein is turned toward the aqueous media, whereas the nonpolar face binds and surrounds the insoluble triglyceride and cholesteryl esters. Apoproteins are designated by capital letters (A, B, and so forth) and roman numerals (I, II, and so forth) or numbers. The functions of the apoproteins and their plasma concentrations are listed in Table 4.4, and their distribution in each lipoprotein class is shown in Table 4.5.

The apoproteins also function as ligands, or binding proteins, that permit lipoproteins to bind to specific cell-surface receptors. In this regard, apoprotein B-100 and apoprotein E are specific recognition proteins. These apoproteins direct or determine the sites to which lipoproteins are delivered, as do the specific cell-surface receptors to which they bind. Abnormalities in the binding properties of apoproteins or their receptors will lead to dyslipidemia. Apoproteins also have the capacity to activate as well as to inhibit a number of enzymes that are critical to normal lipid metabolism. Lipoprotein lipase (LPL), the enzyme system situated on the surface of endothelial cells, is activated by apoprotein C-II and inhibited by apoprotein C-III. A deficiency of apoprotein C-II leads to a defect in the activation of lipoprotein lipase and, consequently, defective hydrolysis of chylomicrons and VLDL triglyceride. Apoprotein A-I is not only a major

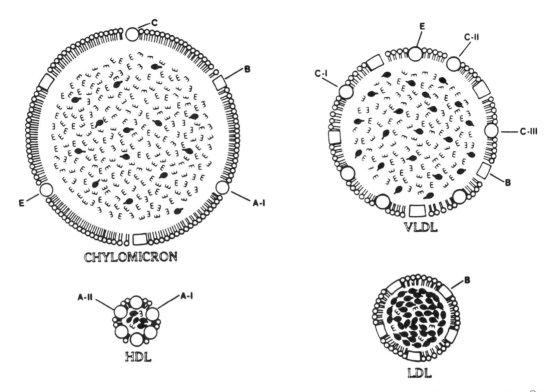

Figure 4.4. Schematic diagram of the structure and composition of four major plasma lipoprotein species: (phospholipid symbol), phospholipid and free cholesterol monolayer; (○) exchangeable apolipoproteins A, C, and E; (□) apolipoprotein B; (●) cholesteryl ester; and (triglyceride symbol) triglycerides. The relative diameter of the lipoprotein particles are approximately to scale, with the exception of the chylomicron, which should be approximately twice as large as shown. (From: Segrest JP, Chung BH, Cone JT, Hughes TA. Coronary heart disease assessment by plasma lipoprotein profile. Ala J Med Sci 1983;20:76–83.)

Table 4.3. Functions of Apoproteins

Function	Apolipoprotein
Structural	apo B-48, apo B-100, apo A-II
Ligands for receptor binding	apo E, apo B-100
Enzyme activation	apo A-I, apo C-II
Enzyme inhibition	apo C-III
Cholesteryl ester transfer	apo D

structural component of HDL but is also a necessary cofactor for the activation of LCAT. This enzyme is synthesized in the liver and released into the circulation; it must be activated by apoprotein A-I before esterification of cholesterol can occur. A defect in the activity of LCAT, because of a defect in apoprotein A-I or a deficiency of the enzyme, leads to impaired cholesterol esteri-

fication and an abnormality in lipoprotein transport. The transfer of cholesteryl esters between various lipoproteins, and specifically from HDL to chylomicron remnants, VLDL, IDL, and LDL, appears to be regulated by apoprotein D, also known as cholesteryl ester transfer protein (CETP), or lipid transfer protein (LTP).

The location of most of the apoproteins in the human genome have been established. Apoprotein A-II is located on chromosome 1, apoprotein B on chromosome 3, apoproteins A-I, C-II, and A-IV on chromosome 11, and apoproteins E, C-I and C-III, as well as the LDL receptor, are on chromosome 19 (Table 4.6).

Many apoproteins have been isolated, purified, and sequenced. Apoproteins are synthesized as propeptides, thereby creating potential sites for point mutations. Muta-

Table 4.4. Plasma Apoproteins

Apolipoprotein	Approximate Molecular Weight (Daltons)	Approximate Concentration in Plasma (mg/dL)	Reported Functions
A-I	28,000	130	Cofactor for LCAT, stuctural role in HDL, reverse cholesterol transport
A-II	14,000	40	Cofactor for HTGL, structural role in HDL
A-IV	45,000	15	Unknown
Apo A [as Lp(a)]	500,000	0-120	Unknown; resembles plasminogen; may interfere with thrombolysis
B-48	240,000	Trace	Structural role in chylomicrons
B-100	510,000	100	Binding ligand for the LDL receptor; structural role in VLDL and LDL
C-I	6,600	7	Cofactor for LCAT
C-II	8,900	4	Cofactor for LPL
C-III	8,800	13	Inhibitor of LPL, inhibitor of chylomicron-remnant and VLDL uptake by cells
D	19,000	6	Cholesteryl ester transfer; also known as cholesteryl ester transfer protein or lipid transfer protein
E	34,000	5	Binding ligand for the remnant chylomicron and LDL receptors

Table 4.5. Apoprotein Distribution in Plasma Lipoproteins (Percentage by Weight)

	HDL	LDL	IDL	VLDL
Apo A-I	100			
Apo A-II	100			
Apo B-48[a]				
Apo B-100		90	8	2
Apo C-I	97		1	2
Apo C-II	60		10	30
Apo C-III	60	10	10	20
Apo D	100			
Apo E	50	10	20	20

[a] Found almost entirely in the chylomicrons. (Adapted from: Havel RJ, Kane J. Structure and metabolism of plasma lipoproteins. In: Scriver CR, Beaudet AL, Sly WS, et al (eds). Metabolic basis of inherited disease. ed 6. New York: McGraw-Hill, 1989.)

Table 4.6. Location of Apoproteins and the LDL Receptor in Human Genome

Apoproteins/ LDL Receptor	Chromosome
A-II	1
B	3
Apo (a)	6
A-I, A-IV, C-II	11
C-I, C-III, E	19
LDL receptor	19

tions may lead to alterations in lipoprotein metabolism and a predisposition to dyslipidemia and vascular disease. Replacement of defective apoproteins is possible with recombinant DNA synthesized proteins or by the administration of normal plasma. Apoproteins B-48, B-100, C-II, and E also are glycoproteins; this may be important in diabetes mellitus, where apoprotein glycosylation could change their structure and function, thereby contributing to the development of dyslipidemia.

Lipoprotein Receptors

Lipoprotein receptors are located on cell surfaces and mediate the removal of certain lipoproteins from the circulation. The apoprotein content usually determines which receptor will bind and internalize (endocytosis) the lipoprotein. The lipoprotein and its respective receptor have a lock-and-key arrangement, so that a defect in the receptor or in the binding portion of the apoprotein may result in a poor fit and defective binding and removal, with subsequent accumulation of the lipoprotein and development of dyslipidemia. The functions and locations of more well-established lipoprotein receptors are listed in Table 4.7.

Chylomicrons

Chylomicrons are synthesized in the intestine and transport dietary triglyceride and cholesterol to the body. They are large particles that have light scattering properties. When plasma or serum is examined after a meal, lactescence is usually observed because of the presence of chylomicrons. In patients who have defects in chylomicron metabolism, such lactescence can persist even after an overnight fast of 12 to 14 hours. Chylomicron density is less than that of water, and chylomicrons, thus, float to the top when plasma is placed in a refrigerator overnight or when it is centrifuged. The consequent formation of a cream layer on top of the tube overlying a clear (Type I) or cloudy (Type V) plasma infranatant can be used as a diagnostic test for the presence of chylomicrons (See Fig. 9.4).

Dietary triglyceride is hydrolyzed in the intestinal lumen, where fatty acids and monoglycerides are taken up by the GI mucosal cell. The triglyceride is reassembled and combined with specific apoproteins inside these cells and then is released into the lymphatics of the intestine. The chylomicron particle consists primarily of a triglyceride core and an envelope of apoprotein and phospholipid. The amount of triglyceride transported by chylomicrons varies with the amount and type of fat consumed. With low-fat diets, chylomicron formation diminishes and triglyceride transport takes place on VLDL particles synthesized by the liver.

The chylomicrons that are released into the lymphatics from the intestine are incomplete (nascent) and contain only apoproteins A and B-48. When nascent chylomicrons reach the systemic circulation they interact with circulating HDL and receive additional apoproteins. The HDL serves as a reservoir for apoproteins C and E and transfers them to the nascent chylomicron; in exchange, HDL receives apoproteins A-I and A-II from the chylomicrons. Apoprotein E acts as a binding ligand and permits removal of chylomicron remnants by the liver. The HDL, therefore, serves as an important reservoir for critical apoproteins that are required for the normal metabolism of chylomicrons.

The hydrolysis of chylomicron triglyceride by LPL releases FFA into the circulation; simultaneously there is a progressive reduction in the size of the chylomicron. The protein and phospholipid surface elements become redundant and dissociate spontaneously, leading to the formation of new HDL particles. As the chylomicron becomes smaller, apoproteins C and E also are returned to HDL for reuse later.

When all the apoprotein C has been removed from the chylomicron, the residual lipoprotein particle is known as a chylomicron remnant; it now contains only apoprotein B-48 and E on its surface. The chylomicron remnant, which is now enriched with cholesteryl ester, binds to a specific hepatic receptor and is removed from the circulation. It appears that apoprotein C prevents

Table 4.7. Function and Location of Lipoprotein Receptors

Lipoprotein Receptor	Function	Location
LDL	Binds lipoproteins containing apo B and apo B plus apo E: internalizes beta-VLDL, IDL, and LDL	All cells; 75% located in the liver
E	Binds lipoproteins containing apo E: primarily internalizes chylomicron remnants but also VLDL and apo E-containing HDL	Liver
Scavenger	Binds oxidized LDL	Macrophages

premature binding of the chylomicron remnant to the receptor. Transfer of apoprotein C back to HDL is a critical step that occurs when lipolysis is optimum and necessary before the remnant can bind. In contrast to the LDL receptor, the receptor that removes the chylomicron remnant is not under metabolic regulation and, consequently, is not susceptible to downregulation or decreased expression by the liver.

The half-life ($t_{1/2}$) of chylomicrons is short (approximately 5 to 30 minutes), indicating rapid metabolism. This accounts for the wide fluctuations in blood triglyceride concentration after a meal. The VLDL particles also are metabolized by LPL, but at a much slower rate, suggesting that apoprotein content or compositional differences induce conformational changes that retard their binding to LPL, thereby making them a less efficient substrate for metabolism.

During postprandial lipemia, the chylomicron content of free cholesterol relative to that of phospholipid is reduced. This compositional change favors the transfer of cholesteryl esters from other lipoproteins to the chylomicrons. Furthermore, the transfer of cholesterol into cells is decreased, whereas the efflux of cholesterol from cells is increased. As a consequence, there is an increase in HDL-C esterification and an increase in cholesteryl ester transfer to chylomicrons, VLDL, and LDL. The redistribution of cholesteryl esters from HDL to other lipoproteins within blood is a result of a specific cholesteryl ester transfer protein (CETP). All of the remnants become enriched with cholesteryl esters. It is likely that postprandial changes are very important in regulating CETP, and persons who have persistent hypertriglyceridemia are likely to have cholesteryl ester enrichment of chylomicron remnants, VLDL, and VLDL remnants, which may be important with regard to the pathogenesis of atherosclerosis. Some persons who have only a modestly prolonged phase of postprandial hypertriglyceridemia may not have fasting hypertriglyceridemia because the chylomicrons can be cleared within 12 to 14 hours. In such persons, there would be increased cholesteryl ester transfer to chylomicron and VLDL remnants, to the degree that compositional abnormalities of chylomicron remnants and VLDL could exist in the absence of fasting hypertriglyceridemia. Such cholesteryl ester enriched lipoproteins could contribute to the process of atherosclerosis. The duration of postprandial elevations in triglyceride-rich lipoproteins is an important and emergent area of investigation.

Chylomicrons are important with regard to energy balance. Free fatty acids from triglyceride are delivered to peripheral tissues to satisfy energy requirements or are used for storage. Because fat is an efficient way to store excess calories, adipose tissue can satisfy basic energy needs for prolonged periods. The cholesterol that is transported on the chylomicron remnant is simply a passenger, because it is not extracted by peripheral tissues. As the chylomicron remnant becomes smaller and its triglyceride content diminishes, the particle becomes relatively more enriched with cholesterol and cholesteryl esters. The remnant particle is removed by the liver and its cholesterol is incorporated into the hepatic cholesterol pool. In this way, there is a direct relation between dietary cholesterol and hepatic cholesterol content.

A diet high in cholesterol and saturated fat increases the delivery of cholesterol to the liver, therefore increasing hepatic cholesterol content. As described in detail later, hepatic cholesterol inhibits the synthesis and expression of the LDL receptor, thus reducing the fractional catabolic removal of endogenously synthesized LDL-C and leading to the development of hypercholesterolemia.

Very Low Density Lipoprotein

Very low density lipoproteins (VLDL) are synthesized in the liver and released into the circulation. The half-lives of VLDL and IDL are relatively short (approx. 12 hours) but considerably longer than those for chylomicrons and their remnants. The VLDL transports free fatty acids from the liver to peripheral tissues to satisfy energy requirements during periods of reduced availability of dietary substrate. Particles range in size from quite large (similar to chylomicron remnants) at one end of the continuum to smaller particles, similar in size and composition to LDL, at the other end. The VLDL particles contain apoproteins C and E in addition to B-100. Nonetheless, VLDL also interacts with HDL in the circulation, much in the same way as chylomicrons, and additional apoprotein C and E are added to the nascent VLDL to form the mature VLDL particle.

The VLDL is metabolized by LPL on the surface of capillary endothelial cells. The size of the particle becomes progressively smaller and ultimately becomes an intermediate density lipoprotein (IDL). The IDL has a density of 1.006 to 1.019 (Sf 12 to 20), contains only apoproteins B-100 and E on its surface, and binds with high affinity to the LDL receptor in the liver and other tissues. During the hydrolysis of VLDL to IDL, phospholipids, apoprotein C, and some apoprotein E are transferred back to HDL. This reemphasizes the important reservoir function of HDL for apoproteins C and E. In the fasting state, more than 50% of these apoproteins are transported by HDL.

Intermediate density lipoprotein can be taken up by the liver or it can be converted into LDL. As the VLDL particle is reduced in size, apoprotein B assumes a more important role. Size appears to alter the surface expression of apoprotein B and apoprotein E, favoring binding to the LDL receptor. Approximately 50% of the IDL generated from VLDL is removed from the circulation by binding to the LDL receptor in the liver. The other 50% of IDL is converted to cholesterol-rich LDL. The conversion of IDL to LDL is caused by the action of hepatic triglyceride lipase (HTGL). This enzyme is located in the vascular endothelium of the liver, and its action results in the additional loss of triglyceride and apoprotein E from

IDL. The HTGL is reduced in chronic liver disease, hypothyroidism, uremia, and, rarely, as a consequence of a genetic defect. Any acquired or inherited abnormality in HTGL will lead to higher than normal circulating levels of IDL. Such conditions cannot be distinguished from dysbetalipoproteinemia (Type III dyslipidemia), in which persistent elevation of IDL is caused by an abnormality of apoprotein E, with consequent defective binding of IDL to the LDL receptor.

The VLDL particles are heterogeneous and are secreted into different flotation intervals, ranging from Sf 20 to Sf 400. They maintain their pedigree and identity even as they undergo hydrolysis. Only 10% of the large, triglyceride-rich VLDL particles (Sf interval 100 to 400) are converted to LDL, whereas 50% of the smaller and less triglyceride-enriched particles (Sf range 20 to 60) are converted to LDL. Approximately 25% to 60% of apoprotein B in LDL is derived from VLDL particles secreted in the Sf interval of 20 to 60. It has been suggested that the very large VLDL particles do not predispose to the development of vascular disease because they do not contribute substantially to the LDL pool. Large VLDL particles are removed from the circulation without additional metabolism, but the mechanisms for their removal are unclear.

Ordinarily, 75% to 80% of the VLDL particles in normal as well as in hypertriglyceridemic persons are in the Sf 20 to 60 range. In patients who have hypertriglyceridemia, at least 70% of the increase in blood triglyceride concentration is caused by increased numbers of triglyceride-containing particles, whereas 30% is attributable to increased particle size (an increased amount of triglyceride per particle). The size of the particles depends on the defect responsible for the hypertriglyceridemia (see Chapter 7). Patients who have familial hypertriglyceridemia also have large triglyceride-enriched particles, most of which are removed directly from the circulation and very few of which are converted to LDL. Patients who have familial combined dyslipidemia, however, secrete increased amounts of particles in the Sf 20 to 60 range, where approximately 50% of the particles are converted to LDL. This difference may be related to the dissimilar risk for atherosclerosis imposed by these disorders.

An abnormality in the structure of apoprotein E can interfere with the binding of IDL to the LDL receptor and, consequently, can lead to the retention of IDL and somewhat larger particles (known as beta-VLDL) in the circulation. This disorder, variously known as broad beta disease, dysbetalipoproteinemia, or Type III hyperlipoproteinemia, is caused by this specific defect.

It is of considerable interest to note that the VLDL particle, which is synthesized by an organ (the liver) that plays such an important role in substrate homeostasis, is the precursor of a particle (LDL) that has absolutely no role in energy balance. Although LDL usually is considered to be derived from VLDL, it may be synthesized directly and released by the liver in the postprandial period. This aspect of LDL production is not clearly defined or understood.

Low Density Lipoprotein

In contrast to VLDL and IDL, the half-life of LDL is relatively long (approx. 3 days). Consequently, the concentration of LDL and the cholesterol it transports are relatively stable and not acutely influenced by diet. This is one of the reasons that cholesterol levels can be screened in the nonfasting state. Furthermore, although subtle abnormalities in VLDL synthesis may not increase the blood concentrations of the short-lived VLDL, they can manifest as increased levels of the longer lasting LDL.

Low density lipoprotein transports approximately 65% of the blood cholesterol, which it delivers to cells to satisfy metabolic and structural needs. The LDL particle contains only a single apoprotein, apoprotein B-100. Each LDL particle carries approximately 1500 cholesteryl ester molecules. In contrast, VLDL transports approximately

7000 cholesteryl ester molecules per particle. The LDL particles bind to a specific cell-surface receptor, the LDL receptor, which recognizes only apoprotein E or apoprotein B-100. Apoprotein B-100 allows the LDL particle to bind to the receptor, at which point both the receptor and LDL are internalized and disassembled by lysosomal hydrolysis. The LDL receptor is recycled to the cell surface to be reused in transport of LDL particles, whereas the LDL particle is degraded in the cell and releases amino acids and cholesterol. Approximately two-thirds of the circulating pool of LDL is removed by this pathway, 70% of which occurs in the liver and 30% in extrahepatic tissues. The remaining one-third of the circulating pool of LDL is removed by nonspecific pathways, including scavenger receptors.

There are four major LDL subspecies based on density differences: LDL-I is the largest and least dense and LDL-IV is the smallest and most dense. Subclasses LDL-I and LDL-II predominate in most healthy persons. The more dense LDL subclasses correlate positively with levels of VLDL and IDL triglyceride and are related inversely to HDL-C concentrations. With blood triglyceride concentrations in the range of 150 to 350 mg/dL (1.69 to 3.95 mmol/L), there is an increase in LDL-III, and in more severe hypertriglyceridemia, increased levels of LDL-IV are found. Patients who have a dense LDL pattern (subclass pattern B, Table 4.8), have low levels of HDL-C and high levels of VLDL-C and IDL-C, whereas

Table 4.8. Lipid and Lipoprotein Differences in Persons Who Have LDL Subclass Patterns A and B

	Pattern A	Pattern B
Triglyceride	Lower	Higher
VLDL[a]	Lower	Higher
IDL[a]	Lower	Higher
Small LDL[a]	Lower	Higher
HDL[a]	Higher	Lower
Large LDL	Higher	Lower

[a]VLDL = very low density lipoprotein; IDL = intermediate density lipoprotein; LDL = low density lipoprotein; HDL = high density lipoprotein

patients who have larger LDL (subclass pattern A) have higher HDL-C and lower VLDL-C levels. Family studies indicate that the phenotypic expression of the gene responsible for the subclass pattern B is influenced by both age and hormonal status. This heritable trait is not fully expressed in men until after age 20 and in women until after the menopause. The type B LDL pattern is more closely associated with atherosclerotic disease.

The structure of the LDL receptor and its postulated five functional domains are shown in Figure 4.5. The ligand-binding domain is located on the surface of the cell plasma membrane. It is rich in cysteine and therefore has a negative charge, permitting it to bind with the basic amino acid structures of apoproteins E and B. Another domain is hydrophobic and spans the plasma membrane, whereas yet another projects into the cytoplasm and may anchor the receptor or relate to its movement or clustering within the cell. The LDL receptors migrate to the surface of the cells to special regions called coated pits (Fig. 4.6). The concentration of LDL receptors at these sites is required for efficient removal of LDL particles. The LDL receptors bind to circulating LDL or VLDL remnants, and the resulting receptor-ligand complex is internalized by endocytosis. The LDL receptor is reused and

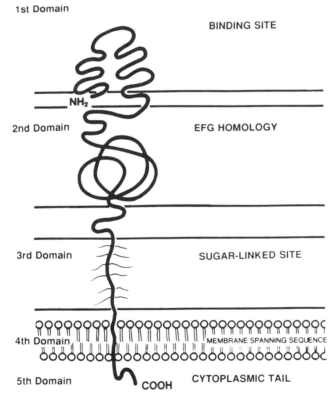

Figure 4.5. The LDL receptor contains more than 800 amino acids. The first domain has nearly 300 amino acids and is the portion of the molecule that binds to LDL. Because of the amino acid repeats in this domain, it may bind to more than one apoprotein at a time and to both apoprotein B and apoprotein E. The functions of the second, third, and fourth domains are largely unknown. Deletion of the fourth domain results, however, in secretion of the LDL receptor from the cell instead of insertion into the membrane. The fifth domain serves to cluster LDL receptors in coated pits. (From: Speroff L, Glass RH, Kase NG. Hormone biosynthesis, metabolism, and mechanism of action. In: Clinical gynecologic endocrinology and infertility. ed 4. Baltimore: Williams & Wilkins, 1989:1–49.)

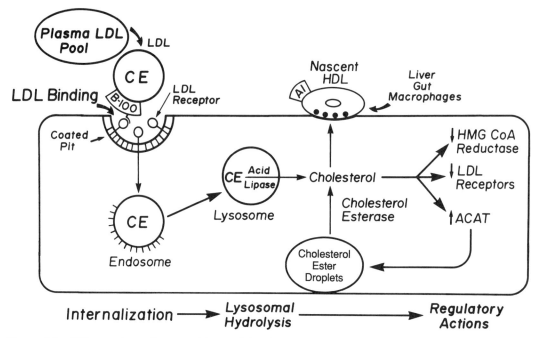

Figure 4.6. LDL receptor pathway and intracellular cholesterol metabolism. Lipoprotein cholesterol in the plasma pool is carried as the esterified form. The LDL particle is removed by the LDL receptor. After endocytosis, the cholesteryl esters are hydrolized into free cholesterol. As cellular free cholesterol rises, HMG CoA reductase and LDL receptors are decreased while ACAT is increased to reesterify free choleterol. Also, free cholesterol is retrieved by nascent HDL. (Adapted from: Bilheimer DW. Lipid metabolism. In: Kelley WN, DeVita VT, DuPont HL (eds): Textbook of internal medicine, vol. 2. Philadelphia: J.B. Lippincott Co., 1989:2138–2144.)

is recycled every 10 minutes for a total of 18 times during its lifetime.

Four mutations have been described in the LDL gene, leading to abnormalities in the LDL receptor that characterize familial hypercholesterolemia. The most common abnormality is a deficiency in LDL receptors, whereas defective function characterizes the other abnormalities. All four genetic abnormalities have identical phenotypes. Consequently, the specific defect cannot be identified without highly sophisticated techniques. In general, the defect will be constant within families as well as within certain ethnic groups. Theoretically, hypercholesterolemia could also be caused by more subtle abnormalities in the receptor, such as point mutations (a single, missubstituted amino acid), which could decrease the efficacy of binding (but not to the degree seen in FH) or an abnormality of a regulatory gene rather than a structural gene, lead-

ing to a slower rate of synthesis of a normal LDL receptor.

The LDL receptor is a high-affinity, low-capacity receptor and, in many regards, is similar to receptors for polypeptide hormones. This receptor binds LDL with such high affinity that only relatively low concentrations of LDL are required to meet the metabolic and structural needs of the cell. In adult animals that do not develop atherosclerosis and in newborn humans, the LDL-C concentration is 25 to 50 mg/dL (0.65 to 1.29 mmol/L). By adulthood, the median LDL-C concentration in humans has increased to approximately 120 mg/dL (3.10 mmol/L) caused primarily by a 30% to 40% reduction in the number of LDL receptors. With increasing age (from 20 to 70 years), there is a progressive reduction in the LDL receptor number, causing a reduction in the fractional catabolic rate for LDL, with a consequent linear increase in the LDL-C concentration of 1.4 mg/dL (0.04 mmol/L) per year. Over a period of 40 to 50 years, the LDL-C concentration can increase from approximately 100 mg/dL (2.59 mmol/L) at age 20, to 160 mg/dL (0.52 to 4.14 mmol/L) at age 60.

The intracellular content of free cholesterol is the primary determinant of the rate of synthesis of cellular cholesterol, as well as the LDL receptor (see Fig. 4.6). When the amount of free or unesterified cholesterol inside the cell is high, there will be inhibition of both LDL receptor synthesis and the synthesis of cholesterol by the enzyme hydroxymethylglutaryl coenzyme A (HMG CoA) reductase. When the free cholesterol content of the liver is low as a result of diet, drugs that inhibit cholesterol synthesis or interventions that increase the conversion of cholesterol to bile acid (such as resins, ileal bypass, or bile drainage), there is increased synthesis and expression of the LDL receptor, as well as increased activity of HMG CoA reductase. It has been suggested that some of the cellular cholesterol is converted into an active "oxy"-sterol form that facilitates entry into the cell's nucleus. This activated cholesterol interacts with regulatory proteins and suppresses the gene encoding for the LDL receptor. When the cellular cholesterol content is too high, events are initiated that prevent additional uptake or synthesis of cholesterol. When the cholesterol content is too low, events are initiated that increase the transport of cholesterol into the cell as well as synthesis of cholesterol by the cell. Cells with LDL receptors, therefore, have the capacity to maintain cellular cholesterol within relatively narrow limits, enabling them to "meter" cholesterol entry and to protect themselves from becoming overfilled with or deficient in cholesterol.

The amount of free cholesterol inside of the cell reflects the balance between cholesterol synthesized or transported into the cell, the rate of conversion of free to esterified cholesterol by the enzyme acyl cholesterol acyl transferase (ACAT), and the rate of cholesterol efflux from the cell. One might imagine that if the ability to esterify cholesterol within the cell was high, then intracellular free cholesterol would be kept relatively low, and cholesterol as well as the LDL receptor would continue to be synthesized. If there was a defect in the activity of ACAT, however, then free cholesterol concentrations inside the cell would be higher, thereby inhibiting HMG CoA reductase as well as synthesis of the LDL receptor. It is possible, therefore, that persons who have relatively inactive ACAT systems might be susceptible to the development of hypercholesterolemia.

In addition to diet and age, hormones also may have an effect on the synthesis and expression of the LDL receptor. In this regard, both thyroxine and estrogen appear to stimulate its synthesis. In hyperthyroidism there are increased numbers of LDL receptors, increased fractional catabolic removal of LDL-C, and a predictably lower cholesterol concentration than in hypothyroidism, in which the reverse would be true. Because estrogens facilitate synthesis of the LDL receptor, part of the increase in LDL-C that occurs with age in women may be caused by a menopausal estrogen deficiency. The administration of estrogens to postmenopausal women decreases the LDL-C concentration by about 10% to 20% because of the enhanced fractional catabolic removal of LDL, presumably a reflection of increased LDL receptor synthesis.

High Density Lipoprotein

High density lipoproteins (HDL) are synthesized in the liver and intestines and, as described, are also products of the metabolism of chylomicrons and VLDL (Fig. 4.7). The major apoproteins of HDL are apoprotein A-I and A-II, which account for about 90% of the HDL protein, with the remainder being derived from apoproteins C and E. As previously discussed, HDL serves as a reservoir for apoproteins C and E and transfers them to chylomicrons and VLDL. Approxi-

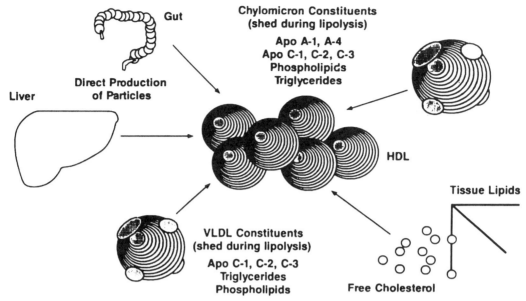

Figure 4.7. Sources of HDL constituents. The components of HDL can be derived from five sources: liver, intestines, chylomicron lipolysis, VLDL lipolysis, or lipid uptake from peripheral cells. The fact that HDL comes from multiple sources reflects the many roles it plays in lipid transport. (From: Coronary heart disease seminar series: Lipid regulation and risk prevention. Parke-Davis, Division of Warner-Lambert Company, 1991.)

mately 50% of circulating apoproteins C and E in the fasting state are found in HDL.

Newly formed (nascent) HDL are derived from chylomicrons and VLDL and as a result of direct secretion by the gut and liver. They appear as bilamellar phospholipid discs that contain apoproteins A-I, A-II, and possibly E. This newly formed HDL readily accepts unesterified cholesterol (Fig. 4.8). As free cholesterol in the HDL membrane is esterified, it moves into the center of the particle and is replaced by another free cholesterol molecule, changing the particle from a disc to a sphere. The process occurs so quickly that few discs can be found in the circulation at any given time.

All esterification of cholesterol in lipoproteins takes place in HDL by the enzyme LCAT. This enzyme is synthesized in the liver and released into the circulation. The optimum activation of LCAT requires its interaction with apoprotein A-I, emphasizing the functional role of this apoprotein (in addition to being the major structural component of HDL).

The HDL particles are the smallest of the lipoproteins and range in size from 75 to 100 angstroms in diameter. They contain nearly equal amounts of lipid and protein. Because of their rapid migration during lipoprotein electrophoresis they also are known as alpha lipoproteins. The HDL particles also are heterogeneous and consist of several subspecies, the most commonly identified forms being HDL_{2a}, HDL_{2b}, and HDL_3, the last being a small spherical form that is cholesterol-poor and represents the newly formed HDL. It is capable of acquiring unesterified cholesterol and converting it into esterified cholesterol. Once optimum esterification of cholesterol has occurred, HDL_3 is transformed into HDL_{2a}, which subsequently is converted to HDL_{2b}. The conversion of HDL_{2a} to HDL_{2b} occurs through CETP, which mediates the exchange of cholesteryl esters for triglyceride with triglyceride-rich lipoproteins. Finally, conversion of HDL_{2b} back to HDL_3 occurs with hydrolysis of triglyceride by HTGL. Under most circumstances, HDL_3 particles outnumber HDL_2 particles by at least 2 to 1.

The HDL particles also can be differentiated by their apoprotein A composition. About two-thirds of HDL particles contain both apoprotein A-I and apoprotein A-II, whereas one-third of particles contain only apoprotein A-I. These apolipoproteins are referred to in the literature as LpA-I:A-II and LpA-I particles. Although men and women have comparable levels of LpA-I:A-II particles, women have significantly higher levels of HDL particles that contain only apoprotein A-I. It also has been noted that CHD patients have fewer LpA-I particles but similar blood concentrations of LpA-I:A-II particles, suggesting that the LpA-I particle is the active protective form.

High blood levels of HDL-C are associated with a lower risk of developing coronary heart disease (see Chapter 2). The mechanisms by which HDL exerts this protective effect are unknown. The major explanation is that HDL participates in the process of reverse cholesterol transport; that is, transporting cholesterol from peripheral tissues to the liver, thereby facilitating its excretion from the body (see Fig. 4.8). The putative mechanism for removal of cholesterol from peripheral tissues involves the transfer of unesterified (free) cholesterol from cell membrane to the nascent HDL, perhaps fa-

cilitated by an HDL receptor; LCAT mediates the esterification of this free cholesterol, creating first HDL_3 and then HDL_{2a}. The HDL_{2a} particle then transfers cholesteryl esters to IDL, LDL, and VLDL through CETP. In this exchange of lipids, HDL acquires triglyceride (HDL_{2b}), which subsequently is hydrolyzed by hepatic triglyceride lipase to regenerate HDL_3. The HDL_3 then can be reused, and the process begins anew. The cholesteryl ester enriched lipoprotein particles (VLDL, IDL, LDL) return the cholesteryl esters to the liver by binding to the appropriate lipoprotein receptors. HDL also deliver cholesteryl esters directly to the liver. The HDL particles that contain apoprotein E (HDL_4) are targeted directly to the liver because of hepatic receptors that recognize apoprotein E. Hepatic cholesterol then is excreted in the bile or converted to bile acids: some cholesteryl esters are transported back to peripheral cells by LDL.

Recent studies in animals demonstrate that intermittent administration of HDL protects against and reverses experimental atherosclerosis without altering HDL-C or LDL-C levels. Consequently, the envisioned concept of reverse cholesterol transport may be too simplistic, and other mechanisms

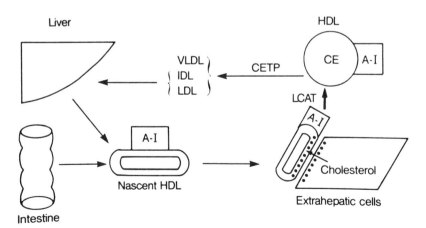

Figure 4.8. Schematic diagram of reverse cholesterol transport. Nascent HDL picks up free cholesterol from the cells where it is converted to cholesteryl esters (CE) by lecithin cholesterol acyl transferase (LCAT) conversion, forming mature HDL. These CEs are then transferred to other lipoproteins (LDL, IDL, VLDL) through cholesteryl ester transfer protein (CETP) and are transported mainly back to the liver. (From: Breslow JL. Apolipoprotein genetic variation and human disease. Physiol Rev 1988;68:85–132.)

may also be operative in the protective effect of HDL against atherosclerosis. It is known that HDL inhibits LDL uptake by tissues and is a major determinant of triglyceride metabolism by providing apoprotein C-II to chylomicrons and to VLDL for activation of LPL. Reduced availability of HDL could be associated with hypertriglyceridemia because of a reduced reservoir of apoprotein C. Alternatively, it has been suggested that the lower HDL in patients who have CHD reflects the accumulation of triglyceride-rich lipoproteins that are atherogenic. Hypertriglyceridemia increases HDL apoprotein A-I catabolism by augmenting HDL triglyceride content while reducing cholesteryl ester in the core of HDL, making it a better substrate for hepatic lipase.

The binding of HDL to tissues is thought to be receptor mediated. The HDL receptor appears to be distinct from the LDL receptor, and an inverse relation between LDL receptors and HDL receptors appears to exist in some tissues. This seems reasonable from the standpoint of cellular cholesterol balance. If there were an increase in the number of LDL receptors as an attempt on the part of the cell to increase its uptake of cholesterol, then it would be prudent for the cell

to reduce the number of HDL receptors, the mechanism by which cholesterol is removed from the cell. If the cell is attempting to rid itself of cholesterol, however, it should downregulate the LDL receptor and upregulate the HDL receptor. The HDL receptor appears to be related positively to cholesterol esterification in the cell and related inversely to cholesterol synthesis and may be regulated by hormones and growth factors. Both insulin and platelet-derived growth factor decrease the number of HDL receptors. These effects tend to stimulate cell growth, in keeping with the need to optimize cholesterol availability for growing cells. Obesity is associated with a reduction in HDL-C concentrations, which may be related to an increase in the total number of HDL receptors and increased HDL removal. In diabetes, there are reduced HDL-C concentrations and an increase in the fractional catabolic rate of HDL.

Lipoprotein (a)

In 1963, Berg described a genetic variation of LDL in men who used heterologous antibodies from rabbits. This subsequently was shown to be a glycoprotein [apoprotein (a)] that is linked covalently by a disulfide bond

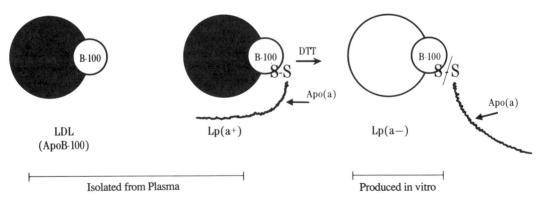

Figure 4.9. Lp(a) consists of an LDL-like particle in which apoB-100 is linked by disulfide bonds to Apo(a) specific to glycoprotein of Apo(a). After action by a reducing agent, dithiothreitol (DTT), a disulfide bridge is cleaved, giving rise to Lp(a) and Apo(a). Both LDL and Lp(a) are components of human plasma. Thus far, Lp(a) and Apo(a) as separate components are documented only in vitro. (From: Scanu AM. Lipoprotein(a): a potential bridge between the fields of atherosclerosis and thrombosis. Arch Pathol Lab Med 1988;112:1045–1047. Copyright 1988, American Medical Association.)

to the LDL apoprotein B-100 and was given the name of Lp(a) (lipoprotein antigen) (Fig. 4.9). There appear to be six genetic isoforms of apoprotein (a) that vary greatly in size. The concentration of Lp(a) in blood is related inversely to the size of apoprotein (a). The majority of apoprotein (a) is synthesized in the liver, but it is not known at what point it becomes attached to LDL. The catabolism of Lp(a) appears at least partly to be controlled by the LDL receptor, and patients who have familial hypercholesterolemia also have higher levels than average. The possibility, however, that other pathways are also involved is being explored.

Apoprotein (a) consists of multiple sequences called kringles, which are structures held together by three internal disulfide bridges (Fig. 4.10). These kringles show a high homology to one of the five kringles (kringle no. 4) found in plasminogen. In addition, human apoprotein (a) contains one kringle with homology to plasminogen kringle 5, and the protease domain of apoprotein (a) has a 94% homology with that of plasminogen. This similarity in structure to plasminogen has raised the possibility that Lp(a) contributes to thrombogenesis by impairing the activation of plasminogen and subsequent fibrinolysis or thrombolysis. Because apoprotein (a) resembles plasminogen, it may compete with it for tissue plasminogen activator and prevent plasminogen from binding to fibrin. In this manner, Lp(a) could prevent normal clot lysis and healing of the vessel wall.

High blood levels of Lp(a) clearly are a marker for an increased risk of developing CHD, and it has been suggested that Lp(a) may be an independent risk factor for CVD, especially among blacks and patients who have familial hypercholesterolemia (FH) and other dyslipidemias (see Chapter 2). Although Lp(a) is associated with an increased risk of premature MI, it appears that it also requires adequate concentrations of apoprotein B-100 and LDL to be contributory.

Notable differences appear in Lp(a) levels and distributions among different ethnic groups. Blood levels of Lp(a) are determined to a large extent by genetic factors, although they may be influenced to some degree by

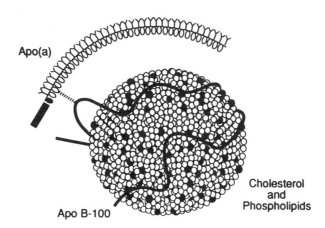

Figure 4.10. Schematic model of Lp(a). The central LDL particle, with its core of cholesteryl esters contains a surface layer of phospholipids and free cholesterol. The apo B-100 molecule, is attached to one molecule of apo (a). The mode of interaction of the LDL component with the apo (a) moiety and the spatial orientation of these components are unknown, as is the exact topology of the postulated (but not proven) disulfide bond (S-S) between apo B-100 and apo (a). Structurally there is a resemblance to plasminogen. (Reproduced, with permission, from: Scanu AM, Lawn RM, Berg K. Lipoprotein(a) and atherosclerosis. Ann Intern Med 1991;115: 209–218.)

nongenetic factors, such as renal disease and hormones. No good evidence exists to show that diet or exercise affects Lp(a) blood levels. The only lipid-lowering drug that has been shown to reduce Lp(a) levels modestly is niacin. There is increasing interest in the effects of drugs that reduce disulfide bonds on Lp(a) levels, but preliminary reports on such effects await further exploration.

Pathways for Removal of Lipoproteins

As already mentioned above, LDL can be removed from the circulation by two receptor-mediated pathways (Fig. 4.11). One is the LDL receptor pathway, which accounts for about 65% of the LDL-C removal. The other is the scavenger pathway, which normally accounts for approximately 35% of the LDL-C removal. Cells that have LDL receptors are protected against undesirably high intracellular levels of cholesterol by the tight metabolic coupling that exists between the free intracellular cholesterol concentration, the activity of HMG CoA reductase, and the synthesis of the LDL receptor. Consequently, as the LDL-C concentration increases a greater portion of it will be forced into the scavenger pathway.

The scavenger pathway is present in monocytes, macrophages, endothelial cells, and vascular smooth muscle cells (see Fig. 2.11). This pathway is capable of binding and internalizing hypertriglyceridemic VLDL and a modified or oxidized form of LDL (OX-LDL). For reasons that are not entirely clear, the receptor of the scavenger pathway is not under metabolic control. The intracellular free cholesterol concentration is not capable of regulating the further internalization of OX-LDL or VLDL cholesterol through this receptor, and the cells continue to take up cholesterol despite increased cellular cholesterol content. As a consequence, they can accommodate progressively more cholesteryl esters and, in doing so, become foam cells. Eventually the cells die, discharging their cholesteryl ester into the surrounding area. It is believed that the

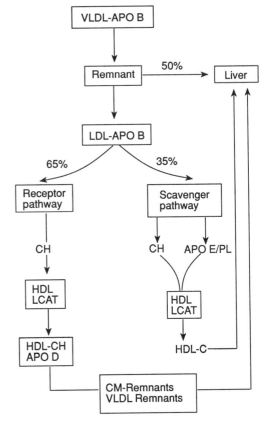

Figure 4.11. About two-thirds of the LDL in the circulation is removed by LDL receptors, but one-third is taken up by scavenger receptors found on macrophages and endothelial cells that do not depend on cellular cholesterol levels. HDL-CH = high density lipoprotein cholesterol; PL = phospholipid. (From: Kreisberg RA. Lipoprotein metabolism and disorders. In: Hershman JM (ed). Endocrine pathophysiology: a patient-oriented approach. Philadelphia: Lea & Febiger, 1988:325–349.)

scavenger pathway plays a central role in the development of atherosclerosis.

Balance between delivery and removal of cholesterol from cells is a critical process. Low density lipoprotein is viewed as delivering cholesterol, HDL as removing it. Under such conditions, persons who have a low HDL-C would be unable to have normal reverse cholesterol transport, even in the presence of normal LDL-C levels. Very high HDL-C concentrations would not only protect against the development of coronary heart disease in patients who have normal

LDL-C concentrations but might also be able to do so in patients who have high LDL-C concentrations because of the greater capacity for reverse cholesterol transport. The Framingham data on CHD risk from HDL and LDL would support this concept (see Chapter 2). Whenever there is an imbalance between cholesterol delivery and cholesterol removal, cholesterol accumulates in the cell and promotes atherogenesis.

Recently it has become clear that LDL-C is not a candidate for uptake by the scavenger receptor until it is oxidized or modified. This modification usually involves peroxidation of the lipid and subsequent alteration of the LDL apoprotein B, which increases its affinity for the scavenger receptor. Peroxidation probably is determined by numerous factors, one of which could be genetic. Consequently, persons who have a high capacity for peroxidation might be at increased susceptibility to atherosclerosis in the absence of obvious lipoprotein disorders (or with minimally increased LDL-C). Peroxidative capacity appears to reside in endothelial cells, macrophages, and mural or muscle cells and could vary from person to person. Drugs that prevent peroxidation theoretically may protect against the development of atherosclerosis. There is accumulating evidence that the antiatherosclerotic effect of probucol (see Chapter 8), at least in animals, is related to this mechanism. Although probucol lowers LDL-C concentrations, the mechanism of protection against CHD may be through other pathways. If peroxidation of LDL is important in this process, then the identification of other drugs that have antioxidant properties could be useful.

Decreased susceptibility of the endothelium to injury and decreased endothelial peroxidation of LDL may be mechanisms by which some patients are less susceptible to atherosclerosis despite the presence of known risk factors. It is also possible that gender differences in susceptibility to atherosclerosis may be mediated at the level of the vascular bed as well as by differences in lipoprotein levels.

Modified VLDL also is removed by the scavenger pathway. Evidence is accumulating to show, however, that hypertriglyceridemic VLDL need not be modified to bind to this receptor. In conditions with increased VLDL and triglyceride concentrations, the apoprotein E content of the VLDL also is increased. The increase in apoprotein E converts VLDL into a particle that is recognized by the scavenger receptor. This could account for lipid accumulation, the formation of foam cells, and the predisposition of certain patients who have hypertriglyceridemia to the development of CHD.

Suggested Readings

Breslow JL. Genetic basis of lipoproteins. Metabolism and relationship to atherosclerosis. J Clin Invest 1990;86:379–384.

Brown MS, Golstein JL. A receptor mediated pathway for cholesterol homeostatsis. Science 1986;232:34–47.

Brown MS, Goldstein JL. How LDL receptors influence cholesterol and atherosclerosis. Sci Am 1984;251:58–66.

Ginsberg HN. Lipoprotein physiology and its relationship to atherogenesis. Endocrinol Metab Clin North Am 1990;19:221–228.

Scanu AM, Lawn RM, Berg K. Lipoprotein(a) and atherosclerosis. Ann Intern Med 1991;115:209–218.

Schaefer EJ, Levy RI. Pathogenesis and management of lipoprotein disorders. N Engl J Med 1985;312:1300–1310.

Tall AR. Plasma high density lipoproteins: metabolism and relationship to atherogenesis. J Clin Invest 1990;86:379–384.

Utermann G. Lipoprotein (a): a genetic risk factor for premature coronary artery disease. Curr Opin Lipidol 1990;1:404–410.

CHAPTER 5

Influence of Growth and Development, Sex, and Age on Lipids

Growth and Development

A substantial amount of data exist to suggest that atherosclerosis begins in childhood and progresses gradually into adulthood. Although the clinical manifestations of CHD rarely appear before the fourth decade, autopsy studies show that early coronary atherosclerosis or precursors of atherosclerosis are often present in younger men. High blood levels of total cholesterol, LDL-C and VLDL-C, and low blood levels of HDL-C have been shown to correlate with the extent of early atherosclerosis in childhood and adolescence.

Population studies reveal that blood cholesterol levels in children also relate to the incidence of adult CHD. In countries that have a low prevalence of CHD, childhood cholesterol levels also are low. The LDL-C values of children from populations that have high rates of CHD escalate more rapidly during childhood than do the LDL-C levels of children from populations that have low CHD rates. Furthermore, lipid levels in children correlate with the subsequent development of fibrous plaques in the coronary arteries of young adults, emphasizing the early initiation of this process. Children of fathers who have survived an MI are likely to have decreased apo A-I levels, increased apo B levels, and increased apo B to apo A-I ratios. Similarly, the frequency of CHD is higher than average in adult members of families of children who have lipoprotein abnormalities. Unlike the situation in white children, data from the Bogalusa study suggest that a history of cardiovascular disease in a black parent is not strongly associated with elevated LDL-C levels in their children. Older black children who have parental history of heart attack, hypertension, or diabetes were about five times more likely to have low levels of HDL-C than those who had no such history.

Lipoprotein abnormalities tend to aggregate in families. In some families the disorder is monogenic (i.e., it occurs as a result of inheritance of a single major gene disorder, such as in familial hypercholesterolemia

[FH] and in familial combined hyperlipidemia [FCH]). The majority of cases are polygenic, however, occurring as a result of expression of a number of genes, each with a small but additive effect combined with shared environmental influences such as diet, obesity, smoking, and exercise habits.

Although the NCEP guidelines for the identification and treatment of elevated cholesterol levels in the adult population do not take age and sex into consideration, it is important to realize that these parameters have a substantial influence on lipoprotein metabolism.

Childhood

Lipid levels at birth are similar in all populations, regardless of CHD rates. The cholesterol concentration in umbilical cord blood and during early infancy is 60 to 90 mg/dL (1.55-2.33 mmol/L), and mean HDL-C levels are approx. 35 to 40 mg/dL (0.91 to 1.03 mmol/L). The most marked changes occur during the first year of life, as solid foods are introduced into the diet, and cholesterol increases between 100 and 150 mg/dL (2.59 and 3.88 mmol/L) appear during the first few weeks of life. Subsequently the serum cholesterol rises slowly to levels of 160 mg/dL (4.14 mmol/L) (50th percentile) at 10 years of age (Figs. 5.1, 5.2). Children who have elevated blood cholesterol levels can be identified by comparison with values in their peer group. The mean and the 5th and 95th percentiles for cholesterol and triglyceride levels at various ages before adulthood are shown in Table 5.1. A total cholesterol of greater than 200 mg/dL (5.17 mmol/L) places boys and girls above the 95th percentile; approximately 5% of American children from age 5 through 18 have plasma cholesterol levels above this level. It is critical, therefore, that appropriate, age-related values be used for assessment of children.

As in adults, the LDL-C in children is affected by the amount of saturated fats and cholesterol in the diet. This is influenced to a large extent by the quality of food eaten away from home. The tendency of young-sters to consume a substantial proportion of their meals and snacks in fast food restaurants has grown in recent years. School food programs also are a major source of meals for children in the United States. A comprehensive approach to modification of nutrition is warranted. It should be emphasized that low blood cholesterol levels in populations are compatible with normal patterns of growth and development, and the concern that has been expressed that diet modification can affect these processes adversely has not been substantiated.

Elevation of the total cholesterol in childhood generally results from increased levels of LDL-C, although some children will have increased HDL-C or VLDL-C levels. As in adults, it is important to determine the lipoprotein profiles of children who have elevated total cholesterol values.

Puberty

The difference in lipid and lipoprotein profiles between men and women begins with the onset of puberty (Fig. 5.3). Impressive changes occur in the metabolism of lipoproteins during adolescence and sexual maturation that are sex- and race-specific.

Boys. From early adolescence to 18 years of age, the mean cholesterol level in white boys declines slightly to 150 mg/dL (3.88 mmol/L) and is caused by a decrease in HDL-C levels, without any change in LDL-C. This 10-mg/dL (0.26 mmol/L) reduction in HDL-C is associated with an increase in plasma testosterone levels. The reduction in the cholesterol transported by HDL is greater than the reduction in apo A-I, reflecting the appearance of a smaller and denser HDL particle. The VLDL-C levels increase in all children during puberty; however, in white boys the values continue to increase during the next decade, while they remain stable in girls and in black boys. These adverse changes may have clinical significance.

Girls. At puberty, the total cholesterol decreases steadily until the age of 19 years in white girls; in contrast to boys, this is

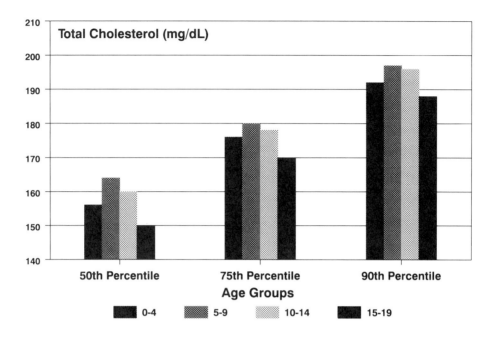

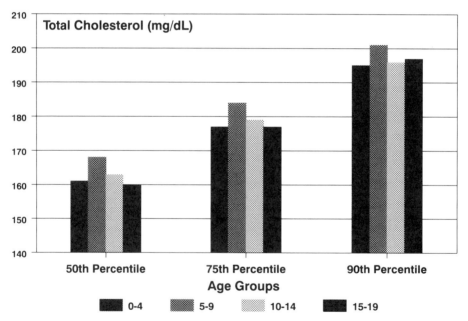

Figures 5.1 and 5.2. Serum total cholesterol levels in male (Fig. 5.1) and female (Fig. 5.2) children and adolescents (ages 0–19) in the LRC Prevalence Study in North America. (Adapted from: U.S. Department of Health and Human Services. The Lipid Research Clinics, Population Studies data book, Vol I-The Prevalence Study. July, 1980 (NIH Pub. No. 80-1527).

Table 5.1. Plasma Lipid Concentrations in the First Two Decades of Life (mg/dL) (By Percentile)[a]

Age (yr)	Cholesterol			Triglycerides		
	5th–	Mean–	95th	5th–	Mean–	95th
0–4						
Males	114	155	203	29	56	99
Females	112	156	200	34	64	112
5–9						
Males	121	160	203	30	56	101
Females	126	164	205	32	60	105
10–14						
Males	119	158	202	32	66	125
Females	124	160	201	37	75	131
15–19						
Males	113	150	197	37	78	148
Females	120	158	203	39	75	132

[a] Based on plasma measurements from 11,219 fasting, white subjects (5749 males, 5470 females), studied in seven North American Lipid Research Clinics. The Prevalence Study. (Adapted from The Lipid Research Clinics Population Data Book, vol. 1. NIH Publication #80-1527. GPO, 1980.)

largely the result of a decrease in LDL-C. Smaller changes are seen in black girls. Triglyceride concentrations, which are higher in white than in black girls, remain relatively constant through the second decade; thereafter, they increase slowly but remain lower than those in men.

Tracking

For most physiologic traits, such as height and weight, everyone tends to maintain relative rank within the population distribution of peers. This phenomenon is called *tracking* and occurs also for the blood lipoproteins. Consequently, concentrations at an early age predict future concentrations. For example, a 31-year-old man who has a total cholesterol of 200 mg/dL (5.17 mmol/L) would be expected to have a cholesterol of slightly more than 230 mg/dL (5.95 mmol/L) at age 60 (Fig. 5.4 curve labeled E). Those persons whose cholesterol is in the upper 25% of all cholesterol values retain this position, and as cholesterol values increase with age, the absolute value of their cholesterol levels also would increase.

Recently developed predictive models for total cholesterol, based on data from the Na-

tional Health and Nutrition Examination Survey 1976 to 1980, describe the association between age and cholesterol levels for men and women aged 20 to 57 (Figs. 5.4 and 5.5). These models allow adult cholesterol levels to be predicted from cholesterol measurements performed at an earlier age. In addition, the age at which persons could be expected to reach borderline-high or high blood cholesterol levels, in the absence of any intervention, can be anticipated.

This observation also applies to children. Young children tend to maintain their same percentile ranking for various lipid parameters, so that those children who are in the upper percentiles for cholesterol tend to have higher adult cholesterol values. Such children subsequently would be expected to be at greater risk of developing CHD. In the Bogalusa Heart Study, children examined initially in 1973 to 1974 and again in 1984 to 1986 showed correlation coefficients ranging from 0.4 to 0.7 for total cholesterol. Tracking also was noted for LDL-C levels in children. It has been estimated that over 70% of those children who have high lipid values remained at the upper end of the distribution as young adults. In the Muscatine Study, 2367 children (aged 8-18 years) were examined on several occasions and were followed up to ages 20 to 30. Among the children who had cholesterol levels exceeding the 90th percentile on two occasions, 43% of girls and 70% of boys had cholesterol levels as adults that qualified for some kind of intervention according to the NCEP guidelines.

The relative cholesterol rank of a person appears to be established as early as 9 to 10 years of age. Such data indicate that children and young adults who are in the upper percentile in the lipid distribution curve should be maintained under careful surveillance, and that their parents should be advised to consider their long-term health habits, particularly consumption of a prudent diet, appropriate levels of exercise, weight control, prohibition of cigarette smoking, and other practices that might influence lipid levels or cardiovascular risk.

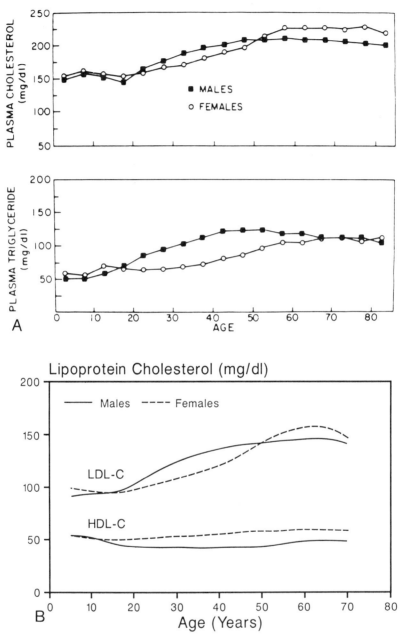

Figure 5.3. A.   A plasma total cholesterol and triglyceride for white males and females not taking sex hormones (median values for 5-year age groups). (From Hazzard WR. Disorders of lipoprotein metabolism. In: Andres R, Bierman EL, Hazzard WR, eds. Principles of geriatric medicine. New York: McGraw-Hill, 1985:764–775. "Reproduced with permission of McGraw-Hill, Inc.") **B.** Plasma LDL-C and HDL-C for white males and females not taking sex hormones (median values for 5-year age groups). (Adapted from: U.S. Department of Health and Human Services, Public Health Service, National Institutes of Health. The lipid research clinics: population studies data book. Volume I. The prevalence study. NIH Publication No. 80-1527, July 1980.)

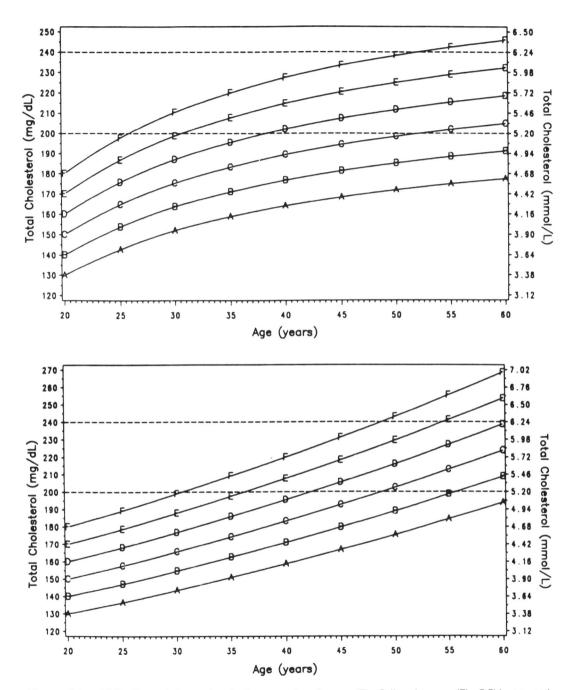

Figures 5.4 and 5.5.   Total cholesterol projections are given for men (Fig. 5.4) and women (Fig. 5.5) by age and initial cholesterol level. By knowing the age and total cholesterol level, the appropriate curve can be extended to a later age, thereby delineating the future cholesterol level anticipated. Ideal (200 mg/dL) and borderline (240 mg/dL) are levels depicted with the dotted lines. (Adapted from: CDC: Morbidity and Mortality Weekly Report 38:(2)365, May 1989.)

Controversy still exists about whether screening for elevated cholesterol should be done in all children (*universal screening*) or only in those children who have a strong family history of dyslipidemia or premature CHD (*selective screening*). Proponents for screening of all children cite studies that show that selective screening misses at least 50% of the children who have elevated cholesterol levels. Proponents of more selective screening emphasize, however, that tracking is imperfect, and that many children who have high cholesterol will not have elevated cholesterol levels in adulthood. They also claim that massive screening would be overly expensive, could lead to harmful labeling of children as patients with a disease, and to family conflicts over implementations of interventions. Other concerns include the lack of sufficient evidence of long-term safety and efficacy of therapy in this age group.

The NCEP Report of the Expert Panel on Blood Cholesterol Levels in Children and Adolescents has adopted the selective approach to screening (see Chapter 9) and recommends screening of all children and adolescents who have a family history of premature CVD or at least one parent who had high blood cholesterol. This approach is aimed at identifying those youngsters who have a familial form of dyslipidemia, which presumably implies a higher cardiovascular risk than nonfamilial dyslipidemia.

Sex

Men

The total cholesterol concentration essentially is constant in men until the age of 20. After age 20, it increases progressively at an annual rate of approx. 2 mg/dL (0.05 mmol/L), until age 30, primarily because of an increase in the LDL-C concentration (Fig. 5.6). Thereafter the rate of increase slows to approx. 1 mg/dL per year (0.03 mmol/L) through middle age. Overall, the LDL-C concentration increases from a mean of 105 mg/dL (2.72 mmol/L) at age 20 to 140 mg/dL (3.62 mmol/L) at age 50. There-

after, no appreciable increase occurs in either the total cholesterol or the LDL-C concentration. The plasma triglyceride concentration also increases over the same interval because of an increase in the triglyceride-rich lipoprotein, VLDL. The increase in the cholesterol transported on VLDL accounts for the other major component of the increase in the total cholesterol concentration that occurs with age.

Women

The age-related changes in cholesterol and triglyceride concentrations in women differ from those in men and appear to depend on their hormonal status (see Fig. 5.6). Because the hormone levels in women change continuously during the menstrual cycle, pregnancies, and menopause, as well as by the administration of exogenous hormones in the form of oral contraceptives and estrogen replacement therapy, the lipoprotein levels in women fluctuate more than in men.

The annual increment in total cholesterol in women is about 1.5 mg/dL (0.04 mmol/L) per year from age 20 to 40; however, the annual increase in cholesterol from ages 40 to 60 is larger than in men and is approximately 2 mg/dL (0.05 mmol/L). After menopause, women have a mean total cholesterol concentration that is approx. 40 mg/dL (1.03 mmol/L) greater than those of comparably aged men. Women who receiving estrogenic hormones have a higher mean cholesterol concentration at age 20 than women who do not receive hormones; however, the difference is only about 10 mg/dL (0.26 mmol/L). Women treated with estrogens also have higher triglyceride concentrations. Although the initial cholesterol levels of women who do not receive estrogens is lower, it increases more rapidly with age and surpasses that of estrogen-treated women by age 50. Analysis of the various lipoprotein profiles indicates that increases in LDL-C are responsible for most of the increases in the total cholesterol; it also has been suggested that the LDL particles become smaller and denser and, therefore,

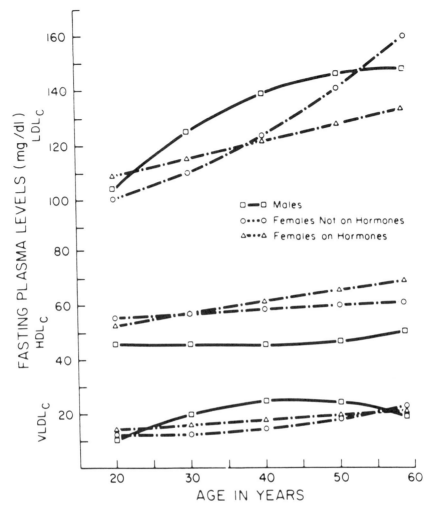

Figure 5.6. Regression estimates of mean plasma lipoprotein-cholesterol values by age for males, females not taking sex hormones, and females on sex hormone preparations. (From: Hazzard WR: Disorders of lipoprotein metabolism. In: Andres R, Bierman EL, Hazzard WR (eds). Principles of geriatric medicine. New York: McGraw-Hill, Inc., 1985:764–775.) "Reproduced with permission of McGraw-Hill, Inc."

more atherogenic after menopause. Estrogen replacement therapy inhibits the normal increase in LDL-C that occurs with age. One study indicates women have higher Lp(a) levels than men, particularly during and after middle age.

Women also have HDL-C concentrations that are approx. 10 mg/dL (0.26 mmol/L) *higher* than those of men. This difference occurs during puberty and persists throughout the lifespan, even with the development of estrogen deficiency at menopause. The HDL-C concentration in estrogen-treated women increases over time, so that their mean HDL-C at 60 years of age is approximately 5 mg/dL (0.13 mmol/L) higher than the levels of women who do not receive estrogen replacement. The higher levels of HDL-C in women may allow them to tolerate higher levels of LDL-C than men; it has been suggested that every increase of 5 mg/dL (0.13 mmol/L) in HDL-C can counterbalance an equivalent increase in LDL-C of approximately 15 mg/dL (0.26 mmol/L) toward risk of CHD.

Menstrual Cycle. Because endogenous hormone levels change during the menstrual cycle, lipid and lipoprotein levels might also be expected to vary in accordance with the relation of the time of blood sampling to the cycle. Although the exact nature of such changes is not completely defined, they seem to be relatively minor and to affect mainly LDL and VLDL levels, with little or no changes in HDL-C levels. The reason for the insignificant variations in HDL-C levels may be partly caused by the relatively modest androgenicity of natural progesterone, compared to synthetic progestins or the longer half-life of HDL in plasma.

Pregnancy. In contrast to the minor effects of the menstrual cycle on lipoprotein metabolism, the large increases in estrogen and progesterone levels during pregnancy are associated with significant changes in lipid and lipoprotein concentrations (Fig. 5.7). Total cholesterol levels increase by approx. 75%, to a mean of about 315 mg/dL (8.15 mmol/L) during the course of pregnancy. This increase is caused primarily by an elevated LDL-C, which increases by about 70%. The HDL-C increases by approximately 40% by midgestation but then declines to levels 15% above nonpregnant levels. The relative increase that occurs in

triglyceride (330%) is more impressive than the increase in cholesterol, from a mean of approx. 95 mg/dL (1.07 mmol/L) before pregnancy to approx. 312 mg/dL (3.52 mmol/L) at the time of delivery. Although VLDL synthesis is increased in pregnancy, VLDL accounts for only approximately 50% of the increase in triglyceride concentration. Therefore, the amount of triglyceride transported by LDL and HDL is also substantially increased. Women who have hypertension also have higher triglyceride concentrations during pregnancy than normotensive women. This is consistent with the current concept that some patients who have hypertension also have mild to moderate insulin resistance and are prone to hypertriglyceridemia.

Following delivery, triglyceride concentrations decrease abruptly and usually are back to or near baseline by 6 to 8 weeks. The HDL-C concentrations also usually revert to prepregnancy levels by the 20th week postpartum. In contrast, the total cholesterol, LDL-C, and apoprotein B concentrations fall more slowly and may still be above the prepregnancy values as late as 20 weeks postpartum. It therefore seems reasonable to

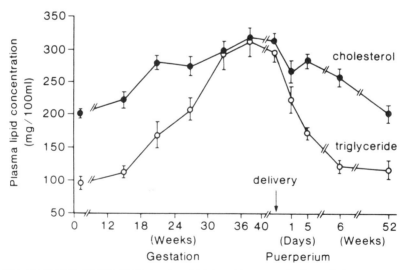

Figure 5.7. The changes in plasma cholesterol and triglyceride concentrations during pregnancy and in the puerperium. Fasting lipid concentrations were measured serially throughout pregnancy, at delivery, in the puerperium, and at 12 months postpartum. The results are the mean ± S.E.M. and include normal and complicated pregnancies. (From: Miller VT: Dyslipoproteinemia in women. Endocrinol Metab Clin North Am 1990;19: 381–398.)

postpone cholesterol measurements for at least 6 months after childbirth.

Estrogens are known to increase the synthesis of VLDL and occasionally to cause hypertriglyceridemia. This is often attributed to unmasking of an underlying defect in VLDL transport. It is likely that the increase in VLDL during pregnancy occurs for similar reasons, although the mechanism probably is more complex because LDL concentrations also are increased. An excessive increase in triglyceride concentrations during pregnancy may identify women who have subtle, previously undetected familial or acquired abnormalities in lipoprotein transport and calls for a careful lipoprotein evaluation postpartum.

Although routine screening for lipids during the course of pregnancy is not recommended, multiphasic screening often is used at some time during prepartum evaluation, and higher than expected values should be noted for future evaluation and treatment. Massive hypertriglyceridemia in pregnancy can be associated with recurrent episodes of abdominal pain or even acute pancreatitis. Intervention with diet clearly would be appropriate. There is little experience with the use of hypolipidemic agents in pregnancy, and they are not recommended except under the most unusual life-threatening circumstances.

Menopause. Menopause is accompanied by an increase in LDL-C levels of approx. 6 mg/dL (0.16 mmol/L) and reductions in HDL-C of approx. 3 to 5 mg/dL (0.08 to 0.13 mmol/L). Surgical menopause with bilateral oophorectomy appears to affect lipoprotein levels to a greater degree than natural menopause; the reasons for this discrepancy are unknown. These changes are abolished by hormone replacement therapy (Fig. 5.8). In addition to the quantitative changes in LDL levels that occur at the time of menopause, there also is an increase in the smaller and denser subspecies of LDL that is more likely to be associated with the development of CHD. The differences in total cholesterol, LDL-C, HDL-C, and triglycerides that occur as a conse-

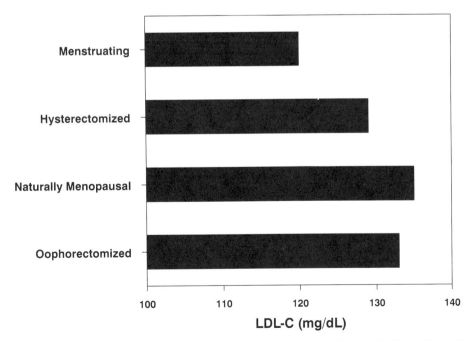

Figure 5.8.   Effect of ovarian function on low density lipoprotein cholesterol concentrations. (From: Bush TL, Fried LP, Barrett-Connor E. Cholesterol, lipoproteins, and coronary heart disease in women. Clin Chem 1988;34:8(suppl B):B60–B70.)

quence of aging and estrogen deficiency are shown in Fig. 5.6.

Increasing evidence shows that estrogen replacement therapy in menopausal women may reduce the risk of developing CHD and, perhaps, even stroke. The reduction in CHD is estimated at 50% to 65%, affecting CHD mortality and morbidity alike. This reduction in risk is caused in part by hormone-induced changes in lipoproteins. Estrogens may also influence vascular reactivity and vessel wall properties that similarly reduce susceptibility to the atherosclerotic process. Estrogen replacement therapy also has beneficial effects on other associated risk factors. It decreases the body mass index in menopausal women by reducing adipose tissue mass and increasing lean body mass. It also decreases the systolic as well as the diastolic blood pressures and reduces the fasting plasma glucose concentration. Although short-term studies of certain progestins indicate that they may reduce HDL-C levels, the overall benefit of hormone replacement therapy during menopause does not appear to be affected adversely by concomitant progestin therapy. The use of androgenic nortestosterone progestin analogues, however, is not currently recommended in high-risk women (see Chapter 8).

Estrogen deficiency that occurs at menopause has an unfavorable effect on lipid metabolism, which probably contributes to the increased risk of developing CHD. Hormonal replacement therapy may prevent these changes and, unless contraindicated, should be considered in all postmenopausal women.

Age

The increase in LDL-C that occurs with aging is caused primarily by a reduction in the rate of LDL catabolism. In large part, this is because there is a reduction in receptor-mediated LDL removal; however, there may also be some increase in the rate of LDL synthesis. Estrogens partially blunt the increase in LDL-C that occurs with age in women. After age 65, nearly 50% of all women and one-third of the men have blood cholesterol levels above 240 mg/dL (6.2 mmol/L) before treatment with lipid-lowering agents. Occult CHD may be present in up to 50% of all persons in this age range. Periodic measurement of cholesterol is necessary, particularly in women whose initial levels are above the median value for the population, because the age-related changes of such persons will push their total cholesterol concentrations into the borderline or elevated ranges.

Currently, an issue of great interest is the importance of cholesterol as a determinant of CHD risk in the elderly. Most authorities agree that the relation between cholesterol and CHD is present in the elderly but that it is weaker than that in other age groups. In the Framingham Heart Study the relation was reduced by approx. 50%, so that for each 1% increase in cholesterol concentration, the risk of CHD increased by only 1% instead of the usually quoted 2%. However, in the Honolulu Heart Study, the relation between the cholesterol level and CHD risk was just as strong in elderly men as it was in middle-aged men (Fig. 5.9). The HDL-C also remains an independent predictor for CHD in the elderly. Even if we accept the fact that the relative risk of CHD from hypercholesterolemia is reduced among the elderly, the prevalence of hypercholesterolemia is high (Fig. 5.10) and the attributable risk of CHD, that is, the proportion of CHD that can be related to hypercholesterolemia, is also very high (Fig. 5.11). This occurs because more older than younger persons have hypercholesterolemia and because older persons have more CHD events in a given time period. The number of potentially preventable cardiovascular events and death, then, may actually be greater among the elderly than is reported.

Presently, there are insufficient data to prove that treatment of dyslipidemias among the elderly improves the cardiovascular risk, because most of the large clinical trials were conducted among the middle-aged. Benefit from cholesterol lowering was demonstrated in men up to the age of 69 years in the LRC Coronary Primary Preven-

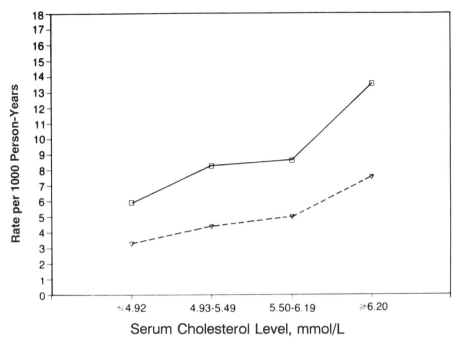

Figure 5.9. Coronary heart disease rates by quartile of total serum cholesterol level in elderly and middle-aged men in the Honolulu (Hawaii) Heart Program. Solid line indicates ages 65 through 74 (relative risk, 1.68); and dashed line, ages 52 through 59 (relative risk, 1.71). (From: Benfante R, Reed D. Is elevated serum cholesterol level a risk factor for coronary heart disease in the elderly? JAMA 1990;263:393–396.) "Copyright 1990, American Medical Association."

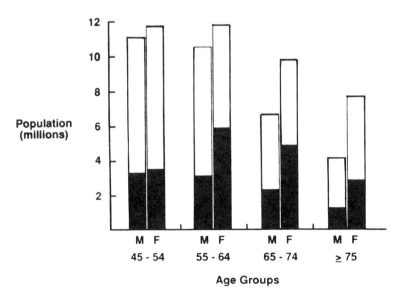

Figure 5.10. Population distribution by age and sex. Solid bar denotes segment of population whose serum cholesterol levels are above 6.20 mmol/L (240 mg/dL). "Reprinted, with permission from" (From: Denke MA, Grundy SM. Hypercholesterolemia in elderly persons:resolving the treatment dilemma. Ann Intern Med 1990;112:780–792.)

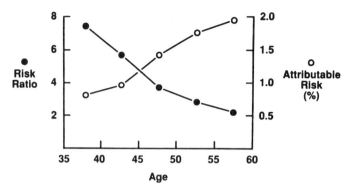

Figure 5.11.   Change in risk ratio (closed circles) and attributable risk (open circles) with age for men in the fifth quintile compared with those in the first quintile of the Multiple Risk Factor Intervention Trial. "Reprinted, with permission from" (From: Denke MA, Grundy SM. Hypercholesterolemia in elderly persons:resolving the treatment dilemma. Ann Intern Med 1990;112:780–792.)

tion Trial, as well as in the Los Angeles Domiciliary Trial that used diet to lower blood lipids. Furthermore, a decrease in CHD mortality has been observed in all age groups in the past decade, including persons 75 years of age and older, presumably because of modifications of CHD risk factors. Elderly persons respond to diet and cholesterol-lowering drugs as well as younger persons. Although some of the elderly may have difficulties in changing dietary and life-style habits acquired in earlier years, many others are willing to make such changes to lower the risk for CVD. It is important to ensure that any new diet is nutritionally balanced and does not aggravate any preexisting nutritional deficiencies. Finally, many of the elderly have other coexisting diseases and require multiple medications, predisposing them to excessive adverse effects from some lipid-lowering drugs; this may be particularly true for nicotinic acid and the resins.

Because the benefit of treating dyslipidemia in some of the clinical trials became apparent only after 1 to 2 years, and because the likelihood of improving the functional status by delaying the development of CHD is low among the elderly whose activity already is limited by other severe chronic diseases, it is essential to select carefully potential candidates for lipid-modifying therapy. Although no definite data exist on

this issue, we believe that the decisions to treat or not to treat should be individualized and that *physiologic* as well as chronologic age should be considered.

Race

Although the major cardiovascular risk factors are universal, the prevalence and relative importance of some risk factors vary among different races. In the United States, blacks have a higher prevalence of hypertension, smoking, obesity, and diabetes mellitus. Although older data had underestimated the role of CHD in black Americans, more recent evidence has dispelled these views, and it is now recognized that for the age groups between 20 and 64, the mortality ratio for CHD is greater in blacks than in whites. Also, the secular trends of decline in CHD mortality experienced in the United States in the past decade are less for black males than for white males.

A population-based study, CARDIA, has provided fasting lipoprotein data on 4858 black and white men and women, aged 18 to 30. The age-adjusted means and standard deviations for total cholesterol, LDL-C, HDL-C, triglyceride, and apoproteins A-I and B for the 18 to 24 years sub-group are given in Table 5.2. Black men had higher HDL-C and apoprotein A-I levels and lower triglyceride concentrations than white men.

Higher HDL-C levels in black men aged 20 through 74 also were found in the Second National Health and Nutrition Examination Survey (NHANES II). In CARDIA, black women had higher LDL-C and apoprotein B levels but similar HDL-C and lower triglyceride concentrations than white women.

Lipoprotein antigen concentrations generally are higher among blacks. In addition, the distribution of Lp(a) is different in the two groups, being bell-shaped in blacks and skewed to the right (more persons who have lower levels) in whites (Fig. 5.12).

Racial differences in lipoprotein levels, especially lower VLDL-C and higher HDL-C concentrations among black children, are established by 9 years of age and are not attributable to environmental factors, although HDL-C levels in blacks seem to be more responsive to physical activity and alcohol than in whites. The levels are greater in men than in women and persist even after stratification for various socioeconomic and behavioral parameters. Age differences in lipoproteins generally are more pronounced among whites than among blacks. To the extent that apo A-I is indicative of HDL particle number, the higher HDL values among blacks also may indicate increased numbers of HDL particles. The higher ratio of cholesterol to apoprotein B in LDL among blacks indicates the presence of larger, less dense, and more cholesterol-enriched LDL particles. This type LDL particle is thought to be less atherogenic than smaller, dense LDL particles.

Overall, blacks have more desirable lipoprotein levels than whites, but because of

Table 5.2. Selected Lipid and Lipoprotein Measures (mg/dL) From Young Adults (18–24 years) in the CARDIA Study by Sex and Race

Lipid/Apoprotein	White Men		White Women		Black Men		Black Women	
Total Cholesterol	168[a]	(32)	174	(30)	168	(32)	176	(34)
LDL-C	105	(29)	105	(28)	102	(30)	109	(32)
HDL-C	46	(10)	54	(13)	53	(13)	55	(12)
Triglycerides	86	(56)	73	(36)	64	(39)	61	(28)
Apo A-I	130	(19)	138	(22)	140	(22)	140	(22)
Apo B	89	(23)	88	(22)	85	(23)	89	(26)

[a] Mean (SD)
(Adapted from Donahue RP, Jacobs DR, Sidney S, Wagenknecht LE, Albers JJ, Hulley SB. Distribution of lipoproteins and apolipoproteins in young adults: the CARDIA study. Arteriosclerosis 1989;9:656–664.)

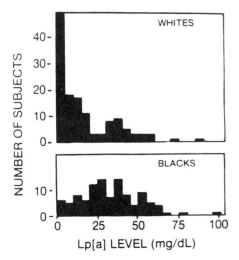

Figure 5.12. Frequency distributions of Lp(a) levels in 134 white (*top*) and 105 black (*bottom*) subjects in Houston. (From: Guyton JR, Dahlen GH, Patsch W, Kautz JA, Gotto AM. Relationship of plasma lipoprotein Lp(a) levels to race and to apolipoprotein B. Arterosclerosis 1985; 5:265–272.)
"By permission of the American Heart Association, Inc."

the higher prevalence and increased severity of nonlipid risk factors, the percentage of persons who require lipoprotein modification is similar among blacks and whites.

Suggested Readings

American Heart Association. Diagnosis and treatment of primary hyperlipidemia in childhood: A joint statement for physicians by the committee on atherosclerosis and hypertension in childhood of the council on cardiovascular disease in the young and the nutrition committee. Circulation 1986;74: 1181A–1188A.

Beufaute R, Reed D. Is elevated serum cholesterol level a risk factor for coronary heart disease in the elderly? JAMA 1990;263:393–396.

Castelli WP, Wilson PWF, Levy D, Anderson K. Cardiovascular risk factors in the elderly. Am J Cardiol 1989;63:12H–19H.

Deuke MA, Grundy SM. Hypercholesterolemia in elderly persons: resolving the treatment dilemma. Ann Intern Med 1990;112:780–792.

Donahue RP, Jacobs DR, Sidney S, Wagenknecht LE, Albers JJ, Hulley JB. Distribution of lipoproteins and apolipoproteins in young adults: The CARDIA study. Arteriosclerosis 1989;9:656–664.

Franklin FA, Brown RF, Franklin CC. Screening, diagnosis, and management of dyslipoprotenemia in children. Endocrinol Metab Clin North Am 1990; 19:399–449.

Garber AM, Littenberg B, Sox HC, et al. Costs and effectiveness of cholesterol screening in the elderly. Washington DC: Office of Technology Assessment, April 1989.

Miller VT. Dyslipoprotenemia in women. Special considerations. Endocrinol Metab Clin North Am 1990;19:381–398.

National Cholesterol Education Program: Report of the Expert Panel on Blood Cholesterol Levels in Children and Adolescents. Pediatrics 1992, in press.

Redmond GP (ed): Lipids and women's health. New York: Springer-Verlag, 1971.

Rubin SM, Sidney S, Black DM, Browner WS, Hulley SB, Cummings SR. High blood cholesterol in elderly men and the excess risk of coronary heart disease. Ann Intern Med 1990;113:916–920.

Sacks FM, Walsh BW. The effects of reproductive hormones on serum lipoproteins: unresolved issues in biology and clinical practice. Ann NY Acad Sci 1990;592:272–285.

SECTION THREE
CLINICAL CONCEPTS

CHAPTER 6

Laboratory Measurements

In dealing with lipid disorders, the performance of the laboratory is critical for all phases of management. For detecting, treating, and following patients with all but the most unusual dyslipidemias, the conventional lipid measures—total cholesterol, triglyceride, HDL-C, and calculated LDL-C—

suffice. As we learn more about lipoprotein metabolism, however, the underlying pathophysiology becomes more complex. As a consequence the number and types of analyses for lipids, lipoproteins, and apoproteins continue to grow. These more detailed and esoteric lipid measurements often must be obtained from special lipid research laboratories or selected commercial laboratories. Routine and special lipid testing for various phases of detection are listed in Table 6.1.

Acceptability of Measurements

Sources of Variability

The accuracy and reproducibility of serum lipid and lipoprotein measurements are affected by analytic *and* biologic variations. Various studies have now examined the issue of single versus multiple lipid samples. The greater the variability of an individual lipid measurement, the more likely that the strength of the relation between that measurement with CHD will be underestimated (Fig. 6.1). Even the average of two measurements will be subject to within-person variability, or "noise," resulting in some bias in the estimation of the true value. For this reason, the average of three measurements or more is even better than two for accuracy.

Biologic Issues. Even with the most accurate analytic techniques, biologic factors

Table 6.1. Phases of Detection

Screening mode
   Total cholesterol
Diagnostic mode
   Total cholesterol
   Triglyceride
   HDL cholesterol
   LDL cholesterol
Special circumstances
   Familial hyperchylomicronemia
      Naked eye appearance
      Lipoprotein electrophoresis
   Dysbetalipoproteinemia
      Lipoprotein electrophoresis
      VLDL cholesterol/triglyceride ratio
      Apo E isoforms
   Familial combined hyperlipidemia/
      hyperapobetalipoproteinemia
      Total apo B
   Severe hyperchylomicronemia
      Lipoprotein lipase estimation
      Apo C-II analysis
   Familial hypercholestrolemia
      LDL receptor analyses
   Early atherosclerosis with no obvious lipid
      abnormality
      LP(a) quantitation

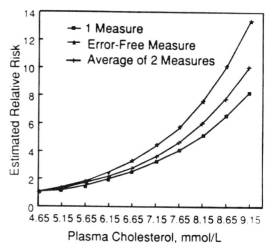

Figure 6.1. Estimates of relative risk for plasma cholesterol based on one measure, the average of two measures, or on an error-free measure. The highest relative risk is associated with the error-free measure and the least relative risk with the single measurement. (From: Davis CE, Rifkind BM, Brenner H, Gordon DJ. A single cholesterol measurement underestimates the risk of coronary heart disease. JAMA 1990;264:3044–3046.)
"Copyright 1990, American Medical Association."

can cause substantial variation of the blood lipid and lipoprotein levels in the same person on different occasions. Such biologic variation typically accounts for more than half of the overall variability. The coefficient of variation (CV), which represents variability, is calculated as the standard deviation (SD) for the measurement divided by the mean value ($\overline{X}$) of the measurement; the number is multiplied by 100 to convert it to a percentage:

$$CV = \frac{SD}{\overline{X}} \times 100$$

A low CV indicates that the measured values cluster around the mean on repeated measurements, whereas a high CV indicates that repeated determinations might deviate considerably from the mean value. For cholesterol, the CV is approx. 6.5%; thus, 95% of the time, the blood cholesterol concentration will vary within a range of about 13%

(twice the CV) above and below the true value. This variability is more than twice what otherwise would be encountered as a result of analytic error in the cholesterol measurement method, where the CV now is expected to be no more than 3%. For this reason, a random single measurement is inadequate to detect fluctuations in cholesterol levels, and several cholesterol measurements should be obtained at intervals of 1 or 2 weeks to accurately establish the patient's cholesterol value. Otherwise, the practicing physician may encounter difficulty in interpreting the results of therapy that reduces blood lipid levels by only 5% to 10%. The dilemma of classifying persons incorrectly concerns mainly those who have lipoprotein values closest to clinical cutpoints for initiating and changing therapy. Treatment decisions must be based on multiple measurements taking major sources of variability into account.

Biologic variability in the triglyceride level is considered to be more important than the variability introduced by technical or analytic factors. Day-to-day variability in the triglyceride level is approx. 20%, compared to variability in total cholesterol, HDL, and LDL-C levels of 5%, 10%, and 8%, respectively.

Seasonal variations in lipid levels are well documented. The cholesterol concentration may be 3% to 5% higher in winter than in the summer, irrespective of the baseline lipid values. Seasonal differences in weight and diet account for less than one-third of this variation. Cholesterol subfractions LDL-C and HDL-C show similar seasonal cycles, but triglyceride levels demonstrate a weaker and somewhat different seasonal pattern, being highest in mid summer and late autumn and lowest in spring. The reasons for these seasonal lipid patterns are largely unknown.

Except for the effect of meals, the time of day has little influence on lipid values and does not need to be considered. Consequently, values obtained in the afternoon are reliable provided there has been a fast of at least 12 to 14 hours. Persons should be on their usual diets for at least two weeks and be at stable body weight before having their lipids measured. Consumption of alcohol can increase the triglyceride concentration transiently, especially in persons who may be sensitive to alcohol. For the measurement of the total cholesterol and perhaps HDL-C, it is not necessary for the sample to be obtained after a 12- to 14-hour fast. This is because little cholesterol is transported by chylomicrons, so that the total cholesterol should be altered minimally, if at all. Chylomicrons are rich in triglyceride, however, and nonfasting samples invariably will be hypertriglyceridemic. The concentrations of HDL-C throughout the day are related inversely to the triglyceride levels. That is, as the triglyceride concentration increases after a meal, the HDL cholesterol concentration decreases, and vice versa. Furthermore, in the nonfasting state, chylomicrons and remnant particles cannot always be isolated clearly in the laboratory procedure. Non-HDL lipoproteins can contaminate the HDL-containing supernatant and produce

misleading HDL-C values. It does not necessarily follow, therefore, that HDL cholesterol concentrations will be uninfluenced by meals, and, in contrast to some authorities, we recommend using a fasting sample if information on HDL-C is desired.

Various other factors that relate to the health status of the patient can influence lipid determinations (Table 6.2). The effects of conditions such as pregnancy, lactation, acute MI, acute infection, stroke, major illness, and surgery will be discussed in greater detail later (see Chapter 8). Other atypical situations also should be considered as potential sources of lipid variability, such as changes in smoking pattern and exercise habits or use of medications.

Analytic Issues. The precision and accuracy of cholesterol and triglyceride measurements are critically important for identifying patients who have lipid abnormalities, as well as for assessing their response to diet and pharmacologic regimens. This is particularly important when the changes are relatively small. Unfortunately, some methods for lipid analysis regularly give imprecise results, whereas others provide accurate values only under seldomly met specific conditions. Furthermore, error can be introduced by the method of collection and storage of the sample. To respond to the national recommendations for managing lipid disorders, physicians must have access to testing methods that are accurate and reproducible.

If lipid concentrations are to be interpreted in the same way as other biochemical

Table 6.2. Nonanalytic Factors Influencing Lipid Measurements

Postprandial state
Recent (post 2 weeks) change in weight, diet, or activity
Acute illness or major injury
MI or CABG[a] within past 12 weeks
Pregnant or less than 12 weeks postpartum
Nonstandard phlebotomy position
Unsatisfactory venipuncture or sample handling
Others (season, menstrual cycle)

[a] Coronary Artery Bypass Grafting

parameters, a standard methodology must be developed and adopted. Different methods for measuring cholesterol can give different results. This is emphasized by a study that was conducted by the College of American Pathologists (CAP), in which an identical blood specimen was sent to more than 5000 laboratories. The laboratories that used the DuPont ACA™ machine reported cholesterol concentrations that ranged from 222 to 270 mg/dL, with a mean of 247 mg/dL. The laboratories that used the Technicon SMAC™ equipment reported values that ranged from 250 to 294 mg/dL, with a mean of 272 mg/dL, whereas laboratories that used a Beckman Astra 4 & 8™ reported values that ranged from 267 to 397 mg/dL, with a mean of 334 mg/dL. For all instruments and for all laboratories that used enzymatic methods, the range of concentrations reported for this single specimen was from 197 to 379 mg/dL. The Centers for Disease Control (CDC) Lipid Standardization Laboratory indicated that the correct cholesterol value was 263 mg/dL.

Obviously, identification of persons at risk and subsequent recommendations for their treatment would differ depending on the technique used. This problem surfaces when a person who is diagnosed as being hypercholesterolemic on a screening test is later found to have an acceptable cholesterol value without a change in diet. Even more difficult for the physician is the patient who has hypercholesterolemia and is given aggressive therapy, only to have the follow-up cholesterol value appear to increase or not change. Because cholesterol can be difficult to measure accurately, caution should be used in interpreting the results. Changes in therapy should be based solely on trends and not on single, isolated determinations. Physicians interested in treating dyslipidemia may need to know more about analytic methods than has been required in the past.

Several analytic issues need to be considered when evaluating a cholesterol testing method:

1. *Accuracy* is defined as the agreement between the measured value and the true value. The most frequently used cholesterol reference method is the modified Abel-Kendall method developed by the CDC. The definitive method for assessment, however, is the isotope dilution mass spectrometry method developed by the National Bureau of Standards (NBS). The measurement of a method's accuracy must be based on the ability to trace the method back to an accepted reference system.

2. *Bias* represents a consistent difference between the method and the reference method. A bias may be positive, which means that the results are consistently higher than the true value, or negative, which means that the values are consistently lower than the true value. A cholesterol testing method with a positive bias will result in classification of patients as being at risk when they are not (false-positive). A method with a negative bias will result in misclassification of patients as being normal when they actually have hypercholesterolemia (false-negative).

3. *Precision* represents the day-to-day reproducibility of measurements. This can be calculated by using the CV. A good laboratory will have a CV of < 3%, whereas an average laboratory will have a CV of 3% to 5%. If there is low precision, it becomes virtually impossible to assess cholesterol values over time.

Desirable Laboratory Features

The choice of a laboratory is important because considerable variability exists in the accuracy with which laboratories measure cholesterol. All clinical recommendations about cholesterol levels assume the use of accurate and reliable measurements. Few clinical laboratories are capable of providing cholesterol measurements with a CV that is ≤ 2% for within-a-day variance and ≤ 3% for day-to-day variation. Physicians should find a laboratory that participates in a suitable standardization program, with adequately documented accuracy and precision.

Careful assessment of the lipid measurements made by the laboratory is essential. The following questions should be asked:

1. Is the accuracy of the values within the specified guidelines of the National Cholesterol Education Program (NCEP) Laboratory Standardization Panel? The laboratory should validate the accuracy of its cholesterol measurements on patient samples by doing split sample analyses with a CDC standardized and certified reference laboratory. A National Reference Method Laboratory Network now is available with regional centers throughout the United States.
2. Has the laboratory met the precision goals recommended by the NCEP Laboratory Standardization Panel? The CV should be no more than 5%.
3. What is the range of values for which the method accurately measures cholesterol? For general purposes, the analysis should demonstrate excellence for values between 150 and 300 mg/dL (3.88 to 7.76 mmol/L), the clinically important range.
4. What percentage of cholesterol measurements performed are inserted in the runs for quality control? An excellent laboratory usually has 20% of its samples for quality control, whereas an average laboratory has only 10%.
5. Does the laboratory participate in a national proficiency testing program such as the College of American Pathologists (CAP) or the American Association of Bioanalysts proficiency testing programs?

To avoid problems in interpreting the results of blood lipid levels, it may be advisable to split a patient's sample periodically and send one part to the regular laboratory and another part to a reference laboratory (such laboratories may be located by contacting the CDC). If the results of the two samples vary more than 5%, then the accuracy of the laboratory should be questioned.

Standardization

As previously stated, cholesterol measurements made by all clinical laboratories in the United States should be standardized so that the cholesterol values are traceable to the CDC reference method or to the NBS definitive method. Laboratories can accomplish this goal and also improve the accuracy and precision of their cholesterol measurements by using certified reference materials currently available from the CDC, NBS, or CAP.

The National Cholesterol Education Program has mandated accuracy and precision standards and has set as an ideal the goal that the CV should be $\leq 5\%$ until 1990 and $\leq 3\%$ thereafter. The CDC has been involved in a Lipid Standardization Program, and, as the NCEP gets underway, it will be important to standardize and verify the accuracy of the methods used by all laboratories to measure cholesterol concentrations. Figure 6.2 indicates the effect of the CV on the commonly used decision point, a borderline total cholesterol value of 240 mg/dL (6.21 mmol/L). At the acceptable precision goal of 5%, the measured values might range from 216 to 264 mg/dL (5.59–6.83 mmol/L). The higher the real concentration of the blood cholesterol, the greater the absolute magnitude of the error. For higher values, the reliability is even less.

Currently there is an effort underway to adopt the *le Système International d'Unités* (SI) for blood cholesterol values. This is an attempt to develop a universally acceptable system of units of measurement and is essentially an outgrowth of the metric system. Although clinicians will be unfamiliar with these values, most medical journals now require their use. To make matters worse, in the SI, the numbers are expressed in decimals. For example, a cholesterol of 250 mg/dL would be equivalent to 6.47 mmol/L. Unfortunately, we may not be able to retain the easily understood traditional system with whole numbers for cholesterol values, and, in clinical practice, we may sub-

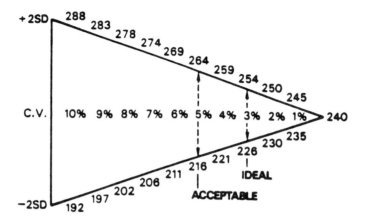

Figure 6.2. The effect of differing degrees of analytic imprecision of cholesterol measurement at a medical decision point of 240 mg/dL (*true value*). An acceptable precision goal is 5% or less from true value; ideal goal is 3% or less from true value.

$$\text{Coefficient of Variation} = \frac{\text{Standard Deviation}}{\text{Mean of Cholesterol Values}}$$

The variability of the cholesterol measurement depends on the magnitude of the true value. The higher the true value, the greater the range of possible values for a given measurement. Multiplying the product of the CV and the actual cholesterol value by two gives the possible true range (with 95% of values falling above and below the measured value). For example, when the CV is 3%, a value of 240 mg/dL would indicate a possible range of 226 to 254 mg/dL (3% × 240 mg/dL × 2) whereas a higher value of 400 mg/dL actually could range from 376 to 424 mg/dL. (Adapted from: Current status of blood cholesterol measurement in clinical laboratories in the United States: a report from the laboratory standardization panel of the National Cholesterol Education Program, U.S. Department of Health and Human Services, Public Health Service, National Institutes of Health. National Institutes of Health, NIH Pub. No. 88-2928, January 1988.)

sequently have to translate the values from the SI system into the more traditional values that patients have come to understand and have learned to accept.

Measurement Techniques

Conventional Methods for Measurement of Cholesterol

Measurement of the total cholesterol estimates both the esterified and the free forms of the sterol. Generally, only two-thirds of the total cholesterol is esterified, with the remainder being free. Color development is of greater intensity with the cholesteryl ester, however, and a large positive bias (i.e., overestimation) can result when the ratio of cholesteryl ester to free cholesterol is increased. Recognition of the limitations of various cholesterol methods can be important to the understanding of laboratory results.

The different methods can be classified by the number of steps involved. The single-step method uses no sample preparation, allowing for a simple and rapid analysis to be carried out. Generally, only one reagent is needed and there is little manipulation of the sample, making this method suitable for automation. Such procedures, however are prone to errors that result from the presence of protein, bilirubin, vitamins A, C, and D, steroid hormones, uric acid, turbidity, and differences in chromogenicity of free and esterified cholesterol. Several automated procedures are based on direct single-step methods, including those adapted to the Technicon SMA-12™ or SMAC-II™ instruments. Such automated methods can have cholesterol values elevated by as much as 30 to 40 mg/dL (0.78–1.03 mmol/L) above the true value because of the above-mentioned interfering factors.

A second level of complexity involves a two-step procedure, with an organic extraction step before measurement of cholesterol and other chemically related steroids. This pretreatment step removes nonspecific chromogens that might interfere with the assay. The deleterious effect of a differential color response from free and esterified cholesterol remains, however, because no attempt is made to split (saponify) the bond between the fatty acid and the cholesterol in the cholesteryl esters. Most routine clinical determinations use this two-step procedure. The values from this technique correspond closely with those of the standard Abel-Kendall procedure, especially when a calibration factor is added to correct for free and ester cholesterol differential color development.

The generally accepted reference measurement of cholesterol, and the one used by the CDC Lipid Standardization Program, is the Abel-Kendall procedure. This technique is somewhat laborious, because, in addition to extraction of cholesterol there is a third step—a saponification step that hydrolyzes the fatty acid moiety from the cholesteryl ester. Only free cholesterol is then measured.

There are a number of kits available for the enzymatic measurement of cholesterol. These commercial kits use the enzyme cholesterol hydrolase to replace the chemical saponification step for converting esterified cholesterol to cholesterol; the cholesterol in the sample then is oxidized with the enzyme cholesterol oxidase to produce hydrogen peroxide in direct proportion to the amount of cholesterol present in the sample. The hydrogen peroxide oxidizes a dye, and the intensity of the color that develops is proportional to the cholesterol concentration.

Lipoprotein Analyses

Multiple techniques, each with its own advantages and disadvantages, can be used to derive lipid and lipoprotein values.

Ultracentrifugation. Ultracentrifugation is the gold standard for quantification of serum or plasma lipoproteins. It is time-consuming, however, and cannot be adapted readily for clinical use. The technique takes advantage of two special properties of lipoproteins: (1) their densities are lower than those of other naturally occurring macromolecules, and (2) the density varies within the lipoprotein classes (*see* Table 4.1). When subjected to prolonged centrifugation at high speed, any lipoprotein that is less dense than the plasma (or serum) will float to the top and can be separated from the other lipoproteins. The plasma density can be adjusted in steps to separate the various lipoproteins differentially.

Analytic Ultracentrifugation. This is the most precise technique for determining the serum or plasma lipoproteins. Its use is limited, however, to a few research centers because expensive equipment is needed and technical difficulties occur in performing the analyses. Its most important application is in patients suspected of having type III dyslipoproteinemia (dysbetalipoproteinemia), which is characterized by accumulation of chylomicron remnants and IDL.

To separate the various lipoprotein classes, a fasting sample initially is adjusted to a density of 1.006 and then is centrifuged for a prolonged period. At this density, the top layer contains VLDL (and chylomicrons, if present) and is separated from the rest of the sample by a clear zone. After removing the top layer, the sample is readjusted to a density of 1.063 and is recentrifuged at high speed. This causes LDL to float to the top and allows for its selective separation. If it is necessary to separate HDL from other lipoproteins, a third density adjustment of 1.210 is made, the sample is centrifuged again, and the HDL lipoprotein fraction at the top of the tube is removed.

Preparative Ultracentrifugation. This procedure can be used to separate VLDL and LDL from HDL. The plasma or serum is overlayed with potassium bromide solution to adjust the density to 1.063 g/mL and is centrifuged. The supernatant solution that contains the VLDL and LDL then is removed, and the infranatant solution is analyzed for the cholesterol concentration. The infranatant fraction, with a density greater than 1.063, contains the HDL cholesterol. This is a tedious but precise method for separating and quantitating the cholesterol in the lipoprotein fractions, and it often is regarded as the reference procedure for estimating the HDL cholesterol content.

Beta-Quantification. For clinical purposes, the most useful technique is a combined ultracentrifugal-precipitation method. Plasma is ultracentrifuged at an adjusted density of 1.006. The cholesterol concentration of the bottom fraction, which contains both LDL and HDL, is determined. The HDL cholesterol then is determined in whole plasma by using a precipitation method, and the cholesterol content of the other lipoproteins is calculated as follows: (1) LDL cholesterol is the cholesterol content of the bottom fraction (density > 1.006) minus the HDL cholesterol; and (2) VLDL cholesterol is the total cholesterol minus the cholesterol concentration in the fraction that contains LDL plus HDL (density > 1.006). It is possible to measure the VLDL-C directly in the fraction whose density is less than

1.006, but this determination will be inaccurate if the triglyceride concentrations are more than 300 mg/dL (3.39 mmol/L) or less than 100 mg/dL (1.13 mmol/L).

Vertical Auto Profile. In this unique method, a small amount of plasma is adjusted to a density of 1.21 g/mL by the addition of potassium bromide, and this is placed in a special test tube. Normal saline is layered on top of the plasma, and the tube is then capped and ultracentrifuged for approximately 45 minutes. After ultracentrifugation, the bottom of the tube is punctured and the effluent is withdrawn continuously into an autoanalyzer that measures cholesterol by an enzymatic method. With each succeeding lipoprotein density, there will be a peak that corresponds to the cholesterol level (Fig. 6.3). In this way, the cholesterol content of all the lipoproteins and many of their subspecies is measured directly and recorded on a strip chart. The area under each peak is quantitated by using a mathematical procedure capable of deconvoluting the profile (Fig. 6.4). The deconvolution program uses a microcomputer that is interfaced with

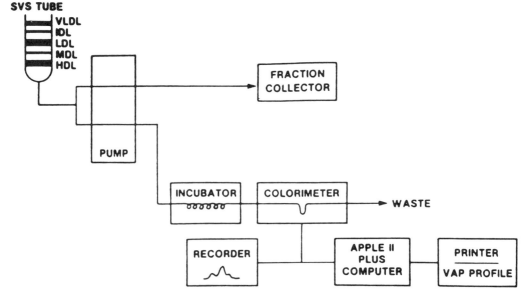

Figure 6.3. Vertical auto profiler (VAP): Simplified flow diagram for computer-assisted VAP. Approximately 20% of the sample is required for on-line, continuous flow cholesterol analysis; thus 80% of the sample can be diverted to a fraction collector for additional protein and lipid analyses. (From: Segrest JP, Chung BH, Cone JT, Hughes TA. Coronary heart disease risk:assessment by plasma lipoprotein profiles. Ala J Med Sci 1983; 20:76–83.)

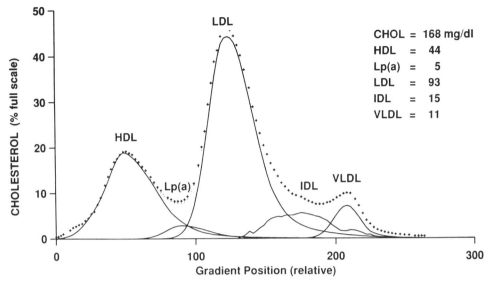

CHOL = 168 mg/dl
HDL = 44
Lp(a) = 5
LDL = 93
IDL = 15
VLDL = 11

Figure 6.4. The vertical auto profile (VAP) of a normolipidemic male. The system directly measures the cholesterol content of the noted lipoproteins and converts the cholesterol into a graphic profile by a microcomputer. Triglyceride level is not directly measured but can be estimated from the levels of lipoprotein cholesterol in VLDL, IDL, and HDL. (From: Lipoprotein Laboratory, Atherosclerosis Research Unit, University of Alabama at Birmingham.)

the analyzer. This method has been validated with the methodology used by the Lipid Research Clinics for measurement of HDL, LDL, and VLDL cholesterol.

The vertical auto profile (VAP) is a rapid and accurate technique of lipoprotein analysis suitable for routine screening of patients who have lipoprotein abnormalities. It is an excellent way to determine the cholesterol concentrations in HDL_2, HDL_3, LDL, IDL, VLDL, and Lp(a). In addition, it also is possible to estimate the concentrations of cholesterol in various LDL subclasses. Its limitation is that triglyceride is not measured directly, although it can be estimated by assuming the percentage of triglyceride in each major lipoprotein class, the major contributors of which are VLDL and IDL. When the triglyceride levels are within normal range, this estimate is reasonably accurate; however, in hypertriglyceridemic samples, it is inaccurate and the triglyceride must be measured directly.

Lipoprotein Electrophoresis. The electric charge of the lipoproteins can be used in their separation by electrophoresis on paper or agarose gel. Although this was once a

popular way to evaluate and type lipoprotein abnormalities, it is time consuming and offers relatively little unique information beyond what is available from routine measurement of the total cholesterol and triglyceride concentrations. The major lipoprotein classes—chylomicrons, VLDL, LDL, and HDL—separate into four distinct bands according to electrophoretic mobility. Chylomicrons remain at the origin of the strip; LDL has beta mobility and moves only slightly from the origin; HDL has alpha mobility and migrates the farthest from the origin, whereas VLDL has intermediate mobility and is located between LDL and HDL. The strip is treated with Sudan Black to stain the lipid in each lipoprotein class, which becomes visible as a discrete band. The strip is then scanned in a way similar to that used for protein electrophoresis, and the area under each peak is quantified. The electrophoretic method is relatively crude and gives no information about the densities or subclasses of the lipoproteins.

HDL Cholesterol. The HDL-C can be measured either by analytic ultracentrifugation or by isolation of HDL and measurement of the particles gravimetrically; however, both methods are laborious and tedious. Instead of measuring the HDL fraction in its entirety, its protein moiety or its lipid moiety is measured as an indirect means for HDL quantification.

The lipid moiety is the one used most often. The lipids found in the HDL molecule include phospholipids, cholesterol, and triglyceride. Because cholesterol is simpler to measure than phospholipids, the HDL-C has prevailed as the indirect means of determining HDL concentrations. Techniques available for HDL-C quantitation require the isolation of HDL, with its subsequent quantitation by one of the many acceptable techniques for measuring cholesterol. High density lipoprotein can be isolated in the clinical laboratory by ultracentrifugation, column chromatography (an experimental method for separating lipoproteins on the basis of size but is of limited clinical value), or electrophoresis. The easiest and least expensive technique consists of precipitation of VLDL and LDL from the sample with polyanion solutions, such as heparin plus magnesium, manganese anions, or phosphotungstate. The cholesterol in the resultant supernatant represents HDL cholesterol.

VLDL and LDL Cholesterol. If the triglyceride value is less than 400 mg/dL (4.52 mmol/L), it is assumed that the ratio of triglyceride to cholesterol in VLDL is constant at 5:1. The triglyceride concentration then is divided by five (or multiplied by 0.2) to estimate the VLDL-C when the values in the formula are in mg/dL (*if in mmol/L, the divisor is 2.18*):

$$\frac{\text{Triglyceride mg/dL}}{5} = \text{VLDL-C (mg/dL)}$$

Because total cholesterol is the sum of HDL-C, VLDL-C, and LDL-C, the latter can be calculated as follows (all quantities are in mg/dL):

$$\text{LDL-C} = \text{total cholesterol} - [\text{HDL-C} + (\text{triglyceride}/5)]$$

If the triglyceride value is above 400 mg/dL (4.52 mmol/L), LDL-C estimation by the above formula becomes less accurate. If necessary, ultracentrifugation in a specialized laboratory can be used to give a more accurate LDL-C level.

The ability to make this calculation is based on the assumption that a constant relation exists between triglyceride and cholesterol in VLDL (at least at blood triglyceride concentrations below 400 to 500 mg/dL [4.52–5.65 mmol/L]). More recently, the appropriate ratio of cholesterol to triglyceride in VLDL has been found to be 0.16 instead of 0.20. Considering the softness of these assumptions, however, 0.20 still is often used. Although this provides a useful approximation of the LDL-C concentration, it should be emphasized that 0.20 is an average ratio established from a group of persons in whom interindividual variability was demonstrated. A calculated LDL-C concentration, therefore, may not always agree with the LDL-C concentration measured by ultracentrifugation.

Because VLDL particles are heterogeneous, the precise relation between triglyceride and cholesterol will depend on the particular VLDL size in an individual patient. When hypertriglyceridemia is caused by an increase in the number of particles without substantial alteration in composition, such as in familial combined hyperlipidemia, the fixed relation between triglyceride and cholesterol may persist beyond the ceiling of 400 to 500 mg/dL (4.52–5.65 mmol/L), perhaps even up to a triglyceride concentration as high as 800 mg/dL (9.03 mmol/L). In states of hypertriglyceridemia that are associated with the synthesis or retention of particles that are abnormal in composition, however, this relation may be totally inaccurate as the triglyceride concentration increases. Consequently, it must be remembered that the calculated LDL-C concentration has limitations, particularly in hypertriglyceridemia.

To avoid this problem, it has been suggested that the non-HDL cholesterol (that is, the total cholesterol minus the HDL-C) be used as a more accurate indicator of risk. This assumes that the cholesterol transported by VLDL also contributes to the atherosclerotic process. Although there is considerable enthusiasm among some researchers for the concept that VLDL is atherogenic, the issue remains controversial.

In the future it may be possible to circumvent some of the above-mentioned problems by measuring the concentration of apoprotein B-100, the major LDL apoprotein, directly. Presently apoprotein B measurements still have major problems and have not yet been standardized to the extent that recommendations for routine use can be made.

Triglyceride Measurement

Triglyceride also can be measured by an automated procedure that depends on the hydrolysis of triglyceride with subsequent enzymatic measurement of free glycerol. Depending on the reagents used for measurement of glycerol, either hydrogen peroxide or NADH is produced, with NADH able to be measured directly, whereas changes in the color of a dye reflect the amount of hydrogen peroxide liberated. In either event, both are proportional to the triglyceride concentration.

Desktop Analyzers

A new generation of simple chemical analyzers are being used increasingly for cholesterol measurements in physicians' offices and at commercial screening sites such as shopping malls. These enzymatic methods were first developed 15 years ago and, until recently, were used almost exclusively in laboratory settings. With the increased demand for rapid cholesterol screening at nonlaboratory sites, enzymatic methods have been adapted to meet these challenges and enjoy increasing popularity.

In these methods, cholesteryl esters are hydrolyzed to free cholesterol that then is oxidized to form cholesterone. The by-product of this reaction, hydrogen peroxide, then reacts with an indicator to produce a color that is measured photometrically. This determination is not completely specific for cholesterol, but cholesterol is by far its largest contributor. Measurements can be performed with desktop analyzers that determine plasma cholesterol directly from whole blood and require only periodic calibration.

Dry Chemistry Analyzers. The term *dry chemistry analyzer* is applied to instruments that use reagents impregnated in small test strips or slides. In the Reflotron™ technique, the sample is applied to these strips and then is incubated for 3 minutes. At the end of the incubation, a light source is reflected from the area of the strip that contains the colored reaction product, and the reflectance of the sample is measured and converted to a digital readout. The use of different test strips allows for the determination of different analytes, such as cholesterol and triglyceride, within a single sample. This approach relies heavily on the manufacturer to calibrate accurately each lot of test strips, because miscalibration will result in inaccurate values.

The Kodak DT-60,™ another dry chemistry analyzer, uses a different technique. Serum or plasma first must be separated from red cells, an inconvenience outside the laboratory setting, and applied by insertion of a pipette into a sample port that is positioned over the test slide. The instrument moves the slide into an incubation chamber for 5 minutes, after which the reflectances of the sample are measured. The Kodak analyzer can perform approximately 50 tests hourly. It requires calibration every 90 days, using lyophilized calibration sera provided by the manufacturer.

Wet Chemistry Analyzers. The Vision Analyzer™ supplied by Abbott Laboratories uses a reagent solution in a small plastic cassette. The plasma sample is inserted into the cassette, mixed with the reagent, and the color development measured. Ten samples can be measured in one batch, taking about 8 minutes. The vision analyzer is calibrated by the operator, using calibration sera provided by the manufacturer.

Accuracy of Measurements. When used properly, desktop cholesterol analyzers can provide accurate and precise cholesterol measurements. Recent studies indicate that these instruments are capable of meeting the current NCEP acceptable level for precision ($\leq \pm 5\%$ CV) and even the 5-year ideal precision goal of 3% CV. The coefficients of variation for these instruments vary from about 2% to 4% as assessed with pooled sera. The measurements correlate well with standardized laboratory techniques but tend to have a negative bias, in the range of 1% to 5%. In other words, the values usually underestimate the true cholesterol obtained with reference laboratory measurements.

Possible sources of inaccuracy include poor calibration of the instrument or erroneous volume measurements; variations in the test strips or reagents; failure to perform appropriate cleaning and maintenance procedures; and deficiencies in the training or experience of the technicians. Users of these desktop instruments should set up a regular system of quality control with a standard laboratory. Quality control techniques could be based on the use of pooled sera or split-sample analysis. Only a qualified technician should operate the instruments, and quality control procedures must be implemented on a regular basis.

As noted earlier, differences in measured values can occur in the same person because of biologic variability. Such considerations apply even more strongly to interpretation of screening measurements in which a subject may be seen only once, especially when the subject's cholesterol concentration is near the cut-off values used for decision making. It must always be remembered that there are several possible sources of variability in measurement between the more conventional laboratory and the newer desktop cholesterol analyzer techniques. Furthermore, it must also be emphasized that an elevated screening total cholesterol concentration should always be followed by a more complete lipoprotein analysis.

Because desktop analyzers are particularly well suited for analysis of small quantities of blood, the fingerstick has been used to obtain capillary samples. Kits now available for total cholesterol allow estimation of a range by comparing color changes (drop of blood on color sensitive material) with a standard color chart. HDL-C and triglyceride determinations also are possible by using similar methodology. There is some controversy, however, about whether capillary blood samples are comparable to cholesterol measurements in venous blood. Generally, measurements of capillary blood samples are slightly lower than measurements in samples obtained by venipuncture.

In any event, current technology makes accurate and reproducible measurement of cholesterol possible in the office and laboratory setting. When the available desktop analyzers are used properly, their accuracy often equals that of clinical laboratories.

Blood Drawing Routine

Procedure

A standardized procedure for drawing blood also is extremely important to ensure reliability of results. Differences in blood collection may complicate the interpretation of the measurements when patients are monitored over time. For fingerstick collection of blood, a deep penetration lancet with a long point should be used. An appropriate stick should provide a free-flowing blood sample; otherwise, manipulation of the finger to obtain a sufficient quantity of blood will introduce sources of error into the measurement. Even with venipuncture, numerous variables, including posture, stress level, tourniquet use, and technician skill, can influence lipid levels.

When a patient assumes the supine position, water is redistributed from the tissues to the plasma. This results in dilution of nondiffusible plasma constituents, including the lipoproteins, resulting in slightly lower blood cholesterol concentration. A change from the standing to sitting position will have a small effect on the cholesterol level. For this reason it is advisable to standardize the procedure and obtain blood while the patient is sitting, the most commonly used position. The patient should be in a sitting position for at least 5 minutes before the blood actually is drawn.

The tourniquet should be applied for as brief a period as possible, because increased lipid values have been reported with prolonged use of a tourniquet. Blood cholesterol values were found to increase an average of 10% to 15% after 5 minutes of occlusion. Increases of up to 5% have been observed after 2 minutes, such as might occur when difficulties arise in obtaining the sample. Removing the tourniquet within 30–60 seconds, therefore, should not affect the lipid determinations.

Choice of Plasma or Serum for Determinations

Generally, either plasma or serum can be used for lipid determinations. Plasma usually is preferred when lipids and lipoproteins are being analyzed. Fibrinogen in an improperly prepared specimen, however, may plug or coat the tubing in continuous flow systems such as autoanalyzers, causing erroneous results. If plasma is used, spontaneous hydrolysis of triglyceride can occur if the samples are left at ambient temperature.

Collection of plasma requires an anticoagulant, usually solid edetic acid (EDTA) (1 mg/mL of blood). The blood cells should be separated as soon as possible, but at least within two hours. If plasma is used, the values should be multiplied by 1.03 to convert the results to the value appropriate for the NCEP cut-points that are based on serum measurements. Both EDTA and heparin cause no detectable change in red-cell volume and decrease total cholesterol concentration by 1% or less. Although EDTA causes slightly larger changes in cholesterol concentrations than heparin, it is preferred for lipoprotein analysis. Edetic acid retards the autooxidation of unsaturated fatty acids and cholesterol by chelating heavy metal ions, such as copper, that promote this change. Also, EDTA inhibits phospholipase-C activity, reducing changes that occur from possible contamination by phospholipase-C-producing bacteria. Certain anticoagulants, such as fluoride, citrate, and oxalate, can bring about large shifts in water between red cells and plasma and should not be used.

Preparation and Storage of Samples

Either serum or plasma left at room temperature for prolonged periods can lead to increased LCAT activity, resulting in an altered lipoprotein composition and movement of cellular cholesterol into the plasma. To prevent this, plasma should be placed on ice and centrifuged as soon as possible. It is best to centrifuge at 4°C (in a refrigerated centrifuge) at $200 \times g$ for about 15 minutes. The plasma should be removed immediately into a clean tube, because exchange of free cholesterol between red-cell membrane and plasma can occur. It is important to prevent any hemolysis of red blood cells. Also, the tube should be sealed tightly to prevent evaporation of the sample. If serum is used, an anticoagulant is not necessary; the sample should be allowed to clot for 30 minutes at room temperature, and the clot should be detached from the tube wall before centrifugation.

It is good practice to perform the lipoprotein analysis as soon as possible, before lipoprotein composition becomes altered by the exchange of cholesteryl esters and triglyceride between HDL and other lipoproteins. Samples kept for future analysis need to be frozen. Studies of samples stored at different temperatures for varying periods, frozen and unfrozen, reveal that levels of all lipoproteins may decrease with storage.

Short-term storage (for a few days) at 4°C probably is acceptable, but some spontaneous hydrolysis of triglyceride takes place that may reduce the measured triglyceride concentration. Samples that require more than short-term storage should be frozen to prevent this spontaneous hydrolysis and oxidation of the unsaturated lipids. The HDL-C reproducibility and resolution of lipoprotein patterns decrease when samples are kept at 4°C for more than 1 week. Freezing at −20°C will keep samples stable for a few months if self-defrosting freezer units are avoided to prevent cyclic thawing and freezing with a breakdown of the lipid and lipoprotein components. Freezing at $-60°C$ may allow for highly reproducible results even for a period of 1 year or more.

Stored samples, whether serum or plasma, must be mixed adequately before analysis, because of layering by density within the sample. Simple inversion of the tube is not sufficient, and the sample should be mixed carefully on a vortex mixer for at least 5 to 10 seconds. If chylomicrons are present at the top of the tube, they must be dispersed evenly throughout the sample before taking an aliquot for lipid analysis. Evaporation of the solvent also can be a problem, unless a good seal is obtained. A well-sealed container or vial also will keep oxygen out of the sample and prevent oxidation of the unsaturated lipids. If samples are to be stored for a long time, it is best to overlay the sample extract with nitrogen before sealing the vial. Screw-capped vials with Teflon-lined caps provide good storage for lipid extracts.

Suggested Readings

Belsey R, Goitein RK, Baer DM. Evaluation of a laboratory system intended for use in physicians' offices. I. Reliability of results produced by trained laboratory technologists. JAMA 1987;258:353–356.

Belsey R, Vanderbark M, Goitein RK, et al. Evaluation of a laboratory system intended for use in physicians' offices. II. Reliability of results produced by health care workers without formal or professional laboratory training. JAMA 1987;258:357–361.

Cardiac risk classification based on lipid screening, [Editorial]. JAMA 1990;263:1250–1252.

College of American Pathologists Comprehensive Chemistry Survey, Participants' Summary, Skokie, Ill: College of American Pathologists, 1985.

Irwig L, Glasziou P, Wilson A, Macaskill P. Estimating an individual's true cholesterol level and response to intervention. JAMA 1991;266:1678–1685.

Kaplan LA, Pesce AJ. Clinical chemistry theory analysis and correlation. In: Nait HK (ed). Methods of analysis. St. Louis: C.V. Mosbey Co., 1985:1194–1212.

Segrest JP, Albers JJ, eds. Methods in enzymology: Volume I. Plasma lipoproteins part A and B—preparation, structure, and molecular biology. Florida: Academic Press, Inc., 1986.

Stein EA. Use of the clinical laboratory for lipid evaluation. In: Redmond GP, ed. Lipids and women's health. New York: Springer-Verlag, 1991:39–47.

U.S. Department of Health and Human Services, Public Health Service. Current status of blood cholesterol measurement in clinical laboratories in the United States: A report from the laboratory standardization panel of the national cholesterol education program. Washington DC: National Institutes of Health, NIH Publication No. 88-2928, January 1988.

Variabilities in serum lipid measurement: do they impede proper diagnosis and treatment of dyslipidemia? [Editorial]. Arch Intern Med 1990;150:1583–1585.

CHAPTER 7

Classification of Primary Lipid Disorders

The first major advance in understanding and communicating about lipid disorders was the classification system proposed by Lees and Fredrickson in 1965. In this system, lipoprotein phenotypes were designated with roman numerals I through V. This subsequently was modified to include six different phenotypes by dividing Type II into Types II$_A$ and II$_B$ (Table 7.1). Persons who have the same phenotype do not necessarily have the same genetic abnormality or biochemical defect, and considerable heterogeneity exists in what phenotypically suggests homogeneity. This classification also does not provide for abnormalities in HDL-C. Despite its shortcomings, however, this classification is still a convenient shorthand for discussing lipoprotein disorders. For this reason, the terminology is preserved in the current system but must be supplemented extensively with description of fundamental disorders and their basic mechanisms. Consequently, the use of lipoprotein phenotyping now has been deemphasized, and a more descriptive classification has been adopted.

To provide an understanding of the heterogeneity that exists within these various lipoprotein phenotypes, some of the different disorders associated with each phenotype will be reviewed (Table 7.2). Because there may be secondary causes of these abnormalities, all patients must be screened carefully before concluding that the defect is primary. In such conditions, treatment of the secondary cause may be all that is necessary for

Table 7.1. Phenotypic Classification of Dyslipidemias

Phenotype	Lipoprotein in Excess	Typical Lipid Range: Total-C	Typical Lipid Range: Triglyceride	Genetic Disorder
I	CM[a]	300–500	5000–6000	LPL deficiency; apoprotein C-II deficiency
II$_A$	LDL	250–400	< 250	FH[e] (heterozygote)
				PH[g]
				FDB[d]
				FCHL
				FCHL[b]
		400–800	< 250	FH (homozygote)
II$_B$	LDL	240–350	250–500	FCHL
	VLDL			PH or FH plus FHT
III	β-VLDL	300–450	300–1000	FD[c]
	IDL			
IV	VLDL	200–240	300-700	FCHL
				FHTG[f]
V (mild)	CM			FCHL plus LPL deficiency
	VLDL	200–300	500–1000	
V (severe)				LPL deficiency
	CM			Apo C-II deficiency
	VLDL	300–1000	2000–6000	

[a]CM = chylomicronemia
[b]FCHL = familial combined hyperlipidemia
[c]FD = familial dysbetalipoproteinemia
[d]FDB = familial defective apoprotein B
[e]FH = familial hypercholesterolemia
[f]FHTG = familial hypertriglyceridemia
[g]PH = polygenic hypercholesterolemia
(Adapted from Grundy SM, Greenland P, Herd A, Huebsch JA, et al. Cardiovascular and risk factor evaluation of healthy American adults: a statement for physicians by an ad hoc committee appointed by the steering committee, American Heart Association. Circulation 1987;75:1340A–2362A.)

Table 7.2. Genetic Hyperlipidemias

Disorder	Blood Lipoprotein Pattern	Estimated Population Frequency
Familial hypercholesterolemia	IIA, IIB	1-2/1,000
Familial defective apoprotein B	IIA	1-2/1,000
Polygenic hypercholesterolemia	IIA, IIB	-
Familial hypertriglyceridemia	IV, V	2/1,000
Familial combined hyperlipidemia	IIA, IIB, IV, V	3-5/1,000
Familial dysbetalipoproteinemia	III	1/10,000
LPL deficiency	I, V	Rare
Apoprotein C-II deficiency	I, V	Rare
LCAT deficiency	-	Rare

(Adapted from: Bierman EL, Glomset JA. Disorders of lipid metabolism. In: Wilson JD, Foster DW (eds). Williams textbook of endocrinology, ed 7. Philadelphia: W.B. Saunders Company, 1985:1108–1136.)

resolution or improvement of the lipoprotein disorder. It is equally important to search for secondary causes in patients who have familial disease, because such disorders may aggravate primary causes. The secondary causes of various forms of hyperlipoproteinemia, manifested by hypercholesterolemia or hypertriglyceridemia, are listed and discussed in Chapter 8.

Hypertriglyceridemia

Triglyceride-rich lipoproteins are transported to the blood stream from the intestines as chylomicrons (exogenous pathway) and from the liver as VLDL (endogenous pathway) (*see* Fig. 4.3). Both of these particles subsequently undergo lipolysis by the endothelial enzyme lipoprotein lipase (LPL) to produce chylomicron remnants or VLDL remnants (IDL); the remnants can be taken up by specific hepatic receptors or (in the case of IDL) can be converted to LDL. Hypertriglyceridemia usually occurs as a result of overproduction of these triglyceride-rich particles or as a result of decreased degradation by LPL (Fig. 7.1). A third mechanism is found in dysbetalipoproteinemia, in which a decreased uptake of chylomicron remnants and IDL by the hepatic receptors results in hypercholesterolemia combined with hypertriglyceridemia.

Hyperchylomicronemia

In the normal state, postprandial chylomicrons are hydrolyzed rapidly by LPL, producing chylomicron remnants that are taken up by the liver; thus, fasting blood does not normally contain chylomicrons. Defects in the catabolism of chylomicrons result in accumulation of these particles in blood even in the fasting state, resulting in elevated blood levels of triglyceride. Such abnormalities may result from primary enzymatic defects (as described later) or from secondary causes. Examples of secondary causes include diabetes, in which there is an acquired defect in the activity of LPL, and paraproteinemia (myeloma, SLE), in which antibodies are directed against LPL or factors that activate LPL, such as heparin.

Chylomicronemia with LPL or Apoprotein C-II Deficiency (Type I). Primary hyperchylomicronemia may occur as a result of absent or defective LPL activity. In patients who have LPL deficiency, chylomicrons may persist for several days, even on a fat-free diet or full fasting. The disorder also may result from a deficiency of apoprotein C-II, which is a necessary cofactor for LPL activation.

Primary abnormalities or deficiencies of LPL and apoprotein C-II are uncommon genetic disorders. Homozygous LPL deficiency occurs in 1 in 1,000,000 persons and is associated with hyperchylomicronemia and hypertriglyceridemia in infants and

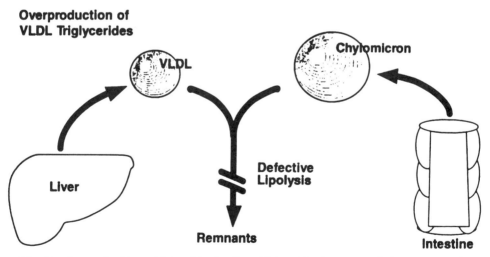

Figure 7.1. Development of hypertriglyceridemia. Hypertriglyceridemia may result from two putative processes: (1) overproduction of VLDL by the liver in response to an increase in free fatty acids flowing to this organ, or (2) a defect in the lysis of VLDL triglycerides and chylomicrons by lipoprotein lipase. When this enzyme is deficient, triglycerides cannot be converted, hydrolyzed, or broken down, and the metabolism of chylomicron and VLDL remnants may be delayed, as in type III hyperlipoproteinemia or dysbetalipoproteinemia. (Source: Coronary heart disease seminar series: lipid regulation and risk prevention. Parke-Davis, Division of Warner-Lambert Company, 1991.)

children. Heterozygous LPL deficiency, in which the affected persons have half the amount of LPL activity as normal persons, occurs in 1 in 500 persons; these patients often exhibit no obvious lipid abnormality and usually are asymptomatic. Recently it has been suggested, however, that heterozygotes may account for approximately one-third of patients who have combined hyperlipidemia.

Hyperchylomicronemia can be associated with recurrent episodes of abdominal pain, attributable to pancreatitis. Cardinal signs of this disorder are eruptive xanthoma and lipemia retinalis (*see* Chapter 9). In addition, in those persons in whom the disorder is primary, hepatosplenomegaly may be seen at an early age as a consequence of the accumulation of triglyceride in reticuloendothelial cells. Lymphadenopathy also has been described recently.

Chylomicronemia with Elevated VLDL (Type V). Severe hypertriglyceridemia can be associated with a constellation of symptoms known as the *chylomicronemia syndrome*. In such patients, a primary genetic disorder of triglyceride metabolism usually is complicated by a secondary (acquired) form of hypertriglyceridemia that is caused by a disease or a drug. These secondary causes can either *drive* the synthesis of VLDL triglyceride, such as occurs with estrogen treatment, alcoholism, obesity, and NIDDM, or they can produce acquired defects in the removal of VLDL and chylomicrons, such as may occur with hypothyroidism or the use of beta-blockers. Consumption of a high-fat diet also may lead to the accumulation of both chylomicrons and VLDL, because both particles compete for limited amounts of LPL. In such patients, the triglyceride concentration may rise above 1000 mg/dL (11.29 mmol/L), and the magnitude of the elevation may be striking.

Presenting clinical manifestations of the chylomicronemia syndrome include abdominal pain, skin rash caused by eruptive xanthomas, memory loss, depression, and dyspnea. Typical physical findings may reveal lipemia retinalis, eruptive xanthoma, and hepatosplenomegaly. Laboratory evaluation reveals (spurious) hyponatremia caused by displacement of water from the plasma or serum sample in which sodium is measured, and decreased pO_2 (also spurious) because of triglyceride interference with the measurement of pO_2.

Familial Combined Hyperlipidemia (Types II_A, II_B, IV)

Familial combined hyperlipidemia (FCHL) is a common lipid disorder that occurs in 1% to 2% of the population and accounts for 15% to 25% of patients who have hypertriglyceridemia. It has a variable phenotype and can be expressed as either Types II_A, II_B, or IV dyslipidemia. Furthermore, the phenotypic expression can change with time within the same person and may vary from person to person within families (Fig. 7.2). The mechanisms responsible for this variable expression are unknown, but environmental factors, such as diet, obesity, and the like may be important in their expression. It is not uncommon for the lipoprotein abnormalities to be absent in adolescents and young adults, only to appear later, in the third or fourth decade. This disorder often can be detected only by periodic screening, because preceding normal levels of lipids may be replaced in time with one or more lipid abnormalities.

In FCHL, the fundamental abnormality appears to be an increased synthesis of apoprotein B; this is accompanied by an increased production of VLDL particles of normal size and composition and the development of hypertriglyceridemia. Variable metabolism of VLDL may explain why some patients have one phenotype and other patients from the same family have a different phenotype, or why some patients can have variable lipid and lipoprotein patterns on different occasions. For instance, if VLDL production is increased with a comparable simultaneous increase in lipolysis, then the only abnormality might be enhanced conversion of VLDL to IDL and LDL and the de-

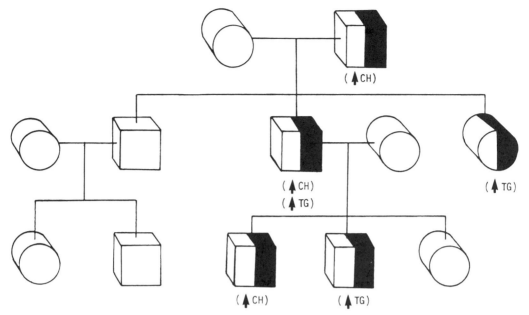

Figure 7.2. Familial combined hyperlipidemia. Inheritance of multiple patterns of dyslipidemia (familial combined hyperlipidemia) as a monogenic-dominant disorder. CH = cholesterol; TG = triglyceride. (Adapted from: Grundy SM. Cholesterol and atherosclerosis:diagnosis and treatment. Philadelphia: J.B. Lippincott Co., 1990.)

velopment of Type II$_A$ hypercholesterolemia. If VLDL production is increased without an increase in VLDL clearance, then hypertriglyceridemia (or Type IV) would develop. A third possibility might be increased VLDL production and increased VLDL clearance, but of a lesser magnitude; this combination of events would lead to increased levels of VLDL and LDL, or the Type II$_B$ phenotype of hypertriglyceridemia and hypercholesterolemia.

If VLDL or LDL clearance varies over time within the same person because of varying metabolic or environmental conditions (such as the development or control of non-insulin-dependent diabetes, the presence or absence of obesity, aging or diet-related changes in LDL-receptor expression), then the variable phenotypes in a single patient could also be explained. It should be pointed out, however, that the Type II$_B$ lipoprotein pattern also could represent a combined heterozygote; that is, a patient who was a heterozygote for familial hypercholesterolemia from one side of the family and a hete-

rozygote for familial hypertriglyceridemia from the other side of the family. Only a carefully obtained family history would permit the identification of these two genetic defects in the same person. Familial combined hyperlipidemia is a serious disorder and may be the most common single disorder to cause dyslipidemia, as well as premature coronary heart disease. Type II$_B$ represented 30% of patients who had lipoprotein abnormalities in the LRC Prevalence Study.

Because of the multiple phenotypes, the diagnosis of FCHL is difficult. Suspicion for the disorder should be high if the hyperlipidemia is mild and the phenotype changes over time, especially in those persons in whom elevated cholesterol or triglyceride level occurs in approx. 50% of first-degree relatives. The diagnosis also should be considered in those patients who have a strong family history of premature CHD characterized by primary hypertriglyceridemia. About one-third of patients who have FCHL have hypercholesterolemia; another one-third have hypertriglyceridemia; and the remaining one-third show elevation of both lipids. When dense LDL predominates, the blood cholesterol concentration may not be elevated; also a low HDL-C may be present. Although no characteristic physical signs of FCHL exist, patients tend to be obese and hypertensive and frequently develop premature atherosclerosis in the fourth or fifth decade of life. The presence of tendon xanthomas in the family of such patients or hypercholesterolemia in a first-degree relative under the age of ten rarely is present in FCHL and should suggest familial hypercholesterolemia.

Familial Hypertriglyceridemia (Type IV)

Familial hypertriglyceridemia (FHTG) is a common autosomal dominant lipid abnormality that occurs in 1% to 2% of the population and accounts for about 15% to 25% of patients who have increased VLDL and hypertriglyceridemia (Fig. 7.3). The abnormality is characterized by elevated triglyceride levels and relatively normal cholesterol levels (*see* Table 7.1). In contrast to

familial combined hyperlipidemia, FHTG is associated with a normal apoprotein B synthesis rate and normal numbers of VLDL particles, although these particles are enriched with triglyceride. Consequently, there is more triglyceride relative to apoprotein B, and the particles are larger and less dense than normal.

Typically, the hypertriglyceridemia develops in early adulthood and is often associated with obesity, glucose intolerance and hyperinsulinemia, hyperuricemia, and hypertension. In some patients no secondary cause can be identified and the disorder is primary. Other manifestations of dyslipidemia are rare, but some patients also may develop hyperchylomicronemia and have eruptive xanthomas and pancreatitis.

Although there is some disagreement over the difference in the lipoprotein abnormalities, susceptibility to development of CHD in persons who have FHTG is lower than that of familial combined dyslipidemia. Although HDL-C is reduced in both syndromes, the apoprotein A-I levels are normal in FHTG, suggesting that the number of

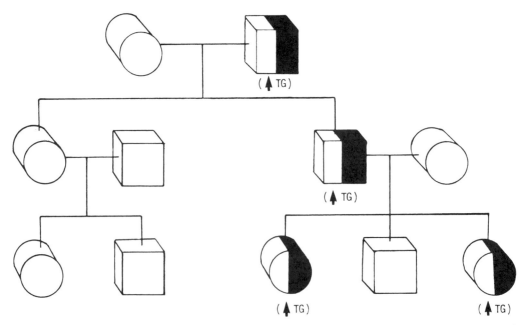

Figure 7.3. Familial hypertriglyceridemia. Inheritance of familial hypertriglyceridemia as a monogenic-dominant disorder. TG = triglyceride. (Adapted from: Grundy SM. Cholesterol and atherosclerosis: diagnosis and treatment. Philadelphia: J.B. Lippincott Co., 1990.)

HDL particles is normal. In contrast, familial combined hyperlipidemia usually is associated with reduced apoprotein A-I levels, and, thus, HDL particles. In addition, only a small fraction (approx. 10%) of the large VLDL particles found in FHTG is converted to LDL, whereas about 40% of normal-sized or small VLDL particles found in familial combined hyperlipidemia are converted to LDL.

The differential diagnosis of isolated hypertriglyceridemia, isolated hypercholesterolemia, and combined hypercholesterolemia plus hypertriglyceridemia is shown in the accompanying diagrams (Fig. 7.4). Family screening is necessary to determine whether the disorder is primary. If so, 50% of all first-degree relatives will have hyperlipidemia. It must be reemphasized that about 50% of hypertriglyceridemia is secondary; therefore, a careful search always should be made for contributory factors such as obesity, excessive alcohol use, estrogens and other medications, high-carbohydrate, low-fat diets, or metabolic defects.

Hypercholesterolemia (Type II$_A$)

From a purely mechanistic standpoint, three basic abnormalities can contribute to hypercholesterolemia: 1) defective clearance of LDL by receptor or nonreceptor-mediated pathways; 2) overproduction of LDL, either through excessive hepatic synthesis of apoprotein B-containing lipoproteins or through increased conversion of VLDL or VLDL remnants to LDL because of decreased hepatic uptake of these lipoproteins; and 3) overloading of LDL particles with cholesteryl esters without any increase in the number of particles.

Primary hypercholesterolemia may be caused by obvious or, at times, not so obvious familial defects. These metabolic defects may be caused by a single gene disorder, such as that in familial hypercholesterolemia, or may be polygenic and result from multiple genes interacting with environmental factors. Hypercholesterolemia also may accompany two other genetic disorders: familial combined hyperlipidemia and dysbetalipoproteinemia. In familial combined

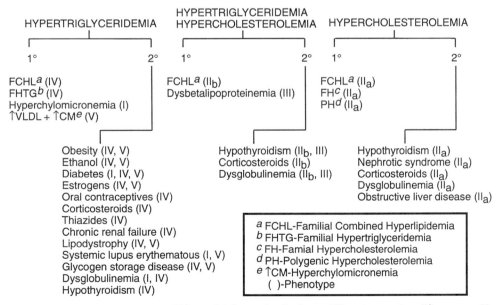

Figure 7.4 Plasma lipid measurement. Differential diagnosis of primary (1°) and secondary (2°) causes of isolated hypercholesterolemia, isolated hypertriglyceridemia, and combined hyperlipidemia. FCHL = familial combined hyperlipidemia; FHTG = familial hypertriglyceridemia; FH = familial hypercholesterolemia; PH = polygenic hypercholesterolemia; ↑CM = hyperchylomicronemia. Parentheses indicate phenotype.

hyperlipidemia, hypercholesterolemia may occur alone (phenotype IIa) or may be accompanied by hypertriglyceridemia (phenotype IIb). This disorder has a variable phenotypic expression, and only some patients (perhaps 25%) have hypercholesterolemia as their only detectable abnormality. In dysbetalipoproteinemia, the accumulation of remnant VLDL (or IDL) leads to hypercholesterolemia and hypertriglyceridemia, resembling the presence of Type IIb; however, the presence of typical physical findings (*see below*) and more sophisticated analyses can lead to the diagnosis of the Type III phenotype. These two disorders will be discussed in more detail in subsequent chapters.

Great variability exists in the relative frequencies of the genetic defects that are associated with hypercholesterolemia. If LDL-C values above the 75th percentile, or 160 mg/dL (4.14 mmol/L), are considered to be abnormal and to increase CHD risk, then 25% of the population would be classified as having hypercholesterolemia. An additional 25% of adults would have borderline LDL-C values in the 50th to 75th percentiles (130 to 159 mg/dL, 3.36 to 4.11 mmol/L). The prevalence rate is influenced by age and sex, so that 30% of men and 40% of women 55 years of age and older would have hypercholesterolemia. The prevalence of the monogenic primary forms of hypercholesterolemia is markedly lower, however, being in the order of 1:1,000,000 for the homozygous form of familial hypercholesterolemia (FH), 1:500 for the heterozygous form of FH, 1:5,000 for clinically manifest dysbetalipoproteinemia, and 1:200 for familial combined hyperlipidemia presenting as isolated hypercholesterolemia, thus giving an aggregate prevalence of hypercholesterolemia of approx. 8:1,000. Because hypercholesterolemia exists in about 25% of the population, over 95% of hypercholesterolemic persons have the polygenic form of the disorder. Thus, the most common cause of hypercholesterolemia is the one that is the least well understood. It is becoming increasingly more clear that patients who have polygenic hypercholestero-

lemia are a heterogeneous group who have differing metabolic abnormalities. Some patients can have marked hypercholesterolemia on a familial or genetic basis because of mild or minimally expressed defects inherited from each parent.

Familial Hypercholesterolemia (FH)

The most carefully studied form of hypercholesterolemia is familial hypercholesterolemia (FH), in which patients have deficient or defective LDL receptors (Fig. 7.5). One gene for the LDL receptor is inherited from each parent, and the inheritance of two defective LDL genes results in the complete absence of functional receptors. The removal of IDL and LDL from the bloodstream thus is severely impaired. This leads to the accumulation of LDL particles in the bloodstream as a result of increased conversion of IDL to LDL, as well as decreased LDL clearance. The blood LDL concentration increases four- to six-fold, resulting in severe hypercholesterolemia. As the blood cholesterol levels increase, more LDL-C must be removed by the scavenger pathway expressed in monocytes and macrophages, with subsequent potential for atherosclerosis. This form of FH, associated with markedly elevated LDL-C and CHD in childhood presents no diagnostic dilemmas.

The inheritance of only one defective gene results in heterozygous FH, in which the number of LDL receptors is reduced by 50% and the cholesterol is only two to three times higher than normal. Although the diagnosis of heterozygous FH is not always apparent, several characteristics should provide clues to its presence. The higher and the less responsive the LDL-C is to diet in any patient, the greater the likelihood of FH. In children, even mild heterozygous FH can be distinguished from other familial forms of hypercholesterolemia, because the hypercholesterolemia associated with FCHL usually is not expressed until the third decade. In young children who have obviously elevated cholesterol values, the likelihood of FH is high because of the rarity of other

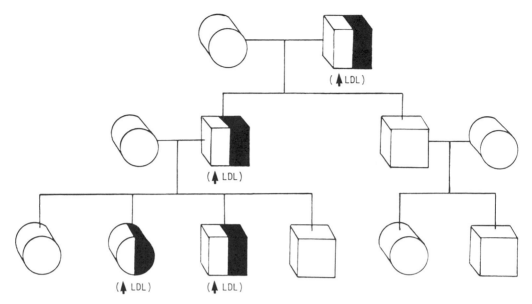

Figure 7.5. Heterozygous familial hypercholesterolemia. Inheritance of familial hypercholesterolemia as a monogenic-dominant disorder. (Adapted from: Grundy SM. Cholesterol and atherosclerosis: diagnosis and treatment. Philadelphia: J.B. Lippincott Co., 1990.)

causes. In patients who have a twofold or greater LDL-C level and tendon xanthomas, the diagnosis of FH is unequivocal.

Although persons who have FH represent only a small percentage of patients who have hypercholesterolemia, they become increasingly important as the cut-point that defines an abnormal cholesterol is increased. Heterozygotes represent about 5% to 10% of young adults hospitalized with acute MI; the highest incidence is in the fifth decade for men and in the sixth decade for women. Consequently, this entity assumes great importance among patients who have myocardial infarction.

Patients who have FH may develop premature corneal arcus, xanthelasma, and tendon xanthomas. Homozygotes develop these abnormalities soon after birth and usually before the age of six years. Tendon xanthomas occur in over half of all heterozygotes, especially older patients. These nodular deposits of cholesterol typically are found in the achilles and patellar tendons and in the tendons of the dorsum of the hand; however, they also may occur elsewhere, such as in the plantar fascia. (Fig.

7.6). Cholesterol deposits also may develop in the skin (tuberous/planar/xanthomas), in the webbing of the fingers (especially between the first and second digits), and in the toes.

Polygenic Hypercholesterolemia

As already discussed, polygenic hypercholesterolemia is the most common cause of hypercholesterolemia and elevated LDL-C. The causes of such abnormalities may be variable, including subtle structural abnormalities of the LDL receptor that cannot be detected with current techniques, abnormalities in the rate of synthesis of the receptor, and defects in the hepatic uptake of VLDL remnants, resulting in increased conversion to LDL. Finally, minor abnormalities in the amino acid sequence of apoprotein B, particularly in regions critical for binding of LDL to the LDL receptor, also can cause hypercholesterolemia. Although such abnormalities are difficult to detect with current technology, their characterization is in progress. It is plausible that unmasking of such subtle genetic abnormalities is caused by other superimposed elements such as

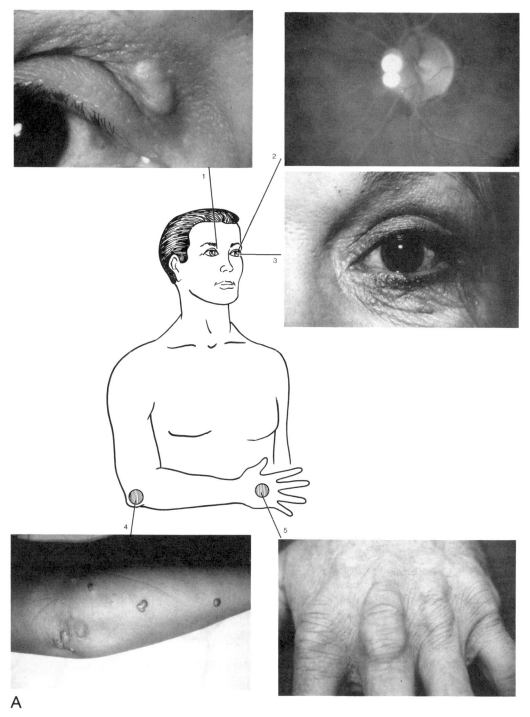

A

Figure 7.6. Characteristic physical findings in dyslipidemic states: Figure 7.6A shows xanthelasma (1), lipemia retinalis (2), arcus cornea and xanthelasma (3), tuberoeruptive xanthoma (elbow) (4), and tendon xanthoma (hand) (5). Figure 7.6B shows tuberoeruptive xanthoma (hand) and striata palmaris (1), tuberoeruptive xanthoma (hand) (2), eruptive xanthomas (3), tendon xanthoma (knee) (4), and tendon xanthoma (Achilles tendon) (5).

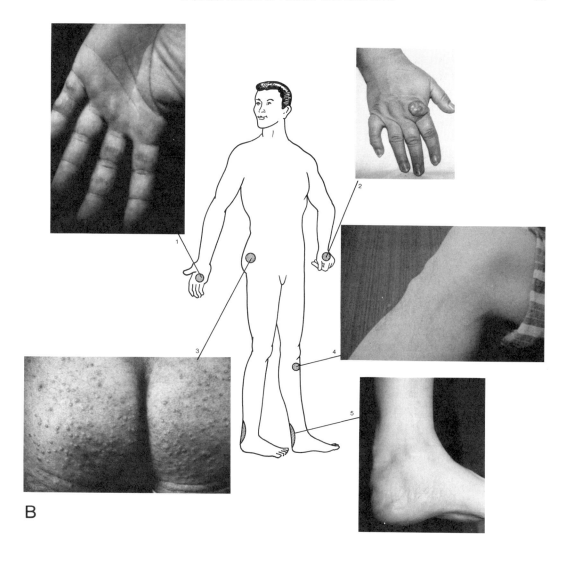

B

aging, dietary factors, exercise level, rates of hepatic conversion of cholesterol to bile acids, and, perhaps, various other environmental factors. Polygenic hypercholesterolemia clearly constitutes the largest and most important group in terms of CHD risk. A better understanding of the mechanisms that contribute to the development of hypercholesterolemia in this group may provide more rational and successful treatment for the majority of patients who are at risk for developing heart disease because of hypercholesterolemia.

The sensitivity of persons who have polygenic hypercholesterolemia to dietary cholesterol and saturated fat is variable. When examined by careful metabolic techniques, protection against the development of hypercholesterolemia appears to be conferred by decreased intestinal absorption of dietary fat and cholesterol, increased excretion of cholesterol as bile acids and perhaps by more efficient inhibition of hepatic cholesterol synthesis. In such persons, the blood cholesterol remains relatively constant. In contrast, persons who absorb more cholesterol, who are less efficient in converting cholesterol to bile acids, or who do not turn off hepatic cholesterol production may become hypercholesterolemic. Up to 35% of persons, both normal and hypercholesterolemic, have this difficulty in absorbing more cholesterol or in regulating cholesterol production (*see* Chapter 10). These observations also may provide insight into why some hypercholesterolemia patients respond to low-fat, low-cholesterol diets and others do not.

Hyperbetalipoproteinemia

Although all patients who have increased LDL-C also have increased levels of apoprotein B-100, the reverse is not always true. A dyslipidemic state called *hyperbetalipoproteinemia* has been described that is characterized by normal cholesterol and LDL-C concentrations but by increased concentrations of apolipoprotein B. The precise mechanism for the increased level of apoprotein B is not understood but could occur through several mechanisms, including increased hepatic secretion of apoprotein B-containing lipoproteins or decreased hepatic removal of VLDL remnants. Both abnormalities could lead to increased conversion of VLDL to LDL and elevated LDL apoprotein B levels.

Patients who have this abnormality appear to have increased numbers of small, dense LDL particles that, as previously discussed (*see* Chapter 4), may be more atherogenic than LDL particles of larger size and normal composition. Because the ratio of cholesterol to apoprotein B in LDL is a reflection of the size of the LDL particle, these patients may display normal LDL cholesterol concentrations, despite their elevated apoprotein B levels. The disorder would not be detected by measurement of the total cholesterol or by calculation of the LDL-C concentration. Whether hyperbetalipoproteinemia constitutes a specific familial defect is unclear at this time, but it is associated with premature CHD and recurrent coronary stenosis following revascularization. Because the disorder is more common in persons who ultimately develop NIDDM, in persons who have hypertriglyceridemia and low HDL-C, and in persons who use beta-blockers, it is not clear whether this is a primary disorder that predisposes to CHD or merely a marker for the presence of other metabolic abnormalities that cause CHD.

Dysbetalipoproteinemia (Type III)

Dysbetalipoproteinemia, or Type III hyperlipoproteinemia, is caused by the accumulation of cholesterol-enriched VLDL remnants (known as beta-VLDL) and IDL. The underlying genetic abnormality usually is the presence of an abnormal form of apoprotein E (E_2). This apoprotein E isotype has a low affinity for the LDL receptor and impedes the removal of VLDL remnants and IDL by the LDL receptor. This defect can account for a spuriously elevated LDL-C concentration because the usual techniques used by clinical laboratories cannot distinguish IDL from LDL. Because beta-VLDL and IDL migrate between LDL and VLDL on electrophoresis, dysbetalipoproteinemia often is referred to as *broad beta disease*.

Intermediate density lipoprotein has some features of VLDL and some features of LDL; it contains apoprotein B and apoprotein E but carries more triglyceride than LDL. As noted earlier, the usual ratio of cholesterol to triglyceride in VLDL is 0.2. Although total cholesterol and triglyceride levels can be elevated comparably in dysbetalipoproteinemia, the ratio of cholesterol to triglyceride in VLDL is greater than 0.35. This feature can be used as one of the diagnostic hallmarks of the disorder.

Knowledge of the molecular basis of dysbetalipoproteinemia and of the functional changes that accompany it provides an important and interesting illustration of how dyslipidemia can develop from genetic abnormalities. Apoprotein E is a glycoprotein with 299 amino acids that is critical for the binding of certain lipoproteins to specific receptors on the hepatic cell surface and for their subsequent removal from the circulation. There are three alleles for apoprotein E that produce three different apoproteins: E_2, E_3, and E_4. The amino acid substitutions at positions 112 and 158 of apoprotein E determine the isoform structure:

	Position 112	Position 158	Gene Frequency
Apoprotein E$_2$	Cysteine	Cysteine	12%
Apoprotein E$_3$	Cysteine	Arginine	77%
Apoprotein E$_4$	Arginine	Arginine	11%

Everyone inherits two alleles, one from each parent. As shown above, the most common (normal) inherited allele is E$_3$, which has arginine at the 158 position. The amino acid located at position 158 influences the capacity of the receptor-binding region of the molecule (positions 140 through 150) to bind to the LDL receptor. Apoprotein E$_2$ has a cysteine residue in this position and, therefore, binds poorly to the receptor, whereas the E$_4$ isoform binds with greater than normal affinity to the receptor.

Normally, approximately 50% of the IDL formed in the lipolytic cascade is removed by the LDL receptor and constitutes an important source of cellular cholesterol. Persons who have the E$_2$/E$_2$ isoform have defective binding of IDL and VLDL remnants to the LDL receptor; this results in the accumulation of these particles in the blood as well as in decreased delivery of cholesterol to the cell by this pathway. As a result, there is an upregulation of LDL receptor synthesis and increased expression of the receptor on the cell surface, with consequent increase in the fractional catabolic rate for LDL and a reduction in its blood concentration.

The apoprotein E phenotype frequencies in the general population are listed below:

$$E_2/E_2 \quad 3\%$$
$$E_2/E_3 \quad 16\%$$
$$E_3/E_3 \quad 62\%$$
$$E_3/E_4 \quad 14\%$$
$$E_4/E_4 \quad 3\%$$
$$E_2/E_4 \quad 2\%$$

One to three percent of the population have the E$_2$/E$_2$ isoform and therefore should be at risk for developing dyslipidemia. Although most of these persons have some abnormalities of IDL composition, even in the presence of normal triglyceride and cholesterol levels, the actual frequency of dysbetalipoproteinemia is only 1:5,000 to 1:10,000. It appears then that the clinical emergence of dyslipidemia requires the presence of other accompanying defects. These defects usually involve the overproduction of VLDL (such as might occur with obesity, noninsulin-dependent diabetes mellitus, familial combined dyslipidemia, familial hypertriglyceridemia, or other secondary causes of hypertriglyceridemia) or a reduction in the number of LDL receptors as a consequence of age, high-fat diets, or postmenopausal estrogen deficiency.

Patients who have these disorders eventually may develop mixed dyslipidemia (elevated cholesterol and triglyceride), with Type III pattern caused by the accumulation of VLDL remnants. The elevation in cholesterol and triglyceride often is comparable; however, if the usual formula is used to calculate LDL-C, it then appears to be elevated. Sophisticated lipoprotein analysis will reveal that the concentration of LDL-C actually is reduced while the concentration of IDL is elevated. Other diagnostic findings include the presence of beta-VLDL and abnormal cholesterol-rich VLDL remnants on electrophoresis. As mentioned, the fact that the VLDL remnants demonstrate beta rather than pre-beta mobility on electrophoresis gives the disorder its name.

Patients who have Type III dyslipidemia also are characterized by two unique physical findings: xanthomas striata palmaris, which are orange/yellow discolorations in the creases of the palms of the hand, and tuberoeruptive xanthomas, which are located typically over the elbows and knees and vary in size from less than a centimeter to quite large (*see* Fig. 7.6). These cutaneous manifestations affect approximately 80% of the patients who have this disorder, and their presence should initiate additional lipoprotein and apoprotein analyses.

Patients who have dysbetalipoproteinemia are prone to develop atherosclerosis of the coronary arteries and peripheral vessels, including the internal carotid arteries and the abdominal aorta. This occurs, however, only after the appearance of dyslipidemia, which usually occurs later in life. Before the appearance of dyslipidemia, low LDL-C may reduce the risk of developing coronary atherosclerosis.

In contrast to the E_2 isoform, IDL that contains the E_4 isoform is removed more efficiently from the circulation than the normal E_3 isoform. This leads to increased intracellular cholesterol, secondary downregulation in LDL receptor synthesis and expression, a subsequent decrease in the LDL fractional catabolic rate, and an increase in the blood concentration of LDL-C. It is estimated that persons who have the E_3/E_4 and E_4/E_4 combinations account for 8% to 10% of the population who have mild to moderate hypercholesterolemia. Postprandial clearance of dietary lipid also is influenced by the apoprotein E isoform, being most rapid with E_4 and slowest with E_2.

Disorders of High Density Lipoprotein Metabolism

Approximately 3% to 5% of persons in the United States have low HDL-C concentrations, defined as < 35 mg/dL (0.91 mmol/L) for men and < 45 mg/dL (1.16 mmol/L) for women, as the predominant lipid abnormality. Although this group has not been characterized extensively, limited studies suggest that the reduction in HDL-C is caused primarily by increased HDL catabolism. Approximately 20% to 30% of patients who have acute MI also have lower than desirable HDL-C concentrations. Asymptomatic persons who have low HDL-C therefore, may be an important group in terms of risk for CHD and other atherosclerotic complications. No currently recognized therapeutic agent exists that selectively increases the HDL-C concentration; also there is no evidence to show that increasing an isolated low HDL-C concentration changes the natural history of CHD. Presently it is unclear what should be done about such patients. Hypolipidemic agents that increase the HDL-C while lowering the LDL-C (nicotinic acid, fibrates, HMG CoA reductase inhibitors) often are used to offset the low HDL-C in high-risk patients.

Although low HDL-C is an important risk factor for development of CHD, it is uncertain whether the specific mechanism that leads to the low HDL-C concentration (decreased synthesis or increased catabolism) selectively affects this risk. For instance, a low HDL-C related to efficient reverse cholesterol transport may have different implications than one caused by decreased synthesis. Some investigators believe that a low HDL-C is important primarily as a marker for other lipoprotein abnormalities that contribute to CHD, such as hypertriglyceridemia and smaller, denser LDL particles (*see* Chapter 4).

Primary High Density Lipoprotein Deficiency Syndromes

Rare genetic defects in the synthesis of apoprotein A-I are associated with premature CHD, corneal opacities, and low HDL-C. Most primary HDL deficiency syndromes involve abnormalities of apoprotein A-I and apoprotein A-II, apoprotein C, or LCAT. Many different genetic mutations of apoprotein A-I synthesis have been described. They can occur alone or (rarely) in combination with apoprotein C-II defects, because the genes for both apoproteins are closely located on the same chromosome (number 11). Heterozygotes for these defects will have HDL levels that are about half the normal concentrations. The underlying defect determines the lipoprotein profile and clinical findings in the primary HDL deficiency syndrome.

Tangier Disease. Tangier disease is a rare lipoprotein disorder named after the first known patients, who were from Tangier Island in the Chesapeake Bay—two siblings who had low blood cholesterol, peculiar appearing tonsils, and almost no HDL. Since then, over 35 patients have been described with this syndrome in the medical literature. These patients have a genetic defect that leads to accelerated HDL catabolism, extremely low HDL-C concentrations, and the accumulation of cholesterol in tissues. Other associated clinical findings include corneal opacities and relapsing polyneuropathy. Despite their low HDL-C, such patients seldom develop premature CHD, perhaps because their LDL-C also is low and insufficient for the development of atherosclerosis. It also is possible that HDL deficiency related to accelerated catabolism has different therapeutic implications than that caused by decreased synthesis.

Familial Apoprotein A-I and C-II Deficiency. Other primary HDL deficiency syndromes include a familial deficiency of apoprotein A-I and apoprotein C-II, with no apparent disorder in other lipoprotein levels. Two separate entities are recognized: (1) reduced synthesis of apoprotein A-I or C-II, with apoprotein A-I levels about 10% of normal, or (2) synthesis of defective apoprotein A-I and no apoprotein C-II. Severe premature CHD is common among homozygotes who have severely depressed HDL, whereas in heterozygotes with HDL levels about two-thirds of normal, CHD usually develops only after age 40.

It is interesting to note that other uncommon syndromes are associated with low (HDL deficiency with planar xanthoma) and moderately low HDL values (Fish-Eye Disease with corneal opacification) without manifestations of CHD.

Familial Hypoalphalipoproteinemia

Reduced levels of HDL-C may occur in families without other accompanying lipid abnormalities (Table 7.3) and predisposes to premature CHD. Although hypo-HDL-emia occurs in only 3% to 5% of the general population, it is found in 25% to 30% of patients hospitalized for acute MI (40% of acute MI

patients have total cholesterol levels within the normal range, of which about two-thirds have low HDL). The significance of these observations is unclear, however, and requires careful prospective verification, because many such patients may have been treated with beta-blocking agents or may have consumed very low-fat diets at the time of their lipoprotein analysis, both of which tend to reduce HDL-C. In addition, hospitalization has been reported to reduce spuriously the level of HDL-C, although the mechanism of this response has not been defined.

In some patients and families, low HDL coexists with mild to moderate elevation of LDL-C, which increases their CHD risk even more. Relatively little is understood about the mechanism of low HDL or its relationship to increased LDL, but studies in a limited number of patients reveal increased catabolism of otherwise structurally normal apoprotein A-I. The HDL-C levels tend to be greater than those found in Tangier disease and are approximately 50% of normal.

High Density Lipoprotein Excess Syndromes

Familial hyperalphalipoproteinemia (hyper per HDL) has been well described and is associated with increased longevity. In addition to familial causes, other disorders and drugs can increase HDL-C (Table 7.4). The mechanisms for the increased level of HDL-C in these patients are not known, and it is possible that they may differ from family to family or ethnic group to ethnic group. In Japan, the most common cause of

Table 7.3. Drugs and Factors Associated with Reduced HDL-C

Nonpharmacologic	Pharmacologic
1. Male sex	1. Probucol
2. Obesity	2. Progestins
3. NIDDM	3. Androgens
4. Cigarette smoking	4. Anabolic agents
5. Low-fat/high-carbohydrate diet	5. Beta-adrenergic blockers
6. Hypertriglyceridemia	6. Retinoids
7. Familial hypoalphalipoproteinemia	
8. Tangier disease	

Table 7.4. Drugs and Factors Associated with Increased HDL-C

1. Alcohol
2. Estrogens
3. Alpha-adrenergic blockers
4. Beta-adrenergic agonists
5. Carbamazepine
6. Diphenylhydantoin
7. Phenobarbital
8. Rifampin
9. Cimetidine
10. Nicotinic acid
11. Corticosteroids
12. Familial

hyper HDL is a deficiency in cholesteryl ester transfer protein (CETP); as described in Chapter 4, this enzyme is thought to mediate the transfer of cholesteryl ester between HDL and other lipoproteins. Despite the putative role of CETP in the reverse cholesterol pathway, these patients also seem to be protected against CHD.

Low Density Lipoprotein Deficiency Syndromes

Hypobetalipoproteinemia

Familial hypobetalipoproteinemia is an autosomal dominant disorder in which the concentrations of the total cholesterol, LDL-C, and apolipoprotein B are abnormally low. Persons who have this beneficial abnormality are heterozygotes for the apoprotein B-100 gene, located on chromosome 2, and have one normal and one abnormal gene. Apolipoprotein B-100 is synthesized by the liver, is a major structural protein of VLDL, and is the sole apoprotein of LDL, where it serves as a ligand for binding of the LDL particle to the LDL receptor.

In the heterozygotic state, patients are asymptomatic and usually have blood LDL-C concentrations that are 25% to 50% of those in normal persons. Such patients are protected from atherosclerotic disease and have increased longevity. It has been estimated that the heterozygous form of familial hypobetalipoproteinemia occurs with

a frequency of 1:1000, so it is likely to be encountered as population screening for cholesterol becomes better established.

Familial hypobetalipoproteinemia is a heterogeneous molecular disorder with similar phenotypes; that is, the patients are similar with regard to low LDL-C concentrations but are different with regard to the underlying molecular defect. Miscoding of an amino acid in hepatic apoprotein B-100 mRNA (point mutation) results in the synthesis of truncated apoprotein B molecules that are 48%, 39%, and 37% of the weight of a normal apolipoprotein B-100. The defect and the size of the apoprotein B molecule are specific for each family. The low LDL-C is caused by decreased synthesis of apoprotein B or accelerated removal of the truncated form of apoprotein B. It is possible that detailed evaluation of patients who have low blood cholesterol levels will identify additional mutations in apolipoprotein B that cause hypobetalipoproteinemia.

Abetalipoproteinemia

Abetalipoproteinemia is an autosomal recessive disorder characterized by the complete absence of apolipoprotein B, with a consequent inability to produce chylomicrons, VLDL, and LDL. The blood levels of triglyceride and cholesterol therefore are extremely low. The clinical features of this syndrome are variable and include malabsorption of fat and fat-soluble vitamins, ataxic central nervous system disease, retinitis pigmentosa, and anemia with acanthocytes. Interestingly, heterozygotes for abetalipoproteinemia have normal blood lipid levels, in contrast to heterozygotes for hypobetalipoproteinemia. The mechanism of abetalipoproteinemia is unknown, but unlike hypobetalipoproteinemia, it is not caused by an abnormality in the apoprotein B gene.

Other Hypocholesterolemic Disorders

Hypocholesterolemia occurs in a number of other medical conditions (Table 7.5) and

Table 7.5. Conditions Associated with Low Blood Cholesterol Levels

1. Abetalipoproteinemia
2. Hypobetalipoproteinemia
3. Hypoalphalipoproteinemia
4. Tangier disease
5. LCAT deficiency
6. Reticuloendothelial hyperfunction
    a. Sarcoidosis
    b. Gaucher's disease
    c. Myeloproliferatieve disorders
    d. Lymphoma
7. Hyperthyroidism
8. Cancer

is interesting more as a curiosity or for what it can teach us about lipoprotein transport than for any pathologic implications. Its association with reticuloendothelial hyperfunction is probably important because it emphasizes the role of monocytes and macrophages in removal of VLDL and LDL. In hyperthyroidism, the low LDL-C is caused primarily by increased synthesis and expression of the cell-surface LDL receptor that accelerates the fractional catabolic removal of LDL and also by decreased synthesis of apoprotein B.

There has always been some concern over the relation between reduced levels of cholesterol and cancer (*see* Chapter 1). The critical question has been whether the cholesterol is reduced because the patient has cancer or whether a low cholesterol predisposes to cancer. These issues are not resolved completely, but the consensus is that a low cholesterol does not predispose to cancer.

Suggested Readings

Breslow JL. Genetic basis for lipoprotein disorders. J Clin Invest 1989;84:373–380.

Dietschy J. LDL cholesterol: its regulation and manipulation. Hosp Pract 1990;25:47–48.

Goldstein JL, Brown MS. The LDL receptor defect in familial hypercholesterolemia: implications for pathogenesis and therapy. Med Clin North Am 1982;66:335.

Grundy SM, Vega GL. Causes of high blood cholesterol. Circulation 1990;81(2):412–427.

Innerarity TL, Mahley RW, Weisgraber KH, et al. Familial defective apolipoprotein B-100: a mutation of apolipoprotein B that causes hypercholesterolemia. J Lipid Res 1990;31:1337–1349.

Mahley RW, Weisgraber KH, Innerarity TL, et al. Genetic defects in lipoprotein caused by impaired catabolism. JAMA 1991;265:78–83.

Schaefer EJ. Diagnosis and management of lipoprotein disorders. In: Rifkind BM, ed. Drug treatment of hyperlipidemia. New York: Marcel Dekker, 1991:17–52.

Schaffer EJ, Levy RI. Pathogenesis and management of lipoprotein disorders. N Engl J Med 1985;312:1300–1310.

Schonfeld G. The genetic dyslipoproteinemias—nosology update 1990. Atherosclerosis 1990;81:81–93.

Scriver CR, Beaudet AL, Sly WS, Valle D (eds). Lipoprotein and lipid metabolism disorders Part 7. In: The metabolic basis of inherited disease, ed 6. New York: McGraw-Hill Information Services Company, 1989:1129–1283.

CHAPTER 8

Secondary Dyslipidemia

Diseases Associated With Abnormal Lipid Profiles

Various clinical disorders can be associated with or can cause dyslipidemia (Table 8.1). Treatment of the underlying condition often can correct the lipid disorder. The possibility, however, of an underlying secondary cause and a primary lipid problem should not be ignored. Typical diagnostic findings for conditions that result in secondary dyslipidemias are given in Table 8.2.

Autoimmune Dyslipidemia

Dyslipidemia associated with monoclonal immunoglobulin abnormalities (IgG, IgA) is rare but well known among lipoprotein specialists. Cutaneous manifestations of the dyslipoproteinemia have been reported to precede the diagnosis of myeloma by as much as 20 years. The lipid abnormality may present as hypercholesterolemia, hypertriglyceridemia, or mixed dyslipidemia. The mechanism usually involves the formation of antibodies that bind to and modify the function of specific lipolytic enzymes (such as LPL and HTGL), apoproteins, and receptors; the antibodies also may interfere with the activation of LPL by binding to heparin.

The most common disease that causes dyslipidemia by autoimmune mechanisms is

Table 8.1. Causes of Secondary Dyslipidemia

Hypercholesterolemia	Hypertriglyceridemia	Combined Hyperlipidemia
Common:		
Hypothyroidism	Diabetes mellitus	Hypothyroidism
Nephrotic syndrome	Obesity	Nephrotic syndrome
Obstructive liver disease	Alcohol	Chronic renal failure
	Chronic renal failure	
Uncommon:		
Acute intermittent porphyria	Myocardial infarction	Liver disease (LP-X)
Pregnancy	Infection (bacterial, viral)	Werner's syndrome
Anorexia nervosa	Systemic lupus erythematous	Acromegaly
	Dysglobulinemia	
	Glycogen storage disease (type I)	
	Lipodystrophy	
	Nephrotic syndrome	
	Bulemia	
	Autoimmune disorders	
	Pregnancy	
Drugs:		
Thiazide diuretics	Beta-blockers	Thiazide diuretics
Retinoids	Retinoids	Glucocorticoids
Glucocorticords	Estrogens	Retinoids
Cyclosporine		
Progestins		
Androgens		

Table 8.2. Diagnostic Findings in Patients Who Have Secondary Hyperlipidemia[a]

Lipid Pattern	Phenotype	Typical Values		Related Conditions
		Cholesterol	Triglyceride	
↑ Chylomicrons	1	300–400	3000–6000	Systemic lupus erythematosus
↑ LDL	2a	300–400	100	Obesity, hypothyroidism, nephrotic syndrome, hepatoma
↑ LDL and VLDL	2b	300–400	250–500	Cushing's syndrome, dysglobulinemia, acute intermittent porphyria, anorexia nervosa, Werner's syndrome
↑ Remnants of VLDL and chylomicron	3	300–500	300–800	Dysglobulinemia
↑ VLDL	4	200–250	300–700	Obesity, diabetes mellitus, estrogen therapy, acromegaly, Cushing's syndrome, uremia, nephrotic syndrome, acute viral hepatitis, dysglobulinemia, alcoholic hyperlipidemia; third-trimester pregnancy
↑ Chylomicrons ↑ VLDL	5	600–800	2000–6000	Poorly controlled diabetes mellitus, estrogen therapy, alcoholic hyperlipidemia

[a](Adapted from: Bilheimer DW: Evaluation of abnormal lipid profiles. In: Kelly WN, ed. Textbook of internal medicine. Philadelphia: J.B. Lippincott Company, 1988:2333.)

multiple myeloma. Dyslipidemia also has been described with systemic lupus erythematosis (SLE), Graves' disease, and idiopathic and thrombocytopenic purpura.

Acute Myocardial Infarction

Plasma lipids are altered in patients who have acute MI and must be interpreted with caution when measurements are made during the convalescent period. The triglyceride concentration reaches a peak within 3 weeks postinfarction and returns to baseline levels by 6 weeks. In contrast, the total cholesterol and LDL-C levels decrease and reach a nadir 1 to 2 weeks after infarction, returning to preinfarction levels by 8 to 12 weeks. The total cholesterol and LDL-C fall by a mean of 20%, but reductions as great as 60% of the preinfarction value have been observed. Recently, acute reductions in HDL-C also have been shown to occur following coronary angiography. The occurrence of similar changes following acute MI could account for, or contribute to, the high prevalence of low HDL observed in such patients. The treatment regimen also may impact on the lipid profile during hospitalization and afterwards. Consequently, lipid measurements that are not obtained promptly (within the first several hours to days after an acute MI), may be misleading and are best postponed for 3 months. The mechanism of the lipid changes during acute MI has not been defined.

Thyroid Disorders

Lipid abnormalities often were used to support a presumptive diagnosis of thyroid dysfunction before the era of sophisticated hormonal measurements. It is well known that hypocholesterolemia is a manifestation of hyperthyroidism and that hypercholesterolemia is a manifestation of hypothyroidism. The LDL receptor appears to play a major role in determining the cholesterol concentration in patients who have thyroid disease (Fig. 8.1). In hyperthyroidism, there is an increased rate of LDL catabolism caused by increased synthesis and expression of the LDL receptor, and the plasma

half-life of LDL is shortened. In contrast, receptor-mediated catabolism is decreased in hypothyroidism, and the LDL half-life is prolonged.

Thyroid hormone may influence the rate of apoprotein B synthesis. In experimental animals, pharmacologic doses of thyroxine suppress the synthesis of apoprotein B-100 by inserting a *stop signal* in hepatic apoprotein B-100 messenger RNA; this prematurely terminates apoprotein B-100 production at the 48th amino acid, producing apoprotein B-48. Apoprotein B-48 normally is synthesized only by the intestine and is the major apoprotein of chylomicrons. Although these observations may explain why hyperthyroidism is associated with decreased apoprotein B-100 concentrations, it does not explain the hypercholesterolemia of hypothyroidism, because apoprotein B-100 synthesis is expressed maximally in the euthyroid state and is not increased any more by thyroxine deficiency.

Hypothyroidism is an important secondary cause of hypercholesterolemia, and because the signs and symptoms may be subtle, thyroid function should be evaluated in all patients who have elevated cholesterol levels, even if thought to be clinically euthyroid. This routine procedure may prevent the unwarranted use of lipid-lowering medications in such conditions.

Thyroid function also influences HDL metabolism, and the blood concentrations of HDL-C and apoprotein A-I are both reduced in patients who have hyperthyroidism. Overall, total HDL-C concentrations are reduced by approximately 20% in patients who have hyperthyroidism, the reduction being confined to the HDL_2 subfraction. Following therapy, HDL_2 levels are restored to normal.

Although abnormalities in the concentration and metabolism of cholesterol are considered to be typical for patients who have thyroid disease, changes also may occur in the concentration and metabolism of VLDL, and hence triglyceride. Hypothyroidism is associated with decreased LPL activity, with consequent decreased clearance of VLDL,

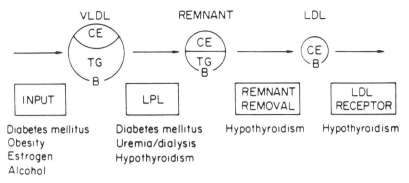

Figure 8.1. Sites of defective lipoprotein metabolism in secondary dyslipoproteinemias. The sites of defects for drugs that affect lipoprotein metabolism (e.g., diuretics, beta-blockers, retinoids) and some of the disorders described in the text are not known with certainty. CE = cholesteryl esters; B = apolipoprotein B; LPL = lipoprotein lipase. (From: Chait A, Brunzell JD. Acquired hyperlipidemia (secondary dyslipoproteinemias). Endocrinol Metab Clin North Am 1990;19:259–278.)

and increased blood triglyceride. Treatment with thyroxine restores LPL activity to normal, thereby reducing the plasma triglyceride concentration. The increase in blood triglyceride that occurs usually is modest, with values in the 200-to-400 mg/dL (2.26 to 4.52 mmol/L) range in patients on a standard diet. Patients who have hypothyroidism and who are on a high-fat diet may develop substantially more hypertriglyceridemia under such circumstances.

In hyperthyroidism, LPL activity is higher than normal and VLDL clearance is accelerated. The concentration of triglyceride remains normal or slightly increased, however, because of increased VLDL synthesis that quantitatively balances or exceeds the increase that occurs in LPL activity.

Diabetes Mellitus

Lipid abnormalities are common in diabetes mellitus, particularly noninsulin-dependent diabetes mellitus (NIDDM). Patients who have NIDDM classically have elevated concentrations of VLDL with an increase in the blood triglyceride level, often accompanied by reduced HDL-C (Table 8.3). The total cholesterol and LDL-C concentrations are usually normal. It should be emphasized, however, that normal concentrations of blood lipids and lipoproteins do not necessarily indicate that lipoprotein production or metabolism is normal, particularly

when diabetes mellitus is associated with obesity. Dynamic alterations in lipoprotein interconversion and in lipoprotein composition may occur in these disorders, even in the absence of evident dyslipidemia (*see* Fig. 8.1). Blood concentrations of apoproteins B and E often are increased, and those of apoprotein A-I are decreased. The presence of smaller and denser LDL particles (*see* Chapter 4) has been demonstrated in hypertriglyceridemic patients who have diabetes or impaired glucose tolerance.

Relation of Dyslipidemia to Insulin and Insulin Resistance. Patients predestined to develop diabetes mellitus and patients who have impaired glucose tolerance or mildly elevated fasting glucose concentrations (150 to 180 mg/dL; 8.32 to 9.99 mmol/L) frequently have increased blood insulin levels because of peripheral insulin resistance. This is associated with increased hepatic synthesis of VLDL, leading to hypertriglyceridemia (Fig. 8.2). There usually is a direct relation between the degree of hyperinsulinemia and the VLDL-triglyceride secretion rate. Although the increase in hepatic VLDL triglyceride synthesis and secretion usually is attributed to the stimulatory effects of insulin, many of these patients actually have resistance to insulin at the level of the liver. It is more likely that the hypertriglyceridemia is caused by increased mobilization of free fatty acids and glycerol from

Table 8.3. Dyslipidemia in NIDDM

Lipid Levels (mg/dL)	LRC Cut-off Value[a]	Nondiabetic (%)	Diabetic (%)
Men			
Total cholesterol	≥260	14	13
Triglycerides	≥235	9	19
VLDL	≥ 40	26	34
LDL	≥190	11	9
HDL	≤ 31	12	21
Women			
Total cholesterol	≥275	21	24
Triglycerides	≥182	8	17
VLDL	≥ 35	31	38
LDL	≥190	16	15
HDL	≤ 41	12	25

[a] Approximate 90th percentile age- and sex-matched values from Lipid Research Clinic (LRC) tables, except for HDL values, which are the 10th percentile.
(Adapted from: Garg A, Grundy SM. Management of dyslipidemia in NIDDM. Diabetes Care 1990;13:153–169.)

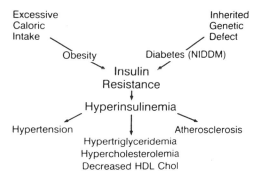

Figure 8.2. Syndrome of insulin resistance. Metabolic cascade that leads from acquired (obesity) or inherited (noninsulin-dependent diabetes mellitus; NIDDM) insulin resistance to hyperinsulinemia and eventually to hypertension, abnormal plasma lipid profile, and atherosclerosis is depicted. (From: DeFronzo RA, Ferrannini E. Insulin resistance: a multifaceted syndrome responsible for NIDDM, obesity, hypertension, dyslipidemia, and atherosclerotic cardiovascular disease. Diabetes Care 1991;14:173–194.) "Reprinted with permission."
"The American Diabetes Association takes no responsibility for the accuracy of the translation from English."

peripheral tissues to the liver, providing more substrates for hepatic triglyceride synthesis.

The blood triglyceride concentration varies directly with hepatic triglyceride secretion rate until the concentration reaches approx. 500 mg/dL (5.65 mmol/L). Additional increases in triglyceride concentrations usually reflect a defect in VLDL

removal. At high triglyceride concentrations, then, there is defective clearance of triglyceride, as well as increased production.

The reduced concentrations of HDL-C and apoprotein A-I that are commonly observed in NIDDM usually are associated with decreased VLDL clearance and increased fractional catabolic disposal of HDL. The HDL in NIDDM often is enriched in triglyceride (*see below*), making it a better substrate for the lipolytic action of hepatic triglyceride lipase. It follows that the hypo-HDL can be caused by decreased synthesis as well as by increased HDL catabolism.

Abnormalities in Lipoprotein Composition. Compositional lipoprotein abnormalities also occur in patients who have diabetes. There is an increased transfer of cholesteryl esters from HDL to VLDL and VLDL remnants, and the free cholesterol content of VLDL and LDL is increased. At the same time, the transfer of triglyceride from VLDL to HDL is increased. Consequently, hypertriglyceridemic patients who have NIDDM have cholesteryl-ester-enriched VLDL and IDL and triglyceride-enriched LDL and HDL particles. In addition, the LDL particles formed are smaller and denser, which tends to make them more atherogenic than the less dense LDL species.

The content of apoprotein E in VLDL is increased in hypertriglyceridemic diabetic patients. This increases the capacity of

VLDL and its remnants to bind to the LDL receptor, leading to its downregulation. The reduced receptor concentration may lead to hypercholesterolemia in some patients. In addition, the increased apoprotein E on VLDL (and its remnants) may increase its affinity to the scavenger receptor, which is thought to be responsible for the development of atherosclerosis. In diabetic patients (particularly women), increased VLDL and triglyceride levels may be important risk factors for the development of vascular disease.

Abnormalities in Lipoprotein Removal. In mild diabetes mellitus, the excessive synthesis of VLDL is associated with an increased conversion of VLDL to LDL; however, there also is an increase in the fractional catabolic removal of LDL, and the LDL-C and apoprotein B levels remain normal despite increased traffic through the pathway.

When diabetes mellitus becomes severe, the increase in VLDL synthesis is no longer accompanied by increased LDL synthesis. This probably reflects impairment in LPL activity associated with the deficiency of insulin. Normal glucose homeostasis and probably insulin concentrations are required for normal function of lipoprotein lipase. In its extreme form, these defects could lead to accumulation of chylomicrons and VLDL, with the development of the chylomicronemia syndrome, which may be associated with triglyceride concentrations of ≥ 1000 mg/dL (11.29 mmol/L) and specific signs and symptoms. Despite the decrease in conversion of VLDL to LDL, a coexistent reduction in the fractional catabolic removal of LDL, caused by increased VLDL apoprotein E (as previously described) and insulin deficiency, leads to an increase in LDL-C levels.

The problems of lipoprotein removal may be aggravated even more in diabetes mellitus by the glycosylation of the apoproteins A-I, A-II, B, C-I, and E. Such glycosylation may interfere with specific functions necessary for normal lipid transport. For example, glycosylation of apoprotein B impairs its binding to the LDL receptor, and this impedes its removal from the bloodstream even more.

Relation to Glycemic Control. In well-controlled, insulin-dependent diabetes (IDDM), lipid and lipoprotein concentrations may be within normal limits. When the diabetes is poorly controlled, the inhibition of lipoprotein lipase results in the development of hypertriglyceridemia, as previously described. Aggressive therapy frequently leads to normalization of lipoprotein lipase activity and a prompt reduction in VLDL, LDL, and apoprotein B levels.

In contrast, restoration of normal lipoprotein lipase activity by insulin in patients who have NIDDM is a slow process, and hypertriglyceridemia and reduced HDL-C levels correct slowly. Nonobese patients who have NIDDM may achieve normal lipid and lipoprotein levels by tight control of their glucose levels. In contrast, dyslipidemia often persists in the obese diabetic despite good control, suggesting that obesity and other coexistent problems contribute to the dyslipidemia.

The presence of markedly elevated triglyceride concentration (greater than 400 to 800 mg/dL [4.52–9.03 mmol/L]) or elevated LDL-C in patients who have noninsulin-dependent diabetes should raise the suspicion of an associated form of familial or secondary dyslipidemia. Hypertriglyceridemia, elevated LDL-C, and reduced levels of HDL-C appear to be more frequent in women than in men in both insulin-dependent and noninsulin-dependent diabetes. The presence of a markedly increased LDL-C in a patient who has NIDDM may also indicate the development of the nephrotic syndrome.

It should be clearly understood that diabetes mellitus should be controlled optimally to reverse any related lipid abnormalities. Nonetheless, because many patients never achieve such glycemic control, persistent lipid abnormalities require treatment in their own right. Such treatment actually may be simpler and more effective than control of hyperglycemia, and should be initiated when indicated. Furthermore, patients who have diabetes mellitus also may have lipid abnormalities unrelated to their diabetes and for which therapy is indicated. If glycemic control is not achieved after 6 months or if lipid abnormalities persist, then specific hypolipidemic therapy should be initiated.

The principles of treatment of dyslipidemia in diabetic patients are discussed in Chapter 12.

Obesity

Obesity can lead to adverse lipoprotein metabolism that results in abnormalities of VLDL, LDL, and HDL. These adverse effects depend not only on the severity of obesity but also on individual genetic regulation of lipoprotein metabolism and distribution of body fat. The predominant effect is an overproduction of VLDL particles related to the insulin resistance and hyperinsulinemia usually associated with obesity, even when glucose tolerance is normal. The increased availability of free fatty acids and glycerol that occur in obesity, especially central obesity, support an increased secretory rate of apoprotein-B-containing lipoproteins (*see* Fig. 8.1).

Overproduction of VLDL particles can result in an elevated triglyceride concentration, the most common lipid abnormality found in obese persons. This overproduction of VLDL frequently is countered, however, by an increase in LPL activity and enhanced lipolysis of VLDL-triglyceride, so that the blood triglyceride level does not become elevated. The conversion of VLDL to VLDL remnants and LDL is augmented, however, and the resultant blood LDL-C levels will depend on the rate of removal of LDL particles. Because obesity usually is accompanied by a compensatory increase of LDL removal, the blood LDL-C concentration often remains normal. Persons who do not have this compensatory mechanism can develop hypercholesterolemia. Another consideration is that the rapid lipolysis of VLDL leads to high cholesterol levels and formation of smaller, more dense LDL particles, which are potentially more atherogenic. Even normal lipoprotein levels in obesity may be associated by abnormal lipoprotein metabolism, emphasizing that measurement

of concentrations under these circumstances can be misleading.

Although there is a direct relation between the degree of obesity and the total cholesterol concentration, this association is not as strong as that which exists between obesity and triglyceride concentrations. Weight loss in obese patients can be expected to lower the cholesterol level, especially when associated with a decrease in the dietary saturated fat and cholesterol content. Peculiarly, some patients seem to have an unexplained increase in total cholesterol when they lose weight. Whether this is caused by increased dietary fat and cholesterol or by some unique metabolic abnormality is unknown.

Obesity is one of the many conditions associated with a low HDL-C concentration. Whether this is caused by the abnormalities in VLDL metabolism and the ensuing hypertriglyceridemia that exists in such patients or whether increased numbers of HDL receptors in adipose tissue reduce the blood levels of HDL is not known. In any case, weight loss should be expected to improve or correct not only the hypertriglyceridemia but also the low HDL-C concentrations (*see* Chapter 10).

Finally, all things being equal, those patients who have upper body (abdominal) obesity have greater hypertriglyceridemia and lower HDL-C levels than similarly obese patients who have lower body (peripheral) obesity. The topography of obesity is important with regard to insulin resistance, metabolic abnormalities, hypertension, and the risk of developing CHD (*see* Fig. 8.2).

Hypogonadism

Men who have hypogonadism and androgen deficiency have increased concentrations of total and LDL-C, with marginal reductions in HDL-C, usually resulting in an increased LDL/HDL-C ratio. The magnitude of the LDL-C elevation is approx. 40 mg/dL (1.03 mmol/L), which could raise its level from a mean of 130 mg/dL (3.36 mmol/L) to the borderline elevated range around 160 mg/dL (4.14 mmol/L). In addition, triglyceride concentrations also are increased. Androgen replacement therapy will optimize the lipoprotein concentrations when used in physiologic doses.

Infections

Hypertriglyceridemia, often marked, may be seen in patients who have severe gram-negative infections, both with and without bacteremia. Gram-positive bacteremia and viral infections also are associated with modest changes in VLDL-C and triglyceride, but such changes seldom produce striking abnormalities. Concentrations of LDL-C decrease promptly with bacterial and viral infection, however, and they remain reduced for variable periods. Cholesterol concentrations can be reduced by 20% to 25% in patients who have gram-positive infections but not in patients who have gram-negative infections. Because of these changes, blood lipid levels should be interpreted cautiously in any patients who have active infection.

Patients who have HIV infection also have higher triglyceride and lower cholesterol concentrations than noninfected patients. Around 50% of HIV-infected patients have triglyceride concentrations in excess of 190 mg/dL (2.15 mmol/L). Although the severity of hypertriglyceridemia is greater in patients who have AIDS than those who have positive HIV serology, the prevalence is the same in both. The mechanism of the hypertriglyceridemia in AIDS is not understood but may be related to the release of various cytokines derived from lymphocytes and macrophages; these have been shown to inhibit triglyceride use and increase its production by the liver. Interestingly, the hypertriglyceridemia in these patients does not appear to correlate with the activity of the disease or with nutritional parameters.

Liver Disease

Of all the organs in the body, the liver probably has the most central role in lipoprotein metabolism. Very low density lipoproteins are synthesized in the liver, secreted into the bloodstream, and converted to IDL and LDL. Although all tissues and cells probably have LDL receptors on their surface, the liver is quantitatively the most important, containing approximately 75% of all LDL receptors in the body. Hepatic triglyceride lipase, localized in hepatic endothelial cells, plays a unique role in the conversion of IDL to LDL and the regulation of HDL subgroup concentrations. In addition, the liver is an important site of HDL synthesis. Because lecithin cholesterol acyltransferasae (LCAT), the circulating enzyme solely responsible for all lipoprotein cholesteryl esterification, is synthesized only in the liver, cholesterol esterification in the HDL particle also is controlled indirectly by hepatic function. Consequently, it is not surprising that hepatic dysfunction leads to numerous lipoprotein and lipid abnormalities.

Acute hepatocellular disease often is associated with hypertriglyceridemia and decreased HDL-C levels. This may result from decreased HTGL activity, with subsequent accumulation of triglyceride-rich IDLs and VLDL remnants. The persistence of these remnants in the circulation facilitates the transfer of triglyceride from the remnants to HDL (as well as transfer of cholesteryl esters in the opposite direction). The resulting triglyceride-enriched HDL is a better substrate for HTGL, leading to lower HDL-C concentrations. The HDL metabolism may be altered even more by decreased production and release of LCAT. This produces a defect in cholesteryl esterification in HDL that may accentuate compositional abnormalities even more.

In contrast, cholestatic liver disease often is associated with hypercholesterolemia. The plasma of such patients contains an abnormal lipoprotein called lipoprotein-X, or LP-X. This lipoprotein appears as a bilamellar disk by electron microscopy and contains phospholipid, free cholesterol, apoprotein C, and minor amounts of triglyceride; it may be responsible for the hypercholesterolemia found in patients who have cholestasis. Unfortunately, lipoprotein-X can occur with intrahepatic and extrahepatic cholestasis and, consequently, cannot be used to distinguish between them.

Renal Disease

Nephrotic Syndrome. Patients who have the nephrotic syndrome also have increased blood cholesterol and triglyceride levels (Fig. 8.3), which appear to be related inversely to the blood albumin. The LDL-C is increased whereas HDL-C often is decreased, accompanied by abnormalities in lipoprotein composition. These lipoprotein abnormalities are thought to increase the susceptibility of such patients to CHD.

The dyslipidemia in patients who have the nephrotic syndrome is not necessarily related to the cause of the nephrotic syndrome and may be associated with all lipoprotein phenotypes, except Type I. The majority of patients have lipid abnormalities–either hypercholesterolemia, hypertriglyceridemia or a mixed dyslipidemia. The dyslipidemia is a result of increased synthesis of apoprotein-B-containing lipoproteins, with increased conversion of apoprotein B remnants to LDL. The elevated triglyceride concentration also may reflect hyperinsulin-

emia, caused by acquired insulin resistance, and increased rates of VLDL secretion. Kinetic studies indicate that the overproduction of lipoproteins returns to normal after recovery.

Chronic Renal Failure. Significant lipid abnormalities often exist in patients who have chronic renal failure. Hypertriglyceridemia characterizes 20% to 70% of dialyzed and undialyzed patients who have chronic renal failure and is the result of a combination of increased VLDL secretion and reduced rates of VLDL removal (*see* Fig. 8.1). The predominant lipoprotein phenotype is hypertriglyceridemia or Type IV. Increased concentrations of LDL-C usually do not occur unless chronic renal failure is complicated by the nephrotic syndrome.

The defects in the lipolytic cascade arise from reductions in LPL and HTGL activities. The extent of the hypertriglyceridemia also relates to the postprandial insulin concentrations, reflecting the acquired insulin resistance of chronic renal failure. These defects lead to increased concentrations of VLDL and VLDL remnants (intermediate density lipoproteins) and to triglyceride en-

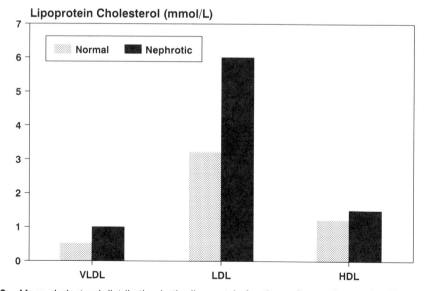

Figure 8.3. Mean cholesterol distribution in the lipoprotein fractions of sera of normal subjects and patients who have the nephrotic syndrome. (Adapted from: Joven J, Villabona C, Vilella E, Masana L, Alberti R, Valles M. Abnormalities of lipoprotein metabolism in patients with the nephrotic syndrome. N Engl J Med 1990; 323:579–584.)

richment of all lipoproteins. Following a meal, there are increased concentrations of chylomicron remnants. These defects are present in undialyzed and dialyzed patients who have nonnephrotic renal failure.

In general, HDL-C concentrations are reduced in quantity and are abnormal in composition in chronic renal failure. These abnormalities persist after institution of chronic hemodialysis (as well as peritoneal dialysis). The ratio of LDL:HDL is increased in these patients, which supposedly predisposes them to an increased risk of developing vascular disease. The function of the HDL particle also is defective in patients on hemodialysis, probably because of a reduction in circulating LCAT activity. Consequently, cholesteryl esterification is reduced by approximately 66%; and the transfer of cholesteryl esters from HDL to VLDL and LDL is reduced by approximately 80%. This defect often reverses to normal in patients who undergo peritoneal dialysis.

Renal Transplant. Renal transplant is being used increasingly to replace long-term dialysis in patients who have endstage renal failure and often results in the correction of many of the metabolic abnormalities associated with uremia and dialysis. The lipid abnormalities often persist, however, and sometimes even worsen. A common pattern observed in many studies is an improvement (but not normalization) in the hypertriglyceridemia accompanied by an increase in blood cholesterol levels and HDL-C. This usually is attributed to the accompanying changes in nutritional, hormonal, and metabolic status of these patients and, perhaps more importantly, to the use of immunosuppresive medications such as corticosteroids and cyclosporine. Lipid elevations often are correlated to the corticosteroid dose and have been reduced by spacing the prednisone dose to alternate days in some, although not all studies.

These data indicate that significant abnormalities of lipoprotein metabolism exist in patients who have the nephrotic syndrome, in patients who have endstage renal disease, and in renal transplant recipients. Even

when triglyceride and cholesterol concentrations are normal in patients who undergo hemodialysis, abnormal lipoprotein composition may predispose to the development of vascular disease. Although the extent to which these lipoprotein abnormalities contribute to the increased incidence of CHD observed in patients who have renal disease is still debated, attempts at minimizing such abnormalities should be contemplated seriously in patients who have chronic renal disease. The abnormalities appear to be less severe in patients who receive peritoneal dialysis, because many of the measured parameters are normal or intermediate between those observed in patients who undergo hemodialysis and in normal subjects. In addition to an increased susceptibility to atherosclerosis, questions are now being raised about whether lipid abnormalities in patients who have underlying renal disease could predispose to, or aggravate, glomerular injury and the development of focal segmental glomerulosclerosis.

Pancreatitis

The relation between hypertriglyceridemia and episodes of abdominal pain is well known and usually is attributed to pancreatitis. In one study approximately 20% of patients who had acute pancreatitis also had lactescent serum with triglyceride concentrations ranging from 490 to 7500 mg/dL (5.53 to 84.68 mmol/L). Pancreatitis occurs with primary as well as with secondary causes of hypertriglyceridemia. Although the older literature suggested that the hypertriglyceridemia was the result of pancreatitis, most recent studies indicate that it actually is the cause of the pancreatitis. Recurrent episodes of pancreatitis, with typical pain and hyperamylasemia, can occur when triglyceride levels are $\geq$ 1000 mg/dL (11.29 mmol/L). Such patients usually have an increase in chylomicrons and VLDL (Type V) or, occasionally, just chylomicrons (Type I). Pancreatitis generally does not result from increased VLDL alone (Type IV). Abdominal pain, similar to but not as severe as that experienced during spontaneous attacks of pancreatitis, often occurs with triglyceride concentrations $\geq$ 600 mg/dL (6.77 mmol /L). At these levels paresthesias and impaired cognitive functioning also have been repeated.

The frequency of pancreatitis is reduced substantially when triglyceride levels are lowered. The relation between chylomicronemia and pancreatitis may sometimes be obscured by the disease itself. For example, triglyceride concentrations may not be elevated impressively in some patients who have pancreatitis in whom hypertriglyceridemia provoked the attack, because the associated pain and nausea prevented food intake and allowed the triglyceride levels to decrease. A high index of suspicion, therefore, is necessary in such patients.

Estrogen supplementation and treatment of skin disorders with oral retinoids have induced pancreatitis in occasional patients who have subtle familial Type V hyperlipoproteinemia. In view of the ability of these drugs to increase VLDL synthesis and cause hypertriglyceridemia, this is not unexpected. Nonetheless, it is additional evidence to support the proposal that hypertriglyceridemia causes pancreatitis. It is also well known that hypertriglyceridemia causes abdominal pain in patients who have preexisting alcoholic pancreatitis when they are placed on a high-fat diet.

Rheumatoid Arthritis and Chronic Inflammatory Arthritides

In patients who have rheumatoid arthritis (RA) and seronegative spondyloarthropathies, the blood concentrations of triglyceride, LDL-C, and HDL-C are reduced. Despite reduced LPL activity, removal of circulating VLDL is more rapid. The fractional triglyceride removal rate correlates directly with the C-reactive protein concentration, suggesting that the degree of inflammation is a direct determinant of lipid levels. The relation between disease activity and the rate of triglyceride removal suggests that VLDL particles may be altered by the inflammatory process or that the scavenger pathway, residing in monocytes/macrophages, is activated and results in an increased rate of VLDL removal. Lipid levels also may be influenced in RA and other connective tissue diseases by concomitant therapy (NSAIDS, chloroquine, corticosteroids), anorexia, and changes in activity.

Systemic Lupus Erythematosus

Dyslipoproteinemia occurs frequently in patients who have SLE. The concentrations of triglyceride, cholesterol, and LDL-C are higher than those of healthy controls as well as of patients who have inflammatory arthritis. These changes probably are attributable to glucocorticoids, because patients who have SLE and who do not receive glucocorticoids have lipid levels similar to those of controls (with the exception of a lower LDL-C). The HDL$_2$ levels are significantly lower than those of controls. Rarely, antibodies may develop and be directed against LPL or the LDL receptor and lead to a striking dyslipidemia.

Drug-Induced Lipid Abnormalities

Drugs used to treat various other medical illnesses can alter plasma lipoprotein levels. The significance of such changes is unknown and controversial. It is not known whether the changes induced by the drugs are qualitatively similar to and have the same implications for CHD as intrinsic lipid abnormalities.

Antihypertensive Agents

Certain antihypertensive agents impact adversely on blood lipoprotein and lipid concentrations. In general, thiazide diuretics, chlorthalidone, and loop diuretics have deleterious effects on plasma lipid concentrations, whereas the effect of potassium-sparing diuretics is unclear. Diuretics increase the total cholesterol, LDL-C, and apoprotein B levels, with variable effects on HDL-C. Triglyceride concentrations generally increase because of the greater secretion of VLDL. The mechanism for the change in VLDL synthesis is not known, but increased insulin levels are associated with increased hepatic synthesis of VLDL.

The dyslipidemic effect of diuretics appears to be unequivocal, although data on the long-term persistence of lipid-induced changes are inconsistent. There appear to be no major differences in response at the usual dosage levels. Dietary modification can alleviate the adverse lipid effects of diuretics and deserves more consideration. Inconclusive data indicate the diuretic, indapamide, increases the total cholesterol and HDL-C concentrations but does not change or reduce the LDL-C. Scant attention has been paid to diuretic drugs unrelated chemically to the thiazides. The effects of diuretics and other drugs thought to influence lipids are given in Table 8.4.

Beta-blockers of the beta-1 (*selective*) and beta-2 (*nonselective*) types often raise the triglyceride and reduce the HDL-C levels. Occasionally, the elevation in triglyceride concentration is striking, but most agents increase triglyceride approximately 15% to 20%. The concomitant reduction of HDL-C is approximately 10%. Beta-blockers usually do not change the concentrations of total cholesterol and LDL-C. It is claimed that cardioselective agents at lower doses have less impact on blood lipids, but little comparative data are available to support these

Table 8.4. Effects of Drugs on Lipids and Lipoproteins

	TC	LDL-C	HDL-C	TG
Thiazide diuretics	↑[a]	↑[b]	→	↑
Loop diuretics	↑	↑	→	↑
Beta-blocker without ISA activity	→[a]	→	↓	↑
Beta-blocker with ISA activity	↓[a]	↓	↑→	→
Terbutaline	→	→	↑	↓
Calcium blockers	→	→	→	→
ACE inhibitors	→	→	→	→
Alpha-blockers	↓	↓→	↑→	↓
Isoretinin	↑	↑	↓	↑
Cyclosporine	↑	↑	→	↑

[a] ↑ = elevated → = no effect ↓ = decrease
[b] Arrows in bold indicate important adverse effects.

statements. Beta-blockers that have intrinsic sympathetic activity (ISA) generally increase the HDL-C concentration and might be considered as alternatives for other beta-blockers in persons who have low HDL-C levels if the clinical indication is appropriate. By way of contrast to beta-blockers, beta-adrenergic agonists (e.g., isoproterenol) increase HDL-C concentrations, although their effects on LDL and triglyceride are variable.

Alpha-blockers (prazosin, terazosin) have a much more favorable effect on lipids, usually being associated with little change or a mild increase in the HDL-C concentration and no change (or a reduction) in triglyceride, total cholesterol, and LDL-C concentrations. Limited information is available on their long-term effects.

Lipoprotein lipase activity is influenced by the balance between alpha- and beta-receptor activity and probably explains, in part, how alpha- and beta-blockers influence blood lipid levels. Stimulation of the beta-receptor activates lipoprotein lipase, whereas stimulation of the alpha-receptor inhibits lipoprotein lipase. Consequently, the use of a beta-blocking drug leaves the alpha-receptor's inhibiting action unopposed, leading to reduced LPL activity and hypertriglyceridemia. Conversely, the use of an alpha-blocking agent leads to unopposed stimulation of the beta-receptor, activation of LPL, and a reduction in triglyceride concentration. The changes in HDL-C levels with these drugs may be caused by the reciprocal nature of the relation between triglyceride and HDL-C levels or the opposing effects of alpha- and beta-receptor stimulation on LPL and HTGL activities.

On the basis of data from multiple controlled trials, it appears that calcium channel blockers and angiotensin converting enzyme inhibitors do not affect lipid metabolism. Consequently, these and alpha-blockers may be the desired drugs in hypertensive patients who have coexistent lipoprotein/lipid abnormalities. Still unresolved is the long-term role of beta-blockers with ISA or alpha-1 adrenergic blocking properties in producing lipo-protein changes.

It has been suggested that the adverse lipid effects of diuretics may be partly re-sponsible for the lack of benefit in reducing CHD mortality in patients who have mild hypertension. In this setting, the reduction in blood pressure produced by the diuretic is relatively less important with regard to CHD risk than the increase that occurs in the con-centration of total cholesterol and LDL-C. As a result, the trade-off for the use of this type drug in patients who have mild hyper-tension, in whom the risk of complications is relatively low, is unfavorable. In population studies, it has been demonstrated that an approx. 5-mm decrease in diastolic blood pressure corresponds to a 14% reduction in CHD, and a 1% decrease in total cholesterol corresponds to a 2% reduction in CHD. In patients who have moderate to severe hy-pertension, however the reduction in risk that occurs as a consequence of blood pres-sure control more than outweighs any in-crease in risk that might occur from unfavor-able changes in lipids or glucose.

Frequently hypertension and dyslipi-demia occur together. As a general rule, it is prudent to consider the effect of the drug on lipids and to prefer the use of lipid-neutral antihypertensive agents when initiating therapy of hypertension; however, cost and adherence considerations may be important determinants of drug choice. Diuretics and reserpine certainly are the least expensive forms of therapy for hypertension, and both have been demonstrated to be beneficial in reducing blood pressure and preventing stroke. Furthermore, in patients with dyslip-idemia who require lipid-lowering therapy the relatively small lipid changes induced by an antihypertensive drug may be inconse-quential. Because of the individual variation in response to drugs, it may be appropriate to temporarily discontinue the antihyper-tensive agent to judge the effect on blood lipoproteins before deciding on a final ther-apeutic regimen.

Various drugs, other than antihyperten-sives, have been reported to alter lipid and lipoprotein levels. These are summarized in Table 8.4.

Hormones

Androgens, estrogens, glucocorticoids, and thyroid hormone can alter lipoprotein levels and the concentrations of cholesterol and triglyceride (Table 8.5).

Androgens and Anabolic Steroids. Ana-bolic steroids and androgens are associated with a reduction of HDL-C, which is caused principally by a reduction in the levels of HDL_2 and apoprotein A-I. These changes are most prominent when androgens and anabolic agents are taken orally and large quantities of these steroids reach the liver but are smaller when androgens are admin-istered parenterally in doses that correct hy-pogonadism. The presence of exceptionally low HDL-C levels in a man, and probably in a woman, should suggest possible use (abuse) of anabolic steroids for the athletic performance-enhancing effect. Hepatic tri-glyceride lipase activity is increased by ana-bolic steroids and is responsible for the reduction in HDL-C concentrations that exist in persons treated with anabolic ste-roids. Such effects are reversible and disap-pear within 5 weeks of the discontinuation of the anabolic steroids. These lipid changes tend to be markedly proatherogenic, how-ever, and the use of such drugs in persons at high risk for atherosclerosis should be discouraged.

Estrogens and Progestins. Estrogens and progestins generally have opposite ef-fects on lipoprotein metabolism; when taken together, the composite effect depends largely on the relative strength of the indi-vidual preparations and their mode of administration.

Estrogens appear to alter the transport of VLDL, LDL, and HDL because of their pro-found effect on the liver, resulting in reduc-tions in LDL-C concentrations and increases in HDL-C and triglyceride concentrations. When the liver is bypassed, as with trans-dermal, intramuscular, or subcutaneous es-trogen administration, only minor changes, if any, may be noted.

Table 8.5. Effects of Hormones on Lipids

	TC	LDL-C	HDL-C	TG
Estrogens				
Transcutaneous	↓(− 3%)[a]	↓(−11%)	↑(+ 4%)	0[b]
Oral	↓(−10%)	↓(− 20%)	↑(+19%)	↑(+14%)[e]
BCP[d] (estrogen dominant)	→	↓	↑	↑
BCP (progestin dominant)	—	—	↓	—[c]
Progestins				
Norethindrone	—	↑	↓	↓
Norethindrone acetate	—	↑	—	→
Levonorgestrel	—	↑	↓	→
Androgens	↑	↑	↓	↓

[a] ↓ = decrease　→ = no effect　↑ = elevated
(　) approx. changes, where known
[b] 0 = no change
[c] — = no information
[d] BCP = birth control pills
[e] Arrows in bold indicate important adverse effects.

Estrogens increase hepatic synthesis of VLDL and inhibit VLDL removal by LPL (*see* Fig. 8.1). Susceptible persons, especially those who have a subtle or asymptomatic defect in VLDL clearance, can develop severe hypertriglyceridemia. The relation between estrogens and VLDL production appears to be dose-dependent. Because VLDL also transports cholesterol, its accumulation may lead to hypercholesterolemia; however, estrogens stimulate the synthesis of the LDL receptor and reduce LDL-C levels by 15% to 20%. Estrogens also increase the concentration of HDL-C, primarily HDL$_2$, by inhibiting hepatic triglyceride lipase. Studies suggest that estrogens increase HDL by reducing its catabolic rate rather than by increasing its synthesis. The effects of estrogens on LDL-C and HDL-C are opposite to those produced by anabolic steroids and certain progestins and may be one mechanism by which postmenopausal women benefit from estrogen therapy.

The progestins used for hormonal treatment are derived from either 19-nortestosterone (19-nor progestins), 17-alpha-hydroxyprogesterone (C-21 progestins), or natural progesterone. Although there are other indications for progestins, the most common use involves their combination with estrogens for contraception and for postmenopausal replacement therapy. Generally, the C-21 progestins show much smaller metabolic effects than the C-19 preparations. Studies with natural progesterone or with the newer progestins do not show any appreciable lipoprotein effect. Synthetic progestins display not only progestational activity but also estrogenic, androgenic, or antiandrogenic activity. The effects of progestins on lipid metabolism are related primarily to their intrinsic androgenicity and, therefore, tend to elevate LDL-C and reduce HDL$_2$-C in a dose-related way.

When a progestin is used in combination with an estrogen, such as in oral contraceptives and in postmenopausal replacement therapy, the overall effect on lipids will depend on the balance between estrogenic and androgenic properties. For that reason, generalizations about the changes that oral contraceptives produce are impossible, because each combination will be different. Those birth control pill combinations that use progestational agents that have strong androgenic properties, such as levonorgestrel, are likely to result in an increase in the LDL-C and a reduction in the HDL-C concentrations. Birth control pills that contain low-dose norethindrone (Modicon, Brevicon, and Ortho-Novum 7/7/7) are associated with the most favorable lipid profiles.

The use of medroxyprogesterone acetate (Provera), a C-21 progestin often used in postmenopausal women to protect the endometrium, may also offset the lipoprotein benefits of estrogen therapy; however, this effect and its significance has been questioned recently. Epidemiologic data suggest that long-term medroxyprogesterone acetate therapy neither increases LDL-C nor lowers HDL-C levels, moreover studies of oral contraceptives in primates suggest that estrogens confer a benefit on the cardiovascular system even when progestational agents produce adverse changes in HDL-C levels.

Both case control and cohort studies indicate that postmenopausal estrogens reduce the relative risk of atherosclerotic vascular disease by about 50%. This protection is associated with lower body weight, lower systolic and diastolic blood pressure, reduced LDL-C, and increased HDL-C. These changes probably are responsible for much of the reduction in cardiovascular disease reported in postmenopausal women who are treated with estrogen.

Glucocorticoids.   In the absence of serious coexistent systemic disease, glucocorticoids increase the levels of all the lipoproteins. Under the usual conditions in which glucocorticoids are often used, however, the resultant increases in VLDL, LDL-C, and HDL-C often are obscured by use of other drugs (such as cyclosporine, azothioprine), conditions (organ transplantations), or diseases (collagen-vascular disease, renal disease, or steroid-induced diabetes mellitus with its own set of lipid abnormalities).

Glucocorticoids increase the synthesis of VLDL by the liver and reduce the usual compensatory increase in LPL activity. The net effect is to produce hypertriglyceridemia. Marked increases in VLDL also may increase the plasma total cholesterol concentration because of the cholesterol transported by VLDL. Because LPL activity also increases, even though it is suboptimal, there is increased conversion of VLDL to IDL and LDL, with consequent elevation in LDL-C levels. The increased HDL levels probably are related to inhibition of hepatic triglyceride lipase activity.

The increase in cholesterol concentration is related directly to the dose of prednisone that is used; antiinflammatory doses of prednisone may increase the total cholesterol by about 15%, LDL-C by about 10%, HDL-C by about 30%, and triglyceride by about 50%. Whether hypertriglyceridemia occurs with glucocorticoid use may depend on other coexistent metabolic abnormalities. A predisposition to hypertriglyceridemia from other medical problems, such as, obesity, insulin resistance, diabetes mellitus, other hereditary abnormalities of VLDL (triglyceride) metabolism, ethanol, and the like may be necessary for such persons to develop more substantial hypertriglyceridemia accentuated by glucocorticoids. The disease process for which the glucocorticoids are being used also may modify the lipoprotein levels or their response to glucocorticoids.

Whether glucocorticoid-induced changes in lipoproteins contribute to the development of atherosclerosis in Cushing disease, in renal transplant patients, and in patients who have connective tissue diseases is unclear, but because these conditions are associated with an increased predisposition to vascular disease, it is not unreasonable to treat identifiable lipid risk factors as would be done in any other patient. The use of alternate-day steroid schedules may have a lesser effect on lipoprotein levels, although the evidence for such benefit is still being debated.

Ethanol

The most common lipid/lipoprotein abnormality associated with the use of ethanol, in the absence of significant liver disease, is hypertriglyceridemia. Although this is caused by both increased synthesis of VLDL and a decreased removal by lipoprotein lipase, the increased VLDL synthesis dominates in drinkers who continue to eat food. This pattern does not prevail in the true alcoholic but rather often characterizes those persons who use liberal amounts of alcohol chronically without intoxication.

Ethanol also increases HDL-C in a dose-dependent fashion, even as it causes hypertriglyceridemia. It is one of the few exceptions to the rule that there is an inverse relation between triglyceride and HDL-C levels. As noted previously, other exceptions include treatment with estrogens or corticosteroids. The selective effect of alcohol on HDL subspecies has attracted much attention, because it has been reported that HDL_3 increases preferentially, and since it has been stated that HDL_2-C is the cardioprotective subspecies, the selective increase in HDL_3-C does not appear to explain the reduced incidence of CHD associated with moderate alcohol drinking. Recent data suggest that HDL_3-C also is cardioprotective; alternatively, it is possible that alcohol reduces CHD by mechanisms other than its effect on lipoproteins. Although this issue is still unresolved, most recent studies suggest that the HDL_3 level increases at mild-moderate doses, whereas the HDL_2 level increases at large doses.

Other Drugs

Patients who have convulsive disorders appear to have a decreased incidence of CHD. Although such findings have been based on nonrandomized studies, the bulk of evidence suggests that antiepileptic drugs, perhaps with the exception of valproate, increase HDL-C and apoprotein A-I. This effect of raising HDL seems to be dose-related. The mechanism for these drug effects on lipoprotein levels has been attributed to induction of the hepatic microsomal enzyme activity, and several studies have shown correlations among the serum HDL-C, liver histology, hepatic cytochrome P-450 concentrations, and the antipyrine clearance in patients treated with antiepileptic drugs. Studies on animals and humans also have shown an increase in bile acid flow and pool size, which may indicate increased conversion of hepatic cholesterol into bile acids.

The effects of oral hypoglycemics on lipid levels have been studied extensively, but the data are inconsistent. Furthermore, the complexity of lipid abnormalities and their mechanism, as well as the nutritional status of diabetics and obese persons, confound interpretation of such studies. Available studies suggest that the biguanides, metformin and phenformin, decrease total LDL-C as well as triglyceride levels in diabetics and nondiabetics alike.

Cyclosporine has been the most studied of the immunosuppressive drugs. It has been associated with an approximate 20% increase in total cholesterol and an approximate 35% increase in LDL-C. Because of the concern for accelerated atherosclerosis and related morbidity and mortality among heart and kidney transplant recipients, these lipid alterations constitute a major problem. The drug is used in complicated clinical situations, however, and conclusions concerning causality are tenuous because of concomitant use of multiple drugs and disease-related changes in metabolism. The mechanism by which cyclosporine affects LDL metabolism is not clear but may be receptor-mediated. Because cyclosporine is not water-soluble, it is transported in the bloodstream by the lipoproteins, especially LDL. It has been speculated that its interaction with the LDL receptor interferes with the LDL removal.

Suggested Readings

Appel G. Lipid abnormalities in renal disease. Kidney Int 39;1991:169–183.

Chait A, Brunzell JD. Acquired hyperlipidemia (secondary dyslipoproteinemias). Endocrinol Metab Clin North Am 1990;19:259–278.

Fahraeus L. The effects of estradiol on blood lipids and lipoproteins in postmenopausal women. Obstet Gynecol 1988;72(suppl 5):18S–22S.

Garg A, Grundy SM. Management of dyslipidemia in NIDDM. Diabetes Care 1990;13:153–169.

Grundy SM. Management of hyperlipidemia of kidney disease. Kidney Int 1990;37:847–853.

Henkin Y, Como JA, Oberman A. Secondary dyslipidemia: inadvertent effects of drugs in clinical practice. JAMA 1992 (In press).

Howard BV. Lipoprotein metabolism in diabetes mellitus. J Lipid Res 1987;28:613–628.

Johnson BF, Danylchuk MA. The relevance of plasma lipid changes with cardiovascular drug therapy. Med Clin North Am 1989;73:449–473.

Joven J, Villabona C, Vilella E, Masana L, Alberti R, Valles M. Abnormalities of lipoprotein metabolism in patients with the nephrotic syndrome. N Engl J Med 1990;323:579–584.

Kaste M, Muuronen A, Kikkila EA, Neuvonen PJ. Increase of low serum concentrations of high-density lipoprotein (HDL) cholesterol in TIA-patients treated with phenytoin. Stroke 1983;14:525–530.

Lardinois CK, Neuman SL. The effects of antihypertensive agents on serum lipids and lipoproteins. Arch Intern Med 1988; 148:1280–1288.

Markell MS, Friedman EA. Hyperlipidemia after organ transplantation. Am J Med 1989;87:5-61N–5-67N.

Marsden J. Hyperlipidemia due to isotretinoin and etretinate: possible mechanisms and consequences. Br J Dermatol 1986;114:401–407.

Wahl PW, Walden CE, Knopp RH, Hoover JJ, Wallace RB, Heiss G, Rifkind BM. Effect of estrogen/progestin potency on lipid/lipoprotein cholesterol. N Engl J Med 1983;308:862–867.

Working Group on Management of Patients with Hypertension and High Blood Cholesterol. National education programs working group report on the management of patients with hypertension and high blood cholesterol. Ann Intern Med 1991; 114:224–237.

CHAPTER 9

The Clinical Evaluation

The purpose of identifying persons who have lipid disorders is to allow institution of therapy that delays or prevents the development of atherosclerotic cardiovascular disease, especially CHD and sudden death. It is now well established that atherosclerosis begins in childhood, progresses over a lifetime, and usually is far advanced when symptoms appear. Thus, preventive therapy is best implemented early in life. The significance of preventing CHD in asymptomatic, "healthy" persons has long been recognized, although much debate surrounds the practical implication. In the past, most of the interest in patients who had established CHD and dyslipidemia centered around the identification of first-degree relatives who would benefit from preventive measures; much less emphasis was placed on active treatment of lipid disorders in such persons. The greater absolute risk of MI (or reinfarction) among these patients, coupled with encouraging data from secondary prevention and arteriographic studies, has led to greater optimism in treating patients who have more advanced lesions.

The purpose of the clinical evaluation is to (1) identify any underlying secondary causes of the dyslipidemia, (2) obtain important family history, (3) collect pertinent information required to determine the patient's risk for CHD, (4) assess historical and physical findings resulting from lipid abnormalities, (5) detect and quantify the extent of CVD so that appropriate management can be implemented, and (6) obtain baseline reference laboratory data for detecting adverse effects of therapy.

Screening for Blood Cholesterol

The National Cholesterol Education Program (NCEP) Expert Panel on Detection, Evaluation, and Treatment of High Blood Cholesterol in Adults has recommended that the total cholesterol level should be measured in all adults at or after 20 years of age at least once every 5 years. Screening usually begins with measurement of the nonfasting serum total cholesterol level, assessment of nonlipid risk factors, and factors likely to influence lipid values, including smoking, obesity, physical activity, stress, blood pressure, alcohol consumption, coffee consumption, and environment (*see* Chapter 2).

Because of biologic and laboratory variability (*see* Chapter 6), the presence of a high cholesterol level must be confirmed with a second test within 1 to 8 weeks and be followed by a complete fasting lipoprotein analysis to provide a more precise estimate of risk before initiating treatment. The standard deviation of repeated measurements in the same person over time is approx. 18 mg/dL (0.47 mmol/L) for total cholesterol and 15 mg/dL (0.39 mmol/L) for LDL-C. This means that two out of three repeated measurements can be expected to vary from about 18 mg/dL (0.47 mmol/L) above to 18 mg/dL (0.47 mmol/L) below the mean total cholesterol value and 15 mg/dL (0.39 mmol/L) above and below for LDL-C. Patients should be asked not to change their eating habits during this series of baseline tests so that usual values can be obtained. If consecutive total cholesterol measurements differ by more than 30 mg/dL (0.78 mmol/L), a third measurement should be made, and an average of the three values should be calculated.

History

A careful history is crucial for the correct assessment of cardiovascular risk and pathogenesis of the lipid disorder. Once a lipid disorder is identified, it is important to document when it was first detected and its mode of presentation (whether by routine screening or by its association with a specific disease or symptoms), as well as the highest documented levels and previous attempts at treatment.

In patients who have no history of known CVD (*primary prevention*), particular attention should be paid to relevant symptoms

such as chest pain, palpitations, syncope, and unusual fatigue. It should be realized, however, that silent ischemia is not uncommon, especially in patients who have diabetes mellitus. Symptoms or signs of atherosclerosis in the cerebral, mesenteric, renal, and peripheral vasculature also should be pursued. Questions should be asked about transient ischemic attacks, flank pain, hypertension, abdominal pain, erectile dysfunction, and hip or calf claudication.

The presence of CHD or other forms of atherosclerotic vascular disease is frequently well established in many patients whom we evaluate (*secondary prevention*). In such patients, the history should be well documented, with particular attention paid to the duration and intensity of symptoms, current medical management, and previous corrective procedures. All these factors impact on management.

Risk Factor Assessment

It is essential to pursue and document the presence of other risk factors, because the presence of these will influence treatment decisions. Of particular importance are cigarette smoking, hypertension, diabetes mellitus, obesity, and physical inactivity.

The most important questions regarding cigarette smoking are the number of cigarettes smoked per day, the number of years the person has been smoking, and (if applicable) when smoking stopped. The type of cigarette, filter, and other characteristics of tobacco consumption are not important. Although duration of smoking and smoking habits are important, the number of cigarettes currently smoked is the best parameter of risk.

A history of hypertension and its treatment is pertinent. Both systolic and diastolic blood pressure levels, regardless of treatment status, should be used to stratify cardiovascular risk in hypertensive patients. The duration of hypertension, degree of elevation, and control by treatment should be evaluated.

It is important to recognize that patients who have a "touch of diabetes" or a "touch of sugar" have impaired glucose tolerance or mild diabetes mellitus. Such persons are clearly at increased risk of developing atherosclerotic vascular disease, even in the absence of frank diabetes mellitus and perhaps the implied or stated unimportance of the problem by previous physicians.

Questions concerning exercise habits should be asked. The type (isotonic versus isometric), intensity, duration, and frequency of exercise should be established, as well as whether the person participates in a conditioning program. Being "busy" or "active" is not good enough.

Secondary Causes of Dyslipidemia

Because lipid disturbances may be either primary or secondary, it is important to exclude carefully other disorders or drugs that cause dyslipidemia or that might interfere with management (see Chapter 8). Specifically, signs and symptoms related to thyroid dysfunction and disease of the gastrointestinal tract, liver, biliary system, kidneys, and previous episodes of pancreatitis should be elicited. Because numerous medications can alter cholesterol and triglyceride concentrations, careful documentation of drug use, both by prescription and over-the-counter drugs, should be made. The current dose, method of administration, and duration of use should be obtained, as well as any drug allergies. Questions that concern the usual frequency and quantity of alcohol consumed are particularly pertinent, especially in patients who have hypertriglyceridemia; the type of alcohol probably makes little difference.

Dietary Assessment

The diet history also is important and should be elicited briefly by the physician and in detail by a trained dietitian (see Chapter 10). Diet is the cornerstone of therapy in many disorders and is particularly useful when the lipid abnormalities are mild to modest. Questions about foods that are high in saturated fat or cholesterol, including consumption of meat, eggs, and dairy products, methods of preparing foods, and

whether labels are read carefully will allow the physician to determine whether the patient is following a prudent diet. Many patients are convinced they are on an appropriate diet, but careful questioning may reveal that they have an incomplete understanding of the principles of appropriate nutrition and, particularly, what foods must be changed. A sample diet questionnaire that can be used by physicians is given in Appendix B.

Weight history should include the current weight, recent weight changes, and the patient's maximum and minimum weights in the past, including the ages at which these occurred. In addition to food intake, it is important to determine the intake of coffee and other caffeine-containing beverages, because they may affect lipoprotein levels. Other lifestyle habits are important as well.

Finally, a thorough social history is desirable in terms of assessing risk factors and ability to comply with dietary guidelines. The number of meals eaten out, peculiarities regarding diet, and even marital status are important in determining a dietary plan. Educational attainment and employment activities also could determine to some extent the likelihood of compliance with the overall regimen.

Family History

Assessment of the history of CVD in first-degree relatives (parents, siblings, and children), especially those aged 55 or younger, is necessary to properly classify CVD risk and to determine whether screening is necessary for children or other relatives. A familial occurrence of lipid abnormalities also can help to classify properly the cause of any dyslipidemia.

Coronary heart disease that occurs at any age aggregates strongly in families, as do many of the risk factors for CHD. The approximate frequencies of concordant abnormalities in two or more siblings who have CHD and are under age 55 are shown in Figure 2.18. The need for collecting information on second-degree relatives depends on the age of the patient being evaluated. For persons in their 50s and 60s, the first-degree relatives should be old enough to have had coronary events, if there is susceptibility for CHD in the family. For younger persons, the history in grandparents and other elderly second-degree relatives may be required to assess such susceptibility. The chance of sharing a common genetic predisposition is around 50% for first-degree relatives and 25% for second-degree relatives. The number and age of the relatives who had CHD also must be considered, because the risk for CHD increases dramatically if more than one family member developed it before age 55 (Table 9.1).

Physical Examination

A comprehensive physical examination, even in asymptomatic and otherwise normal persons, should be performed to find skin changes and other findings related to secondary causes, to detect previously unidentified clues for CHD (Table 9.2) and to establish a baseline from which to make comparisons in the future. The measured blood pressure is more relevant to risk assessment than history or treatment of hypertension. It is important to remember that elevated systolic blood pressure also places persons at greater risk for CHD. Absolute weight can be translated into relative weight with the use of simple tables (*see* Chapter

10) or the body mass index (weight in kg/height in m²) for assessing obesity. In addition to height and weight, patterns of fat distribution refine the risk estimates in assessing the possibility that obesity is influencing lipid values. This can be measured most easily as a waist-to-hip ratio (WHR), which is determined from the minimal waist and the maximal hip girth in the standing position (*see* Chapter 2). Skinfolds also can be used to estimate percent fat but require special calipers and some experience with the technique.

The skin and subcutaneous areas should be examined carefully for the presence of xanthomas, paying attention to the skin overlying tendons, the webs between the fingers, and the creases of the palms. The skin that overlies tendon xanthoma need not appear yellow because the lipid accumulations are deep. The presence of collagen (in addition to the foam cells) initiates a fibrous reaction, so that the lesions feel hard rather than soft. Regression of xanthomas has been described with dyslipidemia therapy; however, because of the fibrous nature of the lesions, complete or even partial resolution cannot be assured.

Tendon xanthomas occur almost exclusively in patients who have familial hypercholesterolemia (FH) and increase in prevalence through middle age. Because of the diagnostic implications for FH, the establishment of the presence of xanthomas is important. These xanthomas are located in extensor tendons of the hand, palm, patellar, Achilles, and plantar areas (*see* Fig. 7.2). The earliest indication of xanthomas generally is on unevenness of the tendinous surface or a thickening of the tendon itself. Inflammation of the tendons, particularly the Achilles, leads to a clinical tenosynovitis. This condition is especially common in persons involved in activities in which the tendon is subjected to trauma and predisposes to rupture of the Achilles tendon. Precipitation of an episode of tenosynovitis, similar to inducement of gout with mobilization of urates, is possible after treatment with lipid-lowering agents. Subperiosteal xanthomas such as those found at the in-

Table 9.1. Relative Risk of Future Coronary Disease According to the Strength of a Positive Family History[a]

Positive Family History		Relative Risk for CHD	
Definition	Frequency (%)	Men (20–39)	Women (40–49)
2 < age 55	2%	12.7	12.9
2 + any age	8%	5.9	4.8
1 < age 55	13%	3.9	2.5
1 + any age	38%	2.9	1.7

[a] Data regarding 94,292 adults from 15,250 Utah families. Definition column indicates number of first-degree relatives who had coronary events. (Adapted from Williams RR, Hopkins PN, Hunt SC. Evaluating family history to prevent early CHD. (Personal Communication).)

Table 9.2. Clinical and Laboratory Features of Dyslipidemia

Type	Lipoprotein Abnormality	Lipid Abnormality		Clinical Features
		TG	CHOL	
I	CM	↑[a]	→	Eruptive xanthoma Lipemia retinalis Hepatosplenomegaly Pancreatitis
II$_A$	LDL	→[a]	↑	Corneal arcus Tendon xanthoma Tuberous xanthoma
II$_B$	LDL, VLDL	↑	↑	Symptoms/signs of CHD
III	IDL	↑	↑	Palmar xanthoma Tuberoeruptive xanthoma Symptoms/signs of CHD
IV	VLDL	↑	→	Pancreatitis Increase glucose and uric acid
V	CM, VLDL	↑	↑	Eruptive xanthoma Lipemia retinalis Hepatosplenomegaly Pancreatitis

[a] ↑ = elevated → = no effect

sertion of the patellar tendon, unlike tendon xanthomas, tend to remain fixed. Thickening in these areas can be detected by careful physical examination and, if necessary, usually can be quantitated with calipers or by CT scanning. Sophisticated radiographic procedures, such as soft-tissue radiography and CT scan, do not supplant careful physical examination in determining the presence of xanthomas. Although tendon xanthomas usually suggest the presence of hypercholesterolemia, they can occur in other conditions, such as cerebrotendinous xanthomatosis and sitosterolemia, in which the blood cholesterol is not elevated.

Persons who have severe hypertriglyceridemia are susceptible to the development of eruptive xanthoma (see Fig. 7.2). These are small, pustular-like lesions with umbilicated tops that occur over the extensor surfaces of the arms and legs and over pressure points. Most of these lesions have a fine erythematous halo at the base, which gives them an inflammatory appearance. They may leave areas of increased pigmentation when they resolve or may lead to depressed scars. The presence of an eruptive xanthoma usually indicates that the triglyceride concentration is or has been above 2000 to 3000 mg/dL (22.58-33.87 mmol/L). Lymphadenopathy also has been described as a complication of hypertriglyceridemia.

Patients who are homozygous for FH and patients who have dysbetalipoproteinemia (Type III dyslipidemia) also may have deposition of cholesterol in the palms and skin (*planar xanthomas*), in soft tissues (*tuberous xanthomas*), and in the webs between the fingers and toes. These sometimes can be mistaken for tumors, tophi, and rheumatoid nodules. The presence of orange-yellow lipid deposits in the creases of the palms (*striae palmaris*) is thought to be pathognomonic for dysbetalipoproteinemia. Patients who have dysbetalipoproteinemia also may have tuberoeruptive xanthoma. Lipid deposits may accumulate as a result of trauma in patients who have this disorder and also have other forms of hypertriglyceridemia (see Fig. 7.2).

Examination of the eyes for the presence of corneal arcus and xanthelasma always should be done in patients suspected of having dyslipidemia or simply for detecting patients who have lipid abnormalities (*see* Fig. 7.2). Most patients who have lipid disorders, however, have no visible ocular manifestations. The presence of a corneal arcus may suggest the presence of an underlying disorder of lipoprotein metabolism. The mechanism of this finding is unknown, however, and some patients may have it without an obvious blood lipoprotein disturbance. In addition, the corneal arcus does not regress with lipid reduction. A true arcus has a clear interval between the arcus and the limbus of the cornea, because the lipid deposition does not extend beyond the termination of Bowman's membrane. Only the peripheral portion of the cornea is affected. Blacks frequently have a pseudo corneal lesion that simulates an arcus but contains no lipid. It is distinguishable clinically from a true arcus because it overlaps the sclera, and there is no rim of iris between it and the sclera as with a true corneal arcus.

Palpebral xanthelasma may be a manifestation of hypercholesterolemia alone or occur in combination with hypertriglyceridemia. Xanthelasmas are present in only a minority of patients, some of whom have only borderline elevation of cholesterol values. Overweight, middle-aged women also seem to have a predilection for xanthelasma without any lipid abnormalities. A familial history of such lesions combined with local factors within the skin may play a role in their development. Some persons who have normal cholesterol and triglyceride concentrations but who have xanthelasma have been demonstrated to have abnormalities in apoprotein B or apoprotein E. In addition, although it has been traditional to teach that patients who have xanthelasma are not at risk for development of CHD in the absence of dyslipidemia, recent studies suggest that this may not be true. Some studies also suggest that a corneal arcus has prognostic significance for CHD, especially in younger persons.

Rare eye findings related to lipid abnormalities also may be found. Lipemia retinalis refers to a salmon or creamy color of the reti-

nal blood vessels secondary to markedly elevated blood triglyceride levels, generally in excess of 2500 mg/dL (28.23 mmol/L), although it may be seen at lower triglyceride levels in patients who have anemia. These findings disappear with lowering of triglyceride levels. Atheromatous emboli that contain cholesterol crystals can be found in retinal vessels and are referred to as *Hollenhorst plaques*. These plaques can lead to transient monocular blindness and most commonly originate from microembolization of larger plaques located in the carotid arteries.

Orange or yellowish-gray discoloration of the tonsils, pharyngeal, or rectal mucosa in a patient who has hypocholesterolemia should raise the suspicion of Tangier disease (*see* Chapter 7). Although it has been claimed that the presence of a diagonal earlobe crease is predictive of the presence of CHD, the relevance of this association and its significance (if any) has yet to be defined.

The heart should be examined for rhythm and evidence of cardiac disease. At the very least, palpation and auscultation should be performed over all arteries, and specific notations should be made about amplitude or the presence of bruits. Specifically, auscultation should be performed over the carotid arteries, in the flanks, at the umbilicus (bifurcation of the aorta), and in the femoral or popliteal areas. Aortic stenosis as a consequence of cholesterol deposits within and adjacent to the valve leaflets has been reported.

The abdomen should be evaluated for epigastric tenderness (pancreatitis) and hepatosplenomegaly, which often is found with severe hypertriglyceridemia, as well as with other rare lipid disorders. Arthritis of uncertain origin that involves the proximal interphalangeal joints, wrists, knees, and ankles may be found in patients who have homozygous FH. In addition, arthritis limited to a single joint or a migratory polyarthritis of large peripheral joints as well as small joints of the hands and feet may occur in patients who have heterozygous FH. Cystic lesions in the proximal femur and tibia also have been described in patients with hypertriglyceridemia.

Laboratory Evaluation

All persons who have lipid disorders should have a thorough laboratory evaluation to identify secondary causes of dyslipidemia and to establish reference baseline values against which adverse effects of pharmacologic therapy can be assessed, if implemented for treatment of dyslipidemia. Most multiphasic screening procedures provide useful information about the secondary causes of dyslipidemia.

A complete blood count should be done because of the possibility of change with drugs. Furthermore, macrocytosis (in the absence of anemia) with associated hyperuricemia and an increased GGT in a patient who has hypertriglyceridemia is strong evidence for heavy ethanol ingestion, even in the absence of obvious alcoholism or chronic liver disease. If there is suspicion of alcohol abuse, this can be confirmed by drawing a morning sample of blood for alcohol levels. Any level of alcohol in such a sample would indicate appreciable consumption the evening before.

Useful biochemical determinations include plasma glucose (treatment decisions, influenced by niacin, potential secondary cause), uric acid (may be altered by niacin), creatinine (decisions on using drugs as well as potential secondary cause), and liver function tests (decisions on using drugs, secondary causes of dyslipidemia, and adverse effects caused by hypolipidemic drugs). Estimation of insulin resistance may be valuable in assessing lipoprotein abnormalities. In nondiabetics the higher the fasting insulin concentration for a given glucose level the greater the insulin resistance. In general a fasting glucose (mg/dL)/insulin (μU/mL) ratio lower than 6 characterizes persons with the constellation of risk factors—abdominal obesity, abnormal glucose tolerance, elevated blood pressure, hypertriglyc-

eridemia and decreased HDL-C. Hypercholesterolemia can be associated with hypothyroidism, and screening for this disorder should be conducted even in the absence of clinical findings because its symptoms may be subtle. The ultrasensitive TSH is the best screening test for primary hypothyroidism, because it is elevated when the T_4 or free T_4 concentration still may be within normal limits in patients who have subclinical hypothyroidism.

It was recommended that patients placed on HMG CoA reductase inhibitors have a baseline slit lamp evaluation for cataracts at initiation of treatment and follow-up at 1 year. The incidence of this complication in humans is extremely low (if any). The FDA has now removed this requirement.

The resting electrocardiogram (ECG) can provide valuable information that would indicate high susceptibility to CHD: nonspecific ST/T-wave changes, left ventricular hypertrophy, and intraventricular conduction defects are particularly useful. However, the exercise electrocardiogram is preferable to the resting ECG. An ischemic response, ST/T-wave changes, or angina with exercise in persons at high risk for CHD show a 5-fold greater likelihood of coronary heart disease, compared to those persons who have a normal exercise test. Results from the Coronary Primary Prevention Trial (CPPT) indicated that the exercise ECG response is an invaluable and independent predictor of the risk of death from CHD in hypercholesterolemic men. The value of exercise testing decreases notably in persons who do not have evidence of CHD or an increased risk for CHD because of an increased frequency of false-positive responses. The increased risk associated with a positive exercise test is greater when ischemia occurs during the early stages of exercise at a low heart rate, or when the ST-segment depression exceeds 2 mm. The exercise test also is invaluable in patients who want to exercise, by providing heart rate criteria for a structured exercise program and, for safety reasons, in ensuring that no ischemic responses will occur at contemplated levels of exercise.

Various other noninvasive techniques, such as echocardiography and radionuclide studies, can provide additional information on left ventricular mass, ischemia, and left ventricular function. Doppler ultrasound studies of carotid and peripheral arteries can demonstrate atherosclerosis in these vessels and consequently provide a valuable indirect indication of CHD risk.

Follow-Up of Initial Lipid Studies
Acceptable Blood Cholesterol

If the screening total cholesterol is less than 200 mg/dL (5.17 mmol/L) and the patient is at low risk for CHD, lipid fractionation usually is not required. General dietary advice should be given and the patient should be reevaluated in approximately 5 years (Fig. 9.1). It must be emphasized that the recommendation for a repeat cholesterol measurement in 5 years is arbitrary, because there are no data to indicate the likelihood of developing hypercholesterolemia at this interval. Nonetheless, because cholesterol generally increases with age and the appearance of hypercholesterolemia in disorders such as familial combined dyslipidemia often is delayed, repeating the measurement periodically is not only reasonable but is also prudent.

We believe that patients who have CHD or persons who have two or more risk factors, even with a total cholesterol of less than 200 mg/dL (5.17 mmol/L), should have lipoprotein fractionation to detect low HDL-C. This is contrary to what is recommended by the NCEP; however, about one-third of high-risk persons who have total cholesterol levels under 200 mg/dL (5.17 mmol/L) have LDL-C above 130 mg/dL (3.36 mmol/L), and two-thirds have low HDL-C, < 35 mg/dL (<0.91) mmol/L). Such persons should be evaluated clinically and treated like other high-risk persons, as described later in this chapter.

Some conditions are specifically associated with a high frequency of low HDL-C levels and require complete lipoprotein fractionation even in persons who have a total cholesterol at or below 200 mg/dL (5.17 mmol/L) (Table 9.3). These include noninsulin-dependent diabetes mellitus, peripheral vascular disease, the presence of high triglyceride values, and a family history of premature CHD or of dyslipidemia. Hypoalphalipoproteinemia also is found in a disproportionate number of cigarette smokers. Various lipoprotein abnormalities also may be found in normocholesterolemic persons who have obesity or hypertension, or in those who take specific medications that influence lipid and lipoprotein levels (beta-blockers, progestins, anabolic steroids, retinoids, and so forth), and the decision about obtaining a lipoprotein profile in such patients should be considered if the CHD risk is high. The rationale is to detect persons who have acceptable total cholesterol levels and who may require active intervention for reduction of CHD risk.

Although the NCEP has not indicated the importance of routine screening for HDL cholesterol, except as a risk modifier in patients who have hypercholesterolemia, the panel eventually may change its position on this issue. Epidemiologic data clearly demonstrate that HDL-C levels are inversely related to the risk of developing CHD in men and even more strongly in women. Furthermore, at any given level of LDL-C, the risk of developing CHD is directly related to the prevailing HDL-C level (*see* Chapter 3). Approximately 3% to 5% of all "normal," free-living persons in the United States will have low HDL-C (< 35 mg/dL or 0.91 mmol/L). For most ages, an HDL-C lower than 35 mg/dL (0.91 mmol/L) in men and lower than 45 mg/dL (1.16 mmol/L) in women places them at or below the 25th percentile. Perhaps 30% to 40% of patients who have CHD also have total cholesterol concentrations lower than 200 mg/dL (5.17 mmol/L), and 70% to 80% of these persons may have low HDL-C. Using the NCEP screening guidelines, patients who have low HDL-C would not necessarily be identified.

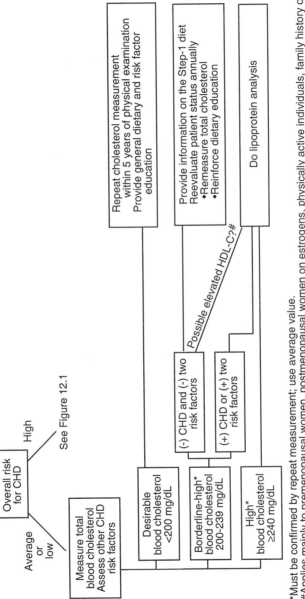

Figure 9.1. Screening and initial evaluation of blood cholesterol levels. (Adapted from: National Cholesterol Education Program, Report of the Expert Panel on Population Strategies for Blood Cholesterol Reduction. National Heart, Lung, and Blood Institute, National Institutes of Health. Arch Intern Med 1988;148:36–69.)

*Must be confirmed by repeat measurement; use average value.
#Applies mainly to premenopausal women, postmenopausal women on estrogens, physically active individuals, family history of high HDL-C, or use of medications that increase HDL-C levels.

Table 9.3. Conditions Frequently Associated with Lipoprotein Abnormalities Even in the Presence of Desirable of Total Cholesterol Levels

NIDDM
Peripheral vascular disease
Hypertriglyceridemia
Family history of premature CHD
Family history of dyslipidemia
Obesity
Hypertension
Medications
　Beta-blockers
　Progestins
　Anabolic steroids
　Retinoids
Cigarette smoking

Although it originally was recommended that cholesterol and LDL cholesterol be stratified by age and sex for evaluation, no concessions currently are made to these factors in interpreting total and LDL-C values. The former concept that age-related increases in cholesterol do not require modification now seems unacceptable because of the age-related increase in the risk of developing CHD. This increased risk is related directly to the associated increase in both LDL and total cholesterol concentrations. In those societies in which the total cholesterol and LDL-C concentrations are low, CHD does not increase with age. It has been stated by some authorities, on the basis of autopsy and population studies, that vascular disease seldom occurs in persons who have cholesterol values at or below 150 mg/dL (3.88 mmol/L). Similarly, studies of established CHD suggest that total cholesterol levels of less than 150 mg/dL (3.88 mmol/L) may be necessary for disease regression. Consequently, the optimal total cholesterol, with regard to atherosclerosis and CHD, may be one that is 150 mg/dL (3.88 mmol/L) corresponding to an LDL-C of 100 mg/dL (2.59 mmol/L).

Elevated Blood Cholesterol

All patients who have blood cholesterol levels 240 mg/dL (6.21 mmol/L) or higher should have lipoprotein fractionation and calculation of the LDL-C level (*see* Fig. 9.2).

Two measurements of LDL-C after an overnight fast are made 1 to 8 weeks apart, and the average is used for treatment decisions. Repeat assessment of LDL-C is important for the same reasons given earlier for total cholesterol. If the time exceeds 8 weeks, then the influence of recently instituted dietary charges will affect the results.

It is possible, however, to save time and effort by making the first LDL-C measurement on the same specimen used for the second total cholesterol test, especially for patients who have high total cholesterol levels on the first test. The LDL-C levels are classified as follows (Fig. 9.2):

desirable < 130 mg/dL (3.36 mmol/L)
borderline high risk 130–159 mg/dL
　　　　　　　　(3.36–4.11 mmol/L)
high risk ≥ 160 mg/dL (4.14 mmol/L).

Those patients who have desirable LDL-C should receive general dietary instructions, whereas those who have borderline or high-risk LDL-C should be evaluated clinically and then given an appropriate LDL-C-lowering treatment program based on risk assessment.

Borderline Blood Cholesterol

Decisions on additional evaluation and treatment of this group (blood cholesterol from 200 to 239 mg/dL [5.17 to 6.18 mmol/L]) are more complicated than those for the previous two groups. Although the blood cholesterol is only mildly elevated, such persons actually comprise the largest group among patients who have premature CHD (*see* Chapter 2), and, thus, deserve appropriate attention. The approach of the NCEP to stratify such patients according to their overall risk for CHD, based on the number of associated risk factors, appears logical and fitting (*see* Fig. 9.1).

Low-Risk Persons. Persons who do not have CHD or more than one risk factor generally are considered to be at low risk and should be given dietary instructions designed to lower their blood cholesterol level, with annual follow-up. It should be reemphasized, however, that a substantial number of women have higher than usual

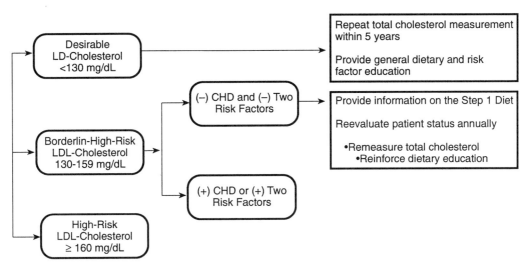

Figure 9.2. Classification based on LDL-cholesterol. (Adapted from: National Cholesterol Education Program, Report of the Expert Panel on Population Strategies for Blood Cholesterol Reduction. National Heart, Lung, and Blood Institute, National Institutes of Health. Arch Intern Med 1988;148:36–69.)

HDL-C concentrations, and, in such women, the total cholesterol often is mildly elevated. Recent studies have suggested that approximately 18% to 20% of women whose screening cholesterol is from 200 to 239 mg/dL (5.17 to 6.18 mmol/L) have an increase in their HDL-C with a normal LDL cholesterol concentration. In cases in which such a condition is suspected, it is important to fractionate the cholesterol and not to assume that the person is a candidate for dietary modification and annual reevaluation. Unfortunately, in men this occurs with considerably less frequency, so that only about 2% of men whose total cholesterol values are 200 to 239 mg/dL (5.17 to 6.18 mmol/L) are likely to have the increase because of a higher than normal HDL-C concentration.

High-Risk Persons. Patients who have CHD or persons who have two or more risk factors require lipoprotein fractionation for additional evaluation of overall risk. Contrary to what is recommended by the NCEP, we believe that *such persons should have lipoprotein fractionation to detect low HDL-C, even if their total cholesterol is less than 200 mg/dL (5.17 mmol/L)* (Fig. 9.3). High-risk persons found to have an LDL-C below 130 mg/dL (3.36 mmol/L) should have appropriate dietary information and then annual follow-up for risk assessment and appropriate therapy. In contrast, those persons who have borderline LDL-C levels (130 to 159 mg/dL or 3.36 to 4.11 mmol/L), together with patients who have higher LDL-C values, should be evaluated clinically and treated accordingly.

Elevated Blood Triglyceride

Like the situation with low HDL-C, the NCEP screening approach does not identify those persons who have hypertriglyceridemia. Because hypertriglyceridemia may be a risk factor for the development of CHD in certain persons, it seems that it is as important to know the triglyceride as it is to know the cholesterol concentration.

Triglyceride concentrations are more variable than cholesterol concentrations. This probably represents greater biologic as well as laboratory variability. There also is less information and more controversy about the relation of the triglyceride level to development of CHD. Consequently, it is more difficult to identify a value above which patients are at risk and below which they are not. Nonetheless, a triglyceride concentration of less than 200 to 250 mg/dL (2.26 to 2.82 mmol/L) is desirable; 250 to 500 mg/dL (2.82-5.65 mmol/L) is borderline and above

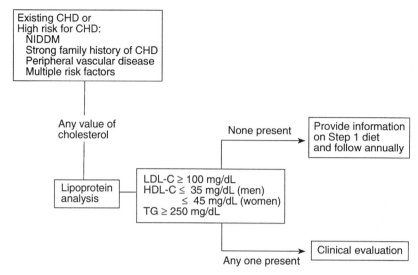

Figure 9.3. Persons who have coronary heart disease or at high risk. Screening for total blood cholesterol and subsequent evaluation in patients at high risk for CHD, irrespective of their total cholesterol values.

Figure 9.4. Visual appearance of plasma left overnight in refrigeration according to lipoprotein phenotype. CM = chylomicron; TG = triglyceride; TC = total cholesterol. (Adapted from: Grundy SM. Cholesterol and atherosclerosis: diagnosis and treatment. Philadelphia: J.B. Lippincott Co., 1990.)

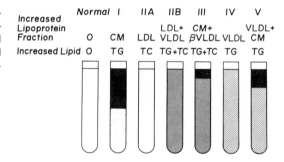

500 mg/dL (5.65 mmol/L) is considered elevated.

In addition to the quantitative measurement of lipids and lipoproteins, useful information often may be obtained from visual inspection of the plasma. Blood obtained in a tube with an anticoagulant, such as EDTA or heparin, can be placed in the refrigerator overnight and inspected the following morning (Fig. 9.4). Lactescence of the plasma, varying from faintly cloudy to milky, may be observed when the triglyceride concentration exceeds 350 to 500 mg/dL (3.95 to 5.65 mmol/L). When lactescence is distributed diffusely throughout the sample without evidence of a cream layer, increased levels of VLDL or IDL are present (Fredrickson Types II-B, III, or IV). A cream

layer on top indicates the presence of chylomicrons. When a cream layer is present and the plasma beneath it is clear, Type I dyslipidemia is present. When there is lactescence as well as a cream layer, indicating the presence of both chylomicrons and VLDL or IDL, then Type III or V is present. Patients who have pure hypercholesterolemia caused by an increase in LDL cholesterol have little or no visual abnormality of their plasma because the LDL particle is small and does not scatter light. Occasionally, the plasma will have a deeper orange color because of carotene-carrying LDL particles.

When the triglyceride concentrations are very high, either because of chylomicrons or large numbers of VLDL particles, drawn whole blood will be seen to be abnormal immediately, with a "cream of tomato soup" appearance. This creamy appearance of blood should suggest the possibility of severe hypertriglyceridemia. In some plasma samples, faint opalescence or lactescence is not associated with hypertriglyceridemia; this has been attributed to the formation of fibrin strands that tend to produce a faint cloudiness of the sample.

Unusual Cases

Although the majority of patients can be classified into one of the groups mentioned above, occasional patients will require an individualized approach. Examples include persons who have premature CHD or a strong family history of CHD in whom no obvious lipoprotein abnormalities or other risk factors are identified, or persons who have physical signs suggestive of Type III dyslipidemia. Such persons should be re-

ferred to specialized clinics for additional assessment of lipoproteins, Lp(a) and possibly hemostatic factors.

Pediatric Considerations

In 1991, after much debate on the issue, the NCEP Expert Panel on Blood Cholesterol Levels in Children and Adolescents published specific guidelines for the detection and treatment of lipid disorders in the younger age group. Similar to the NCEP Adult Treatment Program, this panel recommended two approaches to the problem:

1. *The population approach.* This approach aims at lowering the average levels of blood cholesterol among all American children and adolescents through populationwide changes in nutrient intake and eating habits. The dietary recommendations for children above age 2 are similar to those for adults, and stress the

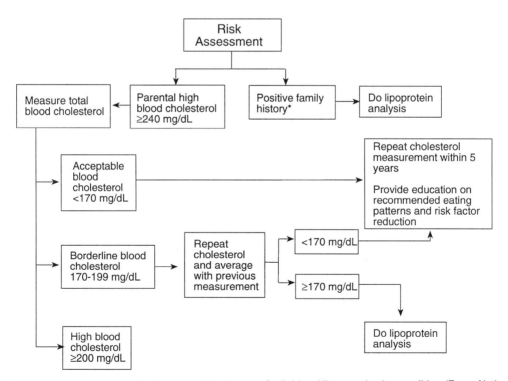

Figure 9.5. Initial screening of children and adolescents for lipid and lipoprotein abnormalities. (From: National Cholesterol Education Program Coordinating Committee. National Cholesterol Education Program Highlights of the Report of the Expert Panel on Blood Cholesterol Levels in Children and Adolescents Pediatrics 1992 (In Press).)

importance of eating a wide variety of foods and of ingesting adequate calories to support growth and development while maintaining desirable body weight. The average total fat should not exceed 30% of the total calories, with saturated fat comprising less than 10% of the total calories, whereas the dietary cholesterol intake should be less than 300 mg/day. Special emphasis is placed on groups that influence eating habits of children and adolescents, such as schools, health professionals, government agencies, the food industry and the mass media.

2. *The individualized approach.* In this aspect, the panel differed in its guidelines from the adult panel by not recommending universal screening of all children. Rather, they endorsed selective screening only in children and adolescents whose parents or grandparents have a history of atherosclerotic CVD before age 55 (proven CHD, peripheral vascular dis-

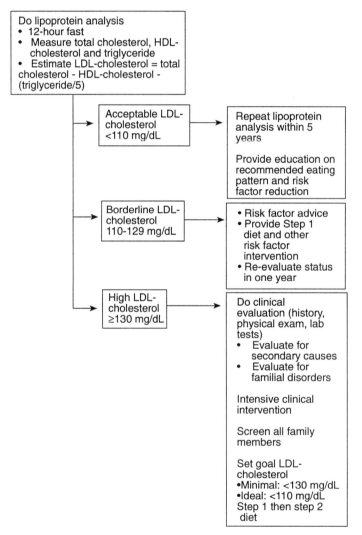

Figure 9.6. Classification, education, and follow-up based on LDL-cholesterol. (From: National Cholesterol Education Program Coordinating Committee. National Cholesterol Education Program Highlights of the Report of the Expert Panel on Blood Cholesterol Levels in Children and Adolescents. Pediatrics 1992 (In Press).)

ease, cerebrovascular disease, or sudden cardiac death), or in those children who have a parent who has elevated blood cholesterol (at least 240 mg/dL or 6.21 mmol/L). For children and adolescents whose parental and grandparental history is unobtainable, as well as for those who are judged to be at higher risk for CHD as a result of smoking, hypertension, obesity, or consumption of excessive amounts of fat and cholesterol, optional cholesterol testing by the practicing physician may be appropriate.

The method of screening varies according to the reason for testing (Fig. 9.5). In those patients tested because of a parental history of hypercholesterolemia, a nonfasting total cholesterol is measured initially and used as a basis for further classification as follows:

Acceptable—< 170 mg/dL (4.40 mmol/L)
Borderline—170–199 mg/dL (4.40-5.15 mmol/L) repeat the measurement; if the average is borderline or high, obtain a lipoprotein analysis
High—≥ 200 mg/dL (5.17 mmol/L); obtain a lipoprotein analysis

In those children tested because of a parental or grandparental history of premature CVD, a lipoprotein analysis is the initial test. In both categories, other decisions on treatment are based on the LDL-C, as shown in Figure 9.6.

The panel's recommendation for selective screening of children and adolescents has caused considerable controversy, because some screening studies indicate that about 50% of children who have elevated cholesterol levels do not have a family history of

dyslipidemia or premature CVD and therefore would not be detected by the above scheme. The panel decided, however, not to recommend universal screening for the following reasons: (1) tracking of the blood cholesterol is not perfect (*see* Chapter 5), and a proportion of the children who have high blood cholesterol will not have high enough levels as adults to qualify for individualized treatment; (2) universal screening could lead to labeling of many young people as patients with "disease," causing an unjustified anxiety in their families; (3) for most children at risk there will be sufficient opportunity to begin cholesterol-lowering therapies when they reach adulthood; and (4) universal screening could lead to overuse of lipid-lowering medications, the efficacy and safety of which have not yet been proved in childhood.

Suggested Readings

Carleton RA, Dwyer J, Finberg L, et al. Report of the expert panel on population strategies for blood cholesterol reduction: a statement from the National Cholesterol Education Program, National Heart, Lung, and Blood Institute, National Institutes of Health. Circulation 1991;83:2154–2232.

Caro JF. Insulin resistance in obese and nonobese man. J Clin Endocrinol Metab 1991;73:691–695.

Cruz PD, East C, Bergstresser PR. Dermal, subcutaneous, and tendon xanthomas: diagnostic markers for specific lipoprotein disorders. J Am Acad Dermatol 1988;19:95–111.

National Cholesterol Education Program, Report of the Expert Panel on Population Strategies for Blood Cholesterol Reduction. National Heart, Lung, and Blood Institute, National Institutes of Health. Arch Intern Med 1988;148:36–69.

Roederer G, Xhignesse M, Davignon J. Eruptive and tubero-eruptive xanthomas of the skin arising on sites of prior injury. JAMA 1988;260:1282–1283.

SECTION FOUR

GUIDELINES FOR TREATMENT

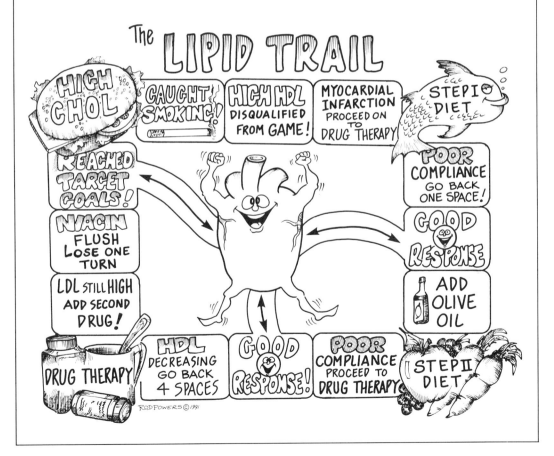

CHAPTER 10

Nondrug Therapies of Lipid Disorders

Of the environmental factors that influence blood lipid levels the most important is the dietary consumption of fat and cholesterol. A vast amount of epidemiologic, clinical, and animal data show that blood lipids increase as dietary cholesterol and saturated fat increase and usually decrease when dyslipidemic persons are put on a low saturated fat, low-cholesterol diet. There is still some uncertainty, however, about what constitutes the ideal diet in specific dyslipidemic states and what the specific effects of several fatty acids are. Doubts also exist about patient predictability of response to certain prescribed diets, the long-term adherence to stringent lipid-lowering diets, and, more importantly, the continued efficacy of such nutritional changes. Also, it appears that dietary modifications produce less striking changes in lipoprotein concentrations among women. Other nondietary approaches, such as exercise and smoking cessation, offer additional possibilities for modifying blood lipids.

Dietary Factors That Affect Blood Lipids

Table 10.1 summarizes the effects of various dietary constituents on the major plasma lipoproteins.

Fat

Dietary fats, or lipids, are substances that are soluble in organic solvents but insoluble in water. They supply essential fatty acids, provide energy sources, and serve as vehicles for absorption of vitamins A, D, E, and K as well as other fat-soluble nutrients. Essential fatty acids are precursors of the prostaglandins and related hormone-like compounds.

The hydrolysis of fats provides fatty acids that can be used immediately for energy or can be stored as triacyglycerols (also called neutral fats or triglyceride) in adipose tissue. Supplying twice the amount of calories per gram as protein or carbohydrate (9 compared to 4), fat represents the most concentrated source of energy available from food.

Most dietary fat is in the form of triglyceride, with a small amount occurring as phospholipid. Although considered a fat, cholesterol really is a sterol, or high-molecular-weight alcohol, with a characteristic cyclic nucleus different from the other

major lipids. Some edible fats and oils contain sterols as well. Phospholipids and cholesterol are derived primarily from ingested fat, but a certain proportion also is reabsorbed from the intestinal lumen after being secreted in bile. To a small extent, the intestine itself contributes to the cholesterol available for absorption through desquamation of intestinal mucosal cells that contain cholesterol.

All fats (triacylglycerol) are made up of fatty acids and glycerol. Fatty acids are classified according to chain lengths, the number of double bonds present, and the position of these double bonds (see Fig. 10.1). These characteristics determine the metabolic activity of each fatty acid. Although the nomenclature of fatty acids is somewhat complicated, understanding of the principles can make it more comprehensible:

1. *Length*—Naturally occurring fatty acids usually contain an even number of carbon atoms. Short-chain fatty acids refers to chain lengths of six or fewer carbons, whereas long-chain fatty acids have 12 or more carbon atoms. The others are called medium-chain fatty acids.
2. *Saturation*—This refers to the number of double bonds. Fatty acids without double bonds are, by definition, *saturated*. Those that have double bonds are *unsaturated* and can be subdivided into *monounsaturated* (one double bond) and *polyunsaturated* (two or more double bonds). The symbol 18:0 denotes a fatty acid with 18 carbons (C_{18}) and no double bonds; 18:2 signifies the same fatty acid with two double bonds. The position of a double bond is given by the symbol Δ followed by a superscript number. For example, Δ^9 means a double bond between carbon atoms 9 and 10.
3. *Omega*—Fatty-acid carbon atoms are numbered, starting at the carboxyl terminus:

Table 10.1. Effect of Dietary Composition on Plasma Lipoproteins

	Lipoprotein		
Treatment	LDL	VLDL	HDL
Saturated fatty acids	↑[a]	↑	↑
Monounsaturates	↓[b]	↓	=
Polyunsaturates	↓	↓	↓
Omega-3 fatty acids	=/↑	↓	=
Cholesterol	↑	=	=
Carbohydrates	=[c]	↑	↓
Protein	=	=	=
Fiber	↓	↓	=
Calories	↑	↑	↓
Alcohol	=	↑	↑

[a] ↑ Increase
[b] ↓ Decrease
[c] = No appreciable change

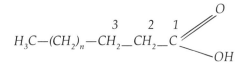

$$H_3C-(CH_2)_n-\overset{3}{CH_2}-\overset{2}{CH_2}-\overset{1}{C}\diagup^{O}_{OH}$$

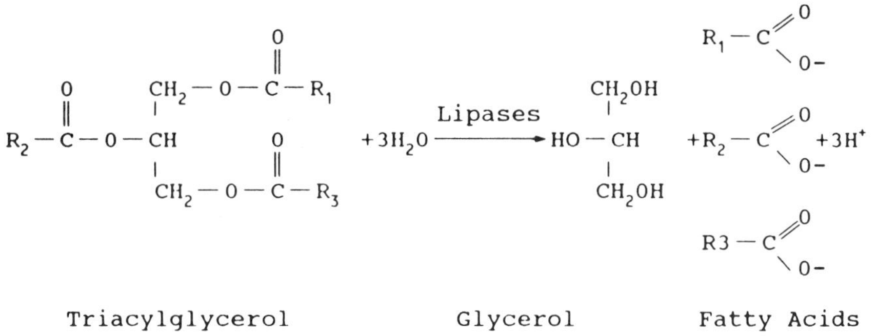

$$CH_2-O-C-R_1$$ ...

Figure with chemical structures:

Triacylglycerol +3H₂O —Lipases→ Glycerol + Fatty Acids

Figure 10.1. Schematic representation of triacyglycerol, glycerol, and free fatty acids. (Adapted from: Stryer L. Biochemistry. ed 3. New York: W.H. Freeman and Company, 1988.)

The methyl carbon atom at the distal (methyl) end of the chain is termed the *omega* carbon. Unsaturated fatty acids can be categorized according to the position of the first double bond in relation to the terminal methyl group. Omega-3 fatty acids, which are commonly found in fish oil, have three carbons between the omega end and the first double bond as shown below:

$$H_3C-CH_2-C{=}C-CH_2-R-COO-$$
$$(1) \quad (2) \quad (3)$$

In contrast, omega-6 fatty acids have six carbon atoms between the omega end and the first double bond. For example, an important omega-6 fatty acid, linoleic acid, has two double bonds and is designated as:

18 : 2 (n-6)

no. of carbon atoms / no. of double bonds / position of first double bond

Oleic acid is the most common fatty acid in the omega-9 family.

4. *Cis* compared to *trans*—The cis configuration of a double bond implies that the hydrogen atoms that are linked by the double bond are both on the same side; this is the configuration found in most naturally occurring unsaturated fatty acids. Trans fatty acids are isomers of the naturally occurring unsaturated fatty acids in which the hydrogen atoms on

Table 10.2. Some Naturally Occurring Fatty Acids[a]

No. of Carbons	No. of Double Bonds	Abbreviation	Common Name
Saturated			
8	0	18:0	Caprylic
10	0	10:0	Caproic
12	0	12:0	Lauric
14	0	14:0	Myristic
16	0	16:0	Palmitic
18	0	18:0	Stearic
20	0	20:0	Arachidic
Unsaturated			
16	1	16:1(n-7)	Palmitoleic
18	1	18:1(n-9)	Oleic
18	2	18:2(n-6)	Linoleic
18	3	18:3(n-3)	Linolenic
20	4	20:4(n-6)	Arachidonic

[a] Adapted from: Stryer L. Biochemistry. 3rd ed. New York: W.H. Freeman and Company, 1988.

carbon atoms linked by double bonds are on opposite sides. They are produced by commercial hydrogenation of cooking oils but also occur in milk fat, beef fat, and lamb fat.

Table 10.2 lists some of the naturally occurring fatty acids found in animals. It is important to remember that *all dietary fat is a mixture of saturated, monounsaturated, and polyunsaturated fatty acids.* The terms saturated, monounsaturated, and polyunsaturated generally refer to the predominant fatty acids in the mixture. Both short-chain length and unsaturation increase the fluidity of fatty acids and their derivatives. Fats that have large amounts of saturated fatty acids are solid at room temperature and usually are animal in origin. Fats that consist of unsaturated fatty acids are liquid and usually are derived from plants. The composition of some common dietary fats is given in Figure 10.2. These dietary oils and fats range from 6% saturated to 92% saturated, with considerable variability in monounsaturated fatty acids as well.

Saturated Fatty Acids. Of the three categories of fatty acids, only the saturated fatty acids are thought to raise the blood cholesterol level. In fact, in humans they increase the blood cholesterol level much more than dietary cholesterol. In the average American diet, approximately 15% of the calories come from saturated fat, and this is thought to be a major determinant for the high frequency of dyslipidemia in the country. Many healthy populations throughout the world exist on diets that contain less than 7% of total calories from saturated fatty acids. At any given level of cholesterol intake, a reduction in the dietary saturated fat will have a salutary effect on lowering the blood cholesterol concentration. This is why some foods that are high in cholesterol content but low in saturated fat (such as shellfish) are not prohibited in cholesterol-lowering diets. Beside their effect on lipoprotein metabolism, limited data suggest an association between the excessive intake of saturated fat and an increased predisposition to thrombosis. Inconclusive data also have linked increased saturated fat with elevated blood pressure and higher fasting insulin concentration.

Animal sources provide most of the saturated fatty acids in our diet: meats, poultry, eggs, and dairy products account for 61% of this fat. Baked goods, cooking oils, snack foods, and other sources, such as mixed dishes (soups, casseroles, and the like) and fried foods, account for the remaining 39%.

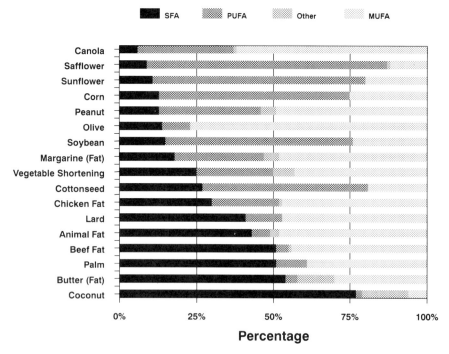

Figure 10.2. Comparison of fatty acids among dietary oils and fats ranked by percent of saturated fat (lowest to highest). Among dietary oils and fats, canola oil has the best fatty acid ratio; safflower, sunflower, corn, and soybean oil appear to be highest in polyunsaturates. Canola and olive oil are highest in monounsaturated fat. The only vegetable oils high in saturated fat are palm oil and coconut oil, but they are among the highest in saturated fat among all the oils and fats (Adapted from: Compare the dietary fats. Educational material produced by Proctor and Gamble, 1990.)

Although most fats of vegetable origin contain only a small proportion of saturated fat, there are a few exceptions to this rule; coconut, palm, and palm kernel oil contain large amounts of lauric and myristic acids, both of which raise the blood cholesterol levels as much or more than the saturates derived from animals (*see* Fig. 10.2). Bakery goods and pastries can derive 50% or more of their calories from highly saturated coconut, palm, palm kernel, or freely hydrogenated vegetable oils. These products often are labeled as "containing no cholesterol" or "made from all vegetable oils," misleading the consumer into buying a product that is high in saturated fat.

Usually, the degree of saturation of a fat determines its cholesterol-raising effect. Saturated fats increase the synthesis of LDL and decrease the LDL fractional catabolic rate, presumably by suppressing the expression of LDL receptors in some poorly understood way. Possible explanations include redistribution of cholesterol within liver cells, thus expanding the metabolically active cholesterol; decreasing the hepatic concentrations of messenger RNA necessary for synthesis of LDL receptor proteins; altering the phospholipid composition of cell membranes, thereby inhibiting normal movement of LDL receptors to the cell surface and clustering in coated pits; or changes in the composition of the LDL particle interfering with binding to the receptors. Most saturated fatty acids downregulate LDL receptor activity and are hypercholesterolemic; however, the relation of saturated fatty acids to atherogenesis is not as simple as originally envisioned. For example, the medium-chain fatty acids (C8 and C10) have little effect on plasma cholesterol, probably because of their conversion to carbohydrate or to their denaturation to monosaturated fatty acids. Stearic acid (C18) also is less hypercholesterolemic than the shorter-than acids—lauric, myristic, and palmitic acids. Some animal research even indicates different effects on blood cholesterol among these fatty acids. Consequently, cocoa butter and other foodstuffs high in stearic acid appear to be less atherogenic than expected. Hydrogenation of vegetable oils for shortenings and margarines is commonly used to convert liquid vegetable oil to solid to protect the fats from oxidation and to add texture to the foods. During this hydrogenation process, transmonounsaturated fatty acids are formed that have potential detrimental effects. Recently, it has been suggested that transmonounsaturated fatty acids, rather than having a neutral effect on cholesterol as the more common cis configuration, actually raise LDL-C levels and reduce HDL-C levels. The concept that the three major classes of fatty acids—saturated, monounsaturated, and polyunsaturated—have distinctly different effects on serum lipids is more complex than previously thought. Some saturated fatty acids raise cholesterol considerably, others to a lesser extent, and some probably minimally, if at all, whereas some unsaturated fatty acids have undesirable metabolic effects. The relation between fatty acid structure and influence on blood lipids and, consequently, atherosclerosis needs additional clarification before specific recommendations within fatty acid classifications can be made.

Monounsaturated Fatty Acids. Monounsaturated fatty acids that contain one double bond occur in all animal and vegetable fats. The major monounsaturate is oleic acid. The usual American diet contains about 15% of calories as monounsaturated fatty acid, most of which is derived from animal fat. Monounsaturated fatty acids make up about 40% of beef fat and up to 45% of chicken fat. They also are a major component of commonly used vegetable oils such as olive oil (77%), canola oil (62%), and peanut oil (49%) (*see* Fig. 10.2). In the past, monounsaturates were considered lipid-neutral, neither raising nor lowering the blood cholesterol levels. Recent evidence suggests, however, that they lower the blood total and LDL cholesterol and thus can be used to replace saturated fatty acids in the diet. This finding has caused much enthusiasm because, in contrast to high-carbohydrate and high-polyunsaturated diets, monounsaturated-enriched diets do not appear to lower the HDL-C. Monounsaturated fat also may protect against oxidation of LDL-C.

For centuries, monounsaturated fatty acids in the form of olive oil have been consumed in large quantities in the Mediterranean area with apparent safety. In these populations, fat intake can reach nearly half of total calories. Yet, CHD in these countries is rare, presumably because much of the fat is monounsaturated. Consequently, it may not be necessary to resort to a high-carbohydrate and low-total-fat diet to reduce LDL; rather, dietary cholesterol and saturated fat can be replaced by monounsaturated fat, restricting the total fat calories to less than 30% of total energy consumption. Such substitution might make it possible to design a diet that will reduce LDL-C, maintain HDL-C and triglyceride levels, and achieve a level of palatability that is absent with very low-fat diets. There also appears to be improved control of hyperglycemia in patients who have noninsulin-dependent diabetes mellitus (NIDDM). Yet, not all studies demonstrate an advantage of monounsaturated over polyunsaturated fat for increasing HDL-C concentrations in patients on low-fat diets, and it is still unclear whether preference of monounsaturates over polyunsaturates and carbohydrates might be more beneficial. In addition, as with the saturates, anomalies may exist. Peanut oil, a monounsaturated fatty acid, has been found in preliminary studies to be unexpectedly atherogenic for monkeys, rabbits, and rats. This may relate to other dietary constituents and requires more documentation about cause and effect.

Polyunsaturated Fatty Acids. The polyunsaturated fatty acids (PUFA) are divided into two primary classes: omega-6 and omega-3 fatty acids, each affecting lipoprotein metabolism differently. As discussed earlier, this nomenclature is based on the position of the first double bond, namely the number of carbon atoms from the methyl (omega) end of the fatty acid chain.

Omega-6 Fatty Acids. Linoleic acid, an essential fatty acid, is the major omega-6 (n-6) variety. Safflower, sunflower seed, soybean, corn oil, and products that contain these oils are all rich in linoleic acid. Substitution of linoleic acid for saturated fat in the diet, as has been known for many years, will lower total cholesterol. Linoleic acid, in addition to being the major arachidonic acid and prostaglandin precursor, is incorporated directly into platelet membrane phospholipids. This process affects membrane fluidity and cell permeability and may account for the reduced platelet aggregation observed with increased dietary linoleic acid.

Polyunsaturated fatty acids decrease the blood levels of the total and LDL cholesterol. This relates primarily to an increase in the concentration of LDL receptors and enhanced removal of LDL-C from the bloodstream, although they also inhibit the synthesis of VLDL in some patients and thereby reduce the formation of LDL particles from VLDL. Other possible mechanisms for the LDL-C-lowering effect include enhanced excretion of cholesterol, redistribution of cholesterol, or a decreased capacity of LDL to transport cholesterol. Replacement of saturated fatty acid with either linoleic (unsaturated) or oleic (monounsaturated) acid produces roughly equivalent reductions in LDL-C. Replacement of dietary saturated fat with polyunsaturated fat also is associated with decreased platelet aggregation and clotting activity. The mechanisms responsible for these effects probably are related to prostaglandin synthesis, but they are poorly understood.

In the past there was great enthusiasm for the use of large amounts of polyunsaturates. Diets high in these fats were seen as alternatives to lowering dietary saturated fats, although there is no strong epidemiologic evidence that such diets prevent CHD. With up to 5 years of follow-up, some clinical studies of persons who consumed diets with 15% of total calories as polyunsaturated fatty acids have shown no ill effects. No large population has ever consumed large quantities of polyunsaturated fatty acids over a prolonged period to demonstrate benefit or safety, although true vegetarians eat an abundance of polyunsaturated fatty acids without known difficulties. Yet, data are sparse, and many investigators are still worried about the possibility of long-term adverse effects. Because of its high caloric value, a large intake of polyunsaturated fat may be undesirable. Also, polyunsaturated fat may predispose to gallstones in some persons. In laboratory animals, diets high in vegetable oils that contain omega-6 polyunsaturates can suppress the immune system and promote tumor development. Because of these concerns and the inconsistent reports of effects on HDL-C, current guidelines do not recommend a polyunsaturated fatty acid intake beyond 10% of total calories. Modest quantities of polyunsaturated fat, 7% to 10% of total calories, favorably affect lipoprotein while minimizing possible untoward effects.

Omega-3 Fatty Acids.   Cold water fish, fish oils, and some plant oils are important sources of the major omega-3 (n-3) polyunsaturated fatty acids, eicosopentaneoic acid (EPA), and docosahexenoic acid (DHA). The role of omega-3 fatty acids, particularly EPA and DHA, in human nutrition is under active investigation. Shortly after it was found that polyunsaturated fatty acids lowered blood cholesterol, investigators realized that marine oils had a similar effect. Little attention was paid to this finding until interest was stimulated by poorly documented reports of low CHD rates among Greenland Eskimos.

Interest also was stimulated by a study of fish intake among middle-aged Dutch men initially free of CHD; at a 20-year follow-up, those men who ate one ounce of fish per day had half the CHD mortality of those who reported little, if any, fish intake. It was concluded that consumption of as little as two fish dishes per week may be cardioprotective. More recently, a randomized trial in England among myocardial infarction survivors indicated that advice to consume fatty fish led to significantly greater survival after a two-year follow-up, compared with similar advice to just reduce dietary fats.

In terms of lipoproteins, the most consistent effect of fish oils at relatively low doses (2 to 5 g/d) has been a reduction in triglyceride and VLDL-C levels, particularly among hypertriglyceridemic persons. This effect is not found with vegetable-derived polyunsaturated fatty acids and is most likely caused by a decrease in the triglyceride content of VLDL particles rather than by a reduction in the number of VLDL particles secreted by the liver. Omega-3 fatty acids can reduce the total blood cholesterol in dyslipidemic persons, but not in normolipidemics. The changes in lipoproteins induced by fish oil supplementation depends in great part on the lipid phenotype (Fig. 10.3). Except for about a 20% decrease in triglyceride, little change occurs in normolipidemics or in patients who have isolated hypercholesterolemia (Type IIa). In contrast, hypertriglyceridemics have large reductions in triglyceride levels, small decreases in total cholesterol, and mild elevations of HDL-C. Most of these patients paradoxically will experience an increase rather than a decrease in the LDL-C. At higher daily intakes of dietary omega-3 fatty acids (15 grams or more per day), a more profound lowering of triglyceride may occur. The larger dose, difficult to take and costly, represents an unrealistic approach in most patients.

Data from multiple sources suggest that the omega-3 fatty acids have important and unique physiologic effects that affect atherogenesis by mechanisms unrelated to the blood lipid levels (Table 10.3). Many of these effects relate to their influence on prostaglandin metabolism. These fatty acids replace arachidonic acid as substrate for the enzyme cyclooxygenase, leading to the preferential synthesis of thromboxane A3, which is a weaker vasoconstrictor and proaggregator prostaglandin than thromboxane A2. It has been shown that consumption of fish or fish oils prolongs the bleeding time and decreases platelet responsiveness. The

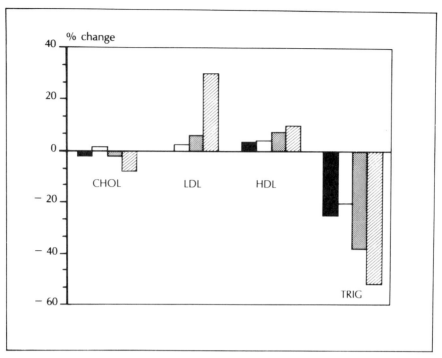

Figure 10.3. Summary of the overall mean percentage change in cholesterol, lipoproteins, and triglyceride in reported fish oil trials varies according to the type of lipid disorder: ■ =normolipidemics; □ = isolated hypercholesterolemia; ▨ = combined hyperlipidemia; ▨ = isolated hypertriglyceridemia; CHOL = total cholesterol; LDL = low density lipoprotein; HDL = high density lipoprotein; TRIG = triglyceride. The changes are most marked for those constituents in the group who had isolated hypertriglyceridemia. (Adapted from: Harris WS. Fish oils and plasma lipid and lipoprotein metabolism in humans: a critical review. J Lipid Res 1989;30:785–807.)

Table 10.3. Potential Mechanisms of Omega-3 Fatty Acids for Inhibiting Atherogenesis

Antiinflammatory component
    Reduced leukotriene B_4

Inhibition of cell growth
    Reduced platelet derived growth factor

Diminished toxic cell metabolites
    Depressed oxygen free radical production

Antithrombotic effects
    Decreased interleukin-1 and tumor necrosis
        factor
    Enhanced production TXA_3
    Decreased fibrinogen

Inhibited vasoconstrictor activity
    Enhanced prostacyclin and endothelial cell
        derived relaxing factor

Other possible effects
    Reduced dysrhythmias
    Decreased blood viscosity
    Decreased blood pressure
    Decreased tissue plasminogen activation

effects on bleeding time and platelet aggregation both are dose-dependent. Omega-3 fatty acids also interfere with arachidonic acid, a precursor of leukotrienes, thus modulating leukotriene synthesis and inhibiting the inflammatory response. These antiinflammatory properties could diminish the vessel wall response to tissue injury and lipid infiltration. Modest reductions in blood pressure, from largely undetermined mechanisms, also have been reported. Fish oil supplements have been reported to inhibit experimental ventricular dysrhythmias in animals. Omega-3 fatty acids may protect against the development of atherosclerosis through multiple mechanisms.

Preliminary data also indicate that omega-3 fatty acids may be beneficial in preventing early restenosis in patients after percutaneous transluminal coronary angioplasty PTCA. The mechanism of this protective effect is not clearly understood, but experimental studies have demonstrated this effect even in the presence of high blood cholesterol concentrations. This suggests that the protective effect is not mediated by lipoprotein alterations but rather by changes of platelet function, reduced leukocyte responsiveness, or some direct effect on vessel wall properties.

Many of the effects of fish oil are dose-related. Fish oil capsules generally contain 300 to 600 mg of omega-3 fatty acids and variable amounts (1 to 10 mg) of cholesterol for each 1-gram capsule. To obtain 5 to 10 g of omega-3 fatty acids per day, 10 to 15 fish oil capsules are required. The use of high-dose fish oil is not devoid of adverse effects. Without other dietary changes, such treatment could add significant calories as well as increased saturated fat and cholesterol intake, leading to weight gain and a higher serum cholesterol. A diet high in fish oil may cause vitamin E deficiency or vitamin A or D toxicity. The antiplatelet activity of low-dose aspirin may also be potentiated by high-dose fish oils. Persons who have bleeding disorders should not take fish oil. Omega-3 fatty acids in patients who have NIDDM may worsen their glycemic control because of an altered tissue responsiveness to insulin and increased hepatic glucose production. The use of large doses of omega-3 fatty acids in patients who have lipid disorders and NIDDM should be undertaken with extreme caution.

Fish oils must be thought of as a drug rather than as a dietary supplement, and insufficient data exist to determine the long-term effects of fish oils on patients who have dyslipidemia. Patients who have elevated triglyceride, particularly with associated increases in chylomicrons and VLDL remnants, may be treated with moderate doses of fish oil under close supervision after a serious attempt at more conventional dietary and drug therapy has proved inadequate. The use of fish oil for otherwise healthy hypertriglyceridemics, especially at the doses recommended by manufacturers (three to six capsules [1 to 2 g of omega-3 fatty acids] per day), is probably neither harmful nor therapeutic.

It seems wise to reserve fish oil for specific therapeutic situations, and to rely generally on increased fish consumption rather than on dietary supplementation with marine oils. Consumption of fish one or more times per week has been associated with a reduced risk of CHD and increased survival after MI in European investigations. A reasonable course of action appears to recommend consumption of two or three fish meals per week until further data are available.

Cholesterol

Cholesterol is an important constituent of cell membranes and is a precursor of adrenal steroids (hydrocortisone and aldosterone), sex hormones (estrogens and androgens), bile acids, and vitamin D. Because all body cells can synthesize cholesterol, there really is *no need to consume exogenous cholesterol to meet metabolic and structural requirements*. In fact, true vegetarians consume less than 10 mg of cholesterol per day without any ill effects.

Dietary cholesterol suppresses the synthesis of LDL receptors; this results in delayed clearance of plasma LDL, as well as lessened uptake of VLDL remnants, both of which lead to elevated blood LDL-C. The hypercholesterolemic effect occurs to a greater extent when dietary cholesterol is consumed with a diet also high in saturated fat.

In addition, dietary cholesterol may influence atherogenesis through mechanisms other than its effect on blood cholesterol, as demonstrated by several prospective population studies (*see* Chapter 2). This independent effect of dietary cholesterol can be accounted for in several ways, all related to the postprandial elevations of blood lipids. Large amounts of dietary cholesterol lead to cholesterol-enriched chylomicron remnants, which are thought to be atherogenic. With lipolysis of chylomicrons by LPL, cholesterol could be displaced in proximity to the arterial wall, facilitating its entry. Enrichment of HDL with dietary cholesterol also could limit the ability of HDL particles to accept cholesterol from peripheral tissues, interfering with the reverse cholesterol transport system. The clinical implications of these postprandial influences could be important. The practice of measuring plasma lipids only in the fasting state may obscure important postprandial lipid changes, and this practice may require modification as our knowledge advances.

Cholesterol is present in muscle and fat from all animal, but not from plant products. Organ meats and egg yolks are particularly rich sources of cholesterol, as are some shellfish (Appendix A). Dairy products that contain butter fat also contribute to cholesterol intake. Recent data on food consumption in this country from the United States Department of Agriculture (USDA) suggest that almost 50% of the cholesterol eaten by adults comes from meat, poultry, and fish. Another one-third comes from milk products, eggs, pastries, and cheese, with the remainder coming from various food sources.

Carbohydrates

Carbohydrates generally are divided into simple (monosaccharide and disaccharide sugars) and complex categories (starches and fibers [polysaccharides]). In the United States, about 45% of the total calories are obtained from carbohydrates, half of which come from simple carbohydrates. The current recommendations to reduce dietary fat dictate an increase in dietary carbohydrate intake: a 10% reduction in fat calories should be accompanied by an increase in carbohydrates calories up to 50% to 60% of the total. Optimally, most of these carbohydrates should be complex to ensure sufficient intake of vitamins, minerals, and fiber. The best sources of complex carbohydrates are grain products, fruits, and vegetables.

Substitution of carbohydrates, polyunsaturated fatty acids, or monounsaturated fatty acids for saturated fatty acids results in similar reductions in LDL-C. A high-carbohydrate intake, however, can stimulate hepatic synthesis of VLDL and triglyceride, the effects being most pronounced with diets that

contain 65% or more of calories as carbohydrate. Patients who have preexisting hypertriglyceridemia, as well as overweight and diabetic persons, have an exaggerated triglyceride response to added carbohydrate. Whether this effect is related more to simple sugars or complex carbohydrates is unresolved, but it can be minimized by a reduction in the total fat content of the diet. Carbohydrate-rich diets also reduce HDL-C levels. This reduction of HDL-C, on the order of 4 mg/dL (0.10 mmol/L) for every 10% of fat calories replaced by carbohydrates, seems to be more sustained than the change in triglyceride. This effect is most pronounced in patients who have NIDDM, who often already have, or are prone to develop, hypertriglyceridemia associated with a low HDL-C. Should any patient develop significant hypertriglyceridemia or substantial lowering of the HDL-C concentration on such a diet, it may be necessary to revise the diet and increase the dietary fat, substituting monounsaturated or polyunsaturated fat for saturated fat.

Protein

The current recommended intake of dietary protein of 15% of total calories provides ample amino acids for synthesis of enzymes and structural proteins and does not need to be increased. Despite suggestions by several experimental studies that certain plant proteins lower cholesterol relative to animal proteins, such findings have never been verified in humans. Isolated clinical studies also report the possibility of a differential effect of protein on blood lipids, but available data have come from short-term metabolic ward studies that used protein in the form of liquid supplements. Most data indicate that diets prepared from mixed sources of protein, at fixed levels of carbohydrates and fat, can vary widely in protein composition with little difference in the plasma cholesterol levels.

Despite the lack of evidence that consumption of large quantities of protein aids muscle building and muscle repair, some persons alter their diets for this purpose. Attempts at high protein intake generally entail high meat consumption and frequently result in undesirably high intakes of saturated fats. High-protein diets also appear to be detrimental to persons who have poor renal function. The daily protein requirement is 0.8 g/kg of body weight, regardless of physical activity level. Although the quality of protein in a low-fat diet may not change, the proportion of vegetable protein is expected to increase because of less protein coming from animal sources. Vegetable sources can provide the necessary protein, but a strict diet without any meat, dairy, or egg products can result in vitamin B_{12} deficiency. Grains, legumes, and vegetables, combined with small amounts of lean meat, fish, or poultry, provide ample protein and vitamins with little fat and calories.

Fiber

Dietary fiber consists of a group of heterogeneous substances that serve as integral components of plant cell walls in fruit, vegetables, and grains. High-fiber foods actually contain a mixture of several types of fiber. Insoluble fiber, such as that found in wheat bran, does not dissolve in water and is derived from cellulose, hemicellulose, and lignins. It adds bulk to the stool and reduces transit time, factors that may prevent colon disorders but have little effect on lipids. Water-soluble fiber includes pectin, certain gums, mucilages (such as citrus fruits, dried beans, and seeds) and psyllium (found primarily in fruits, oat products, and legumes) (Table 10.4). Among the cereals, only oats and barley, which are rich in beta-glucons, contain such fiber. In the intestine, these substances are water-soluble but are not absorbed and, thus, slow down absorption of digestive products.

Intake of soluble fiber at levels of 15 to 25 g per day has been reported to reduce blood cholesterol levels by as much as 10%. The changes that have been most noticeable are decreased LDL-C and triglyceride levels; the effects on HDL-C have been variable. Yet, in most instances, fiber does not produce a major change in blood cholesterol. In addi-

tion, not all soluble fibers are structurally alike, and the amount of soluble fiber present in a food product does not necessarily correspond to its cholesterol-lowering effect. For example, beans seem to have the same effect on blood cholesterol as an equal weight of oat bran, despite less soluble fiber.

The mechanism of blood cholesterol reduction by water-soluble fiber is not fully understood. The fiber may reduce blood cholesterol by binding bile acids, much as the bile acid resins do; by moving absorption of fat more distally; by increasing fecal fat losses; or by altering the size of lipoprotein particles formed by the intestinal mucosa. Some investigators believe that soluble fiber is fermented in the colon to generate short-chain fatty acids; these might be absorbed into the portal vein and, consequently, inhibit cholesterol synthesis in the liver. It also is possible that fiber lowers blood cholesterol in part not by its intrinsic properties but, rather, by replacing dietary saturated fat and cholesterol. Certainly, diets that contain large amounts of fiber contain less fat and cholesterol.

The effect of a particular food on blood cholesterol can be determined only by clinical studies. Guar, pectin, oat bran, and psyllium all have been used in trials with dyslipidemic patients. These studies were short-term, usually only several months in duration, and did not evaluate whether water-soluble fiber as a regular part of the diet differs from supplements given in addition to the regular diet. In healthy volunteers who already are following a low-fat diet meticulously, an additional 35 to 40 g of soluble fiber from oat bran or other foods high in soluble fiber (Table 10.4) usually results in only a small additional reduction in total cholesterol ($< 5\%$). In most studies, the amount of fiber required to demonstrate significant LDL lowering has been 70 to 100 g per day. It is unlikely that most people can achieve such high levels, and increasing dietary fiber to more practical limits of around 40 grams per day may have marginal beneficial effects on lipid concentrations. Most Americans eat between 11 to 23 g of dietary

Table 10.4. Fiber-Rich Foods[a]

	Serving Portion	Total Fiber (gms)	Soluble Fiber (gms)
Bread and Crackers			
Pumpernickel bread	1 slice	4.3	.57
Whole-wheat bread	1 slice	1.4	.30
Whole-meal bread	1 slice	2.1	.47
Graham crackers	2 squares	.7	.33
Grain Products			
Whole-grain cornmeal	2 tbsp	3.0	.78
Whole-meal flour	2 1/2 tbsp	2.1	.35
Brown rice, cooked	1/2 cup	2.4	.20
Oat flakes, fortified	2/3 cup	2.7	1.26
Oats, instant, cooked	3/4 cup	2.8	1.40
Oats, regular, cooked	3/4 cup	2.8	1.40
Fruits			
Apple, raw (with skin)	1 fruit	2.8	.97
Grapefruit, raw	1/2 medium	1.7	.48
Peach, raw	1 medium	1.6	.60
Pear, raw	1/2 medium	2.5	.49
Raisins	1 1/2 tbsp	1.0	.25
Strawberries, raw	3/4 cup	2.0	.74
Vegetables			
Asparagus, cooked	3/4 cup	3.1	.81
Green beans, cooked	1/2 cup	2.1	.50
Beets, cooked	1/2 cup	2.2	.76
Broccoli, cooked	1/2 cup	2.0	.85
Brussels sprouts, cooked	1/2 cup	3.9	1.59
Carrots, raw	1/2 cup	1.3	.61
Cauliflower, cooked	1/2 cup	1.6	.50
Corn, cooked	1/2 cup	3.9	1.74
Okra, cooked	1/2 cup	2.6	.96
Peas, young green, cooked	1/2 cup	4.1	1.12
Potatoes, cooked	1/2 cup	1.6	.78
Sweet potato, cooked	1/2 large	2.1	.81
Legume (dried peas and beans)			
Butter beans, cooked	1/2 cup	4.4	1.19
Kidney beans, cooked	1/2 cup	5.8	2.51
Lima beans, cooked	1/2 cup	4.4	1.19
Pinto beans, cooked	1/2 cup	5.8	1.82
Split peas, cooked	1/2 cup	5.1	1.70

[a]Adapted from: Anderson JW, Plant Fiber in Foods. Lexington, KY: HCS Nutrition Research Foundation, Inc. 1986.

fiber per day—less than the recommended 20 to 35 g per day. The value of fiber rests in its use as an adjunct to an appropriate lipid-lowering diet, but the long-term value of large amounts of fiber, as well as the differential effect of soluble versus nonsoluble fiber, remains to be established.

Fibrous foods that are relatively low in calories provide a rapid and sustained feeling of satiety but should be introduced into the diet slowly. Increasing dietary fiber too quickly can result in bloating, gaseous distension, flatulence, and loose bowel movements. Contrary to common impressions, severe constipation may result from high fiber diets if the liquid intake is inadequate. Other possible untoward effects from high-fiber intake include deficiencies of minerals and trace elements caused by binding of positive ions such as zinc, calcium, and magnesium; and the remote, but reported, possibility of intestinal obstruction from mechanical blockage. Finally, analphylactic reactions from eating psyllium-containing cereals, even without known prior exposure, have been reported among health care workers, presumably sensitized from inhalation.

Calories

In addition to raising the serum cholesterol and triglyceride and lowering HDL-C (*see* Chapter 3), obesity is associated with other major CVD risk factors, such as hypertension and glucose intolerance. Although it previously had been thought that the effect of obesity on CHD was indirect through its influence on these risk factors, long-term data indicate that obesity is an independent risk factor for CHD (*see* Chapter 2). An important consideration, then, in making dietary recommendations for hypercholesterolemia concerns in the need to reduce caloric intake and achieve desirable body weight in overweight persons.

The possible mechanisms responsible for the lipoprotein abnormalities in obesity are discussed in Chapters 3 and 8. The fat distribution, which can be estimated by the waist-to-hip ratio, correlates best with blood lipids. Abdominal (central) obesity especially is associated with hypertriglyceridemia and low HDL-C; however, the precise relation between weight and lipids is unclear. Fortunately, weight reduction will aid in normalizing these lipoprotein abnormalities, especially when accompanied by a reduction in the dietary fat composition. In most persons, these changes occur even without achieving ideal goal weight. Among men, a 10% reduction in relative weight results in an average fall in the serum cholesterol of about 10 mg/dL (0.26 mmol/L). A reduction of one unit of the body mass index (approximately 7 pounds) corresponds to an increase of about 0.7 mg/dL (0.02 mmol/L) in HDL-C levels. As mentioned already the degree of these changes depends to a large extent on accompanying changes in dietary fat composition. There also appear to be sex differences in the lipoprotein response to weight loss. HDL-C levels rise more and LDL-C levels decline more in men. However Lp(a) may decrease with weight loss in premenopausal women more so than men.

For most persons, dietary caloric restriction probably is more helpful than exercise in achieving weight reduction. Nevertheless, physical activity should be an integral component of any program aimed at increasing caloric expenditure and maintaining weight loss.

Alcohol

The lipoprotein effects of alcohol ingestion and the mechanisms responsible for these effects are discussed in Chapter 3. In some persons, regular alcohol ingestion produces clinically significant hypertriglyceridemia because of elevated VLDL and chylomicrons. Excessive alcohol intake is a major contributing factor to hypertriglyceridemia in susceptible persons, such as diabetics or those who have underlying dyslipidemias. Triglyceride levels in excess of 750 mg (8.47 mmol/L) may be high enough to increase the risk of acute pancreatitis. Questions regarding alcohol abuse and attempts at minimizing ingestion of alcohol should be an integral part of the management of every

patient and even more so in hypertriglyceridemic patients.

Although most health-related effects of
alcohol are detrimental, population studies
have shown a controversial increase in
longevity associated with a modest alcohol
intake, compared to complete abstention.
Several explanations are plausible; it has
been suggested that this added longevity
is caused by the effect of alcohol on raising HDL-C (*see* Chapters 3 and 8). Observational studies indicate that those persons
who consume two to three drinks per day
generally have HDL-C values 5- to 10-
mg/dL higher than those who abstain from
alcohol. Consumption of about two drinks
per week, equivalent to an ounce of alcohol,
will raise the HDL-C concentration approximately 1-mg/dL. Whether this involves predominantly HDL_2 or HDL_3 and whether the
initial level of HDL influences the increment
is not clear.

No evidence exists to show that various
alcoholic products differ in this effect. Although less than 2 ounces of alcohol per day
may not be harmful, its therapeutic use for
raising HDL cannot be recommended because of the possibility of habituation, its
potentially adverse effects on behavior,
other medical disorders, and an increased
propensity for accidents.

Other Factors

Coffee. As explained in Chapter 3,
cross-sectional studies suggest that some
brands of coffee, especially when boiled or
decaffeinated, appear to increase the blood
cholesterol levels. Although the quantitative
value and the clinical significance of this effect is still unresolved, it appears appropriate to recommend a reduction in coffee
consumption by hypercholesterolemics who
drink more than 3 to 5 cups daily.

Lecithin. Lecithin belongs to the category of lipids known as phospholipids,
which act as biologic detergents to enhance
lipid solubility in water and in other fats.
Phospholipids are an important component
of cell membranes and the outer structure of
the blood lipoproteins. Claims have been

made that lecithin reduces total cholesterol
concentrations but no well-controlled clinical trial substantiates this effect. The ability
of lecithin to lower blood cholesterol levels is
too insufficiently documented to justify its
use as a cholesterol-lowering agent. Supple-
mental lecithin also is unnecessary because
its breakdown products—choline, phos-
phate, glycerol, and unsaturated fatty
acids—are consumed in common foods.

Pantetheine. Pantothenic acid, a B vitamin, has been used in preliminary studies of
lipid lowering among children. The early results appear to show some benefits with few,
if any, toxic effects.

Determinants and Variability of Dietary Responsiveness

Hyperresponders Compared to Hyporesponders

Dietary fat and cholesterol have a major
influence on blood cholesterol levels and
the risk of developing CHD. International
population studies clearly demonstrate this
relation. Those nations that have a high intake of saturated fat and cholesterol generally have a high prevalence of CHD,
whereas those nations that have a low intake
of dietary saturated fat and cholesterol have
little or no CHD (Fig. 10.4). Even in the presence of other major risk factors, the occurrence of CHD usually is low in countries
without concomitantly high levels of fat
intake.

The 1985 National Cholesterol Consensus Conference concluded that the high risk
of developing CHD in the United States is
caused in large part by high blood cholesterol concentrations that result from excessive intake of saturated fat, cholesterol, and
calories. Yet, even within the United States,
where the diet has been customarily high in
saturated fat and cholesterol, not everyone
develops hypercholesterolemia and CHD.
The importance of cholesterol-elevating
diets is related, in part, to their ability to unmask subtle genetic differences in lipoprotein transport. For reasons that are not well
understood, saturated fat and cholesterol re-

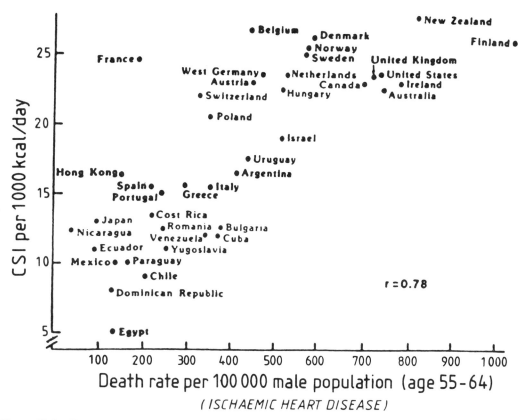

Figure 10.4. The relation between cholesterol-saturated fat index (CSI) and coronary heart disease death rate in men aged 55 to 64 years among major nations. An association exists between the intake of cholesterol and saturated fat with death rates for CHD among middle-aged men in countries throughout the world. Deaths from CHD included 410 to 414, ICD A list, A83, 8th Revision, 1965. The death rate is expressed per 100,000 population in men aged 55 to 64, 1977. The cholesterol (mg/dL) and saturated fat (g/day) were computed from food balance sheets of the Food and Agricultural Organization for 1975 to 1977. The intake of cholesterol and saturated fat is represented by the Cholesterol-Saturated Fat Index (CSI), in which CSI = (1.01 × g saturated fat) + (0.05 × mg C). (From: Connor SL, Gustafson FR, Artaud-Wild SM, Flavell DP, Classick-Kohn, C.J., Hatcher LF, Connor WE. The cholesterol/saturated-fat index: an indication of the hypercholesterolemic and atherogenic potential of food. Lancet 1986;1(8492):1229–1232.)
"Copyright by the Lancet Ltd."

duce the synthesis and expression of the LDL receptor and inhibit the removal of LDL-C from the blood.

For many persons, a diet high in saturated fat and cholesterol does not necessarily elevate the blood cholesterol concentration. In contrast to dietary triglycerides and phospholipids, the digestion and absorption of which are virtually complete, only about 50% of dietary cholesterol is absorbed. Apparently, considerable individual variability in the absorptive capacity of the intestine for cholesterol exists. Other dietary fat probably influences absorption because the bile composition varies with the polyunsaturated to saturated fat (P/S) ratio. The quantity of dietary fat influences the formation of mixed micelles (particles composed of bile salt, fatty acids, and B-monoglycerides) in the gut; some phospholipid monoglycerides and fatty acids are needed for this process. Excess dietary fat also leads to an influx of chylomicron remnants rich in cholesteryl esters into the liver. Dietary cholesterol, in the form of chylomicron remnants, is removed from the circulation by the liver. Cholesterol is structurally different from other major lipids of the body and contains a steroid-ring nucleus (Fig. 10.5). Unlike fat, protein, and carbohydrate, the ring structure of the sterol nucleus cannot be broken down and must be excreted or reabsorbed and stored. Dietary intake of cholesterol usually ranges from 250 to 500 mg (6.47-12.93 mmol) per day, but an additional 600 to 1000 mg (15.6-25.9 mmol) of cholesterol is excreted into the gastrointestinal tract each day in bile. Routes for excretion are limited, and an efficient enterohepatic pathway recaptures much of the cholesterol excreted into the bile. In the absence of bile acids, cholesterol cannot be absorbed efficiently from the gastrointestinal tract and about 50% of intestinal cholesterol is excreted as fecal neutral steroids.

Individual sensitivity to dietary cholesterol relates to differences in cholesterol absorption, the quantity of endogenously synthesized cholesterol, the suppressibility of endogenous cholesterol synthesis by dietary cholesterol and the ability to increase bile acid production from cholesterol. Tight metabolic regulation of these factors permits the blood cholesterol and LDL-C concentration to be maintained within relatively narrow limits. Approximately two-thirds of all persons are capable of regulating their blood cholesterol concentration to compensate for a high dietary intake of cholesterol (Fig. 10.6). On a low-cholesterol (200 mg/day [5.17 mmol]) diet, the contribution of absorbed dietary cholesterol relative to total input (diet and endogenous) of cholesterol is small. As dietary cholesterol intake is augmented, two responses are possible. For the majority of the population, compensation for the increased cholesterol occurs by feedback suppression of endogenous synthesis and maintenance of LDL receptor levels and enhanced secretion of bile acid. However, about one-third of the population (noncompensators) are unable to reduce endogenous synthesis effectively or increase bile acid synthesis, so that cellular cholesterol input continues and apoB/E receptor levels decrease, resulting in an elevated blood cholesterol concentration.

In these vulnerable noncompensators, high intakes of saturated fat and cholesterol increase cholesterol absorption and do not inhibit endogenous cholesterol synthesis efficiently. Because hepatic cholesterol inhibits the expression of the LDL receptor, these persons may possess a physiologic mechanism that allows smaller increments in blood cholesterol to inhibit receptor synthesis (i.e., increased negative feedback sensitivity). These persons experience a much greater increase in the blood cholesterol concentration in response to diet than others. On the typical American diet, such persons are uniquely susceptible to hypercholesterolemia and, consequently, are also the ones who are most likely to benefit from a diet restricted in saturated fat and cholesterol. The mechanism whereby a low-fat diet decreases LDL synthesis is not yet resolved.

Dietary restrictions can be expected to lower total blood cholesterol, on the average, about 15%, depending on initial cholesterol levels the degree of restriction of dietary saturated fat and cholesterol, adher-

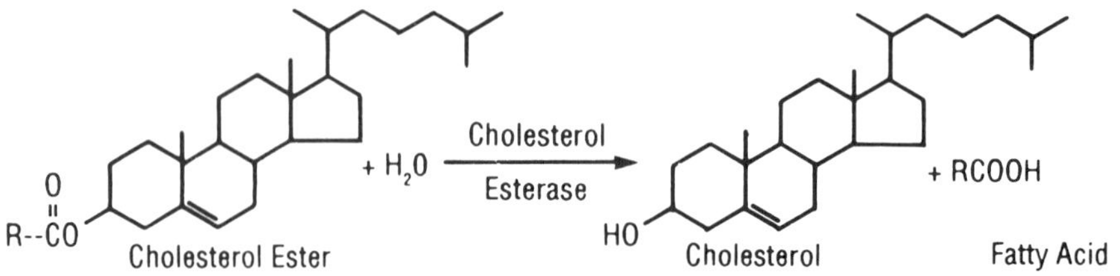

Figure 10.5. Chemical structures of cholesteryl ester and free cholesterol. (From: U.S. Department of Health and Human Services, Public Health Service, National Institutes of Health. Recommendations for improving cholesterol measurement:a report from the Laboratory Standardization Panel of the National Cholesterol Education Program. NIH Pub. No. 90-2964, Feb. 1990.)

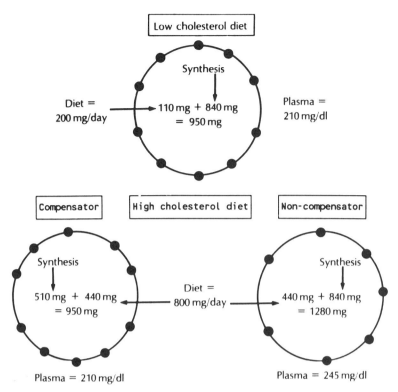

Figure 10.6. Schematic representation of the effects of dietary cholesterol on endogenous cholesterol synthesis and apoB/E receptor levels in man. On a low dietary cholesterol intake about 55% of dietary cholesterol is absorbed, and the majority of cholesterol is derived from endogenous synthesis, allowing apoB/E receptors and plasma cholesterol to be maintained at a steady-state level. With a moderate increase in dietary cholesterol (800 mg per day or approx. 300 mg per day more than the average American consumes), two types of response generally take place: The compensator is able to reduce endogenous cholesterol synthesis (840 mg to 440 mg) to compensate for the increased dietary cholesterol and keep apoB/E receptor levels (represented by dots) and plasma cholesterol concentrations constant. The noncompensators are unable to suppress endogenous cholesterol synthesis. The combined cholesterol input results in diminished apoB/E receptor levels and an increase in plasma cholesterol concentrations. (Adapted from: McNamara DJ: Dietary cholesterol: effects on lipid metabolism. Curr Opin Lipid 1990;1:18–22.)

ence, and inherent individual responsiveness to dietary changes. Age, sex, previous diets, and metabolic status also will influence the response to a restricted diet. It has been reported experimentally that even the number of meals per day could affect blood lipids. In any event, there is a marked variability in response to dietary changes.

Estimation of Dietary Effects on Blood Cholesterol

The average changes in blood cholesterol anticipated from dietary changes can be estimated roughly from data based on controlled clinical studies. It cannot be emphasized too strongly that such theoretical calculations are only group estimates, not necessarily applicable to individual situations, and apply only to a typical range of dietary fat intake. After a series of experiments, Keys and coworkers developed a formula for predicting the change in serum cholesterol from the change in percent of calories derived from saturated fat, polyunsaturated fat, and dietary cholesterol. The Keys equation shows that saturated fatty acids are twice as powerful in raising cholesterol levels as polyunsaturated fatty acids are in lowering them, whereas dietary cholesterol has a lesser effect on the blood cholesterol:

$$CHOL = 1.35\ (\ 2\ S\ -\ P) + 1.52\ Z$$

where

CHOL = estimated change in serum cholesterol in mg/dL; S = change in percentage of daily calories from saturated fat; P = change in percentage of daily calories from polyunsaturated fat; and Z = change in the square root of daily dietary cholesterol in mg/1,000 calories.

Hegsted and coworkers also developed similar regression equations that predict the response of plasma cholesterol concentration to changes in dietary fats. It generally is believed that 100 mg dietary cholesterol per 1000 calories per day will increase the total cholesterol by 8 to 10 mg/dL (0.21-0.26 mmol/L), although changes as low as 2 to 5 mg/dL (0.05-0.13 mmol/L) also have been reported. Conceptually, a minimum of about 100 mg of dietary cholesterol per day would be necessary to produce an increase in the blood cholesterol concentration. Vegetarians and others who routinely consume a diet that contains less than 10% of the calories as fat and less than 100 mg of cholesterol per day have blood cholesterol concentrations around 140 mg/dL (3.62 mmol/L), similar to that of young adults in this country. The effect of dietary cholesterol is clearly expressed among Americans whose intake generally ranges from 250 to 500 mg (6.47 to 12.93 mmol) per day. With more caloric intake and cholesterol consumption, the relative effect of dietary cholesterol diminishes, and it has been established that intakes that exceed 500 mg/day have only small incremental effects on the blood cholesterol levels.

A large segment of the population who have high blood cholesterol values have this problem because of an excessive intake of saturated fat and cholesterol. It is not unreasonable, then, to initiate therapy with nutritional counseling. The magnitude of the change in blood cholesterol depends in large part on factors that are not included in the prediction equations. Consequently, all patients must be given an empirical trial of dietary therapy to identify those who will respond and the extent of the response. It is possible to lower the blood cholesterol levels by as much as one-third by losing weight and restricting saturated fat and cholesterol intake, although 10% to 15% reductions in total cholesterol values are more common. Despite the relatively modest expectations, dietary alterations certainly may be sufficient to convert borderline cholesterol concentrations to acceptable levels, preventing the need for lipid-lowering drugs in some patients and reducing the dose of such drugs in others. Most physicians and their patients would prefer to control the blood cholesterol without the use of drugs and should be given the opportunity to use diet alone. To adequately determine the effectiveness of dietary control, patients must be given a trial of a cholesterol-lowering diet for at least 3 to 6 months. Many hypercholesterolemic patients wrongly believe that they are consuming a low-cholesterol diet and may experience a substantial reduction in their blood cholesterol after they have received appropriate nutritional advice. Also, as noted previously, dietary cholesterol probably affects CHD risk *independent* of its effect on the blood cholesterol.

Diet remains the cornerstone of management for most patients who have lipid disorders. Liberalizing a diet because it was ineffective as sole therapy for a lipid problem usually makes it more difficult to obtain control with drugs. Drug therapy always should be accompanied by diet to minimize the drug dose and related adverse effects.

Dietary Guidelines for Healthy Adults

Recently, specific guidelines for healthy American adults have been proposed by several health agencies: the Public Health Service, the American Heart Association (AHA), and the National Research Council. For adults in the United States, dietary goals formulated by these expert committees are remarkably similar to the guidelines adapted from AHA recommendations, which stipulate that:

1. Total fat intake should be reduced to 30% or less of calories.
2. Saturated fat intake should be less than 10% of calories.
3. Polyunsaturated fat intake should not exceed 10% of calories.
4. Monosaturated fat intake should provide the remainder of fat calories.
5. Cholesterol intake should not exceed 300 mg per day.
6. Carbohydrate intake should constitute 50% or more of calories, with emphasis on complex carbohydrates.
7. Protein intake should provide the remainder of calories but at levels less than twice the Recommended Daily Allowance (RDA) (1.6 g/kg body weight) for adults.
8. Sodium intake should not exceed 3 g per day.
9. Alcohol consumption should not exceed 1 to 2 ounces of ethanol per day.
10. Total calories should be sufficient to maintain recommended body weight (Table 10.5) but not be excessive.
11. A wide variety of foods should be consumed to meet nutritional requirements for vitamins, minerals, and macronutrients.

Despite such recommendations, the dietary habits in the United States have remained ill-suited to maintaining desirable lipid levels. The mean daily intake of nutrients recently has been estimated and related to current recommendations from the AHA in Table 10.6. Based on one-day dietary recalls, the American diet was estimated to contain 37% of daily calories as fat. Carbohydrate and protein account for 46% and 16% of daily calories, respectively, and fiber intake approximates 15 g. Cholesterol intake for men (435 mg) tops that for women (304 mg). The total calorie intake generally exceeds requirements for maintaining desir-

Table 10.5. Desirable Body Weight Ranges[a]

Height Without Shoes	Weight Without Clothes	
	Men (Pounds)	Women (Pounds)
4'10"	—	92–121
4'11"	—	95–124
5'0"	—	98–127
5'1"	105–134	101–130
5'2"	108–137	104–134
5'3"	111–141	107–138
5'4"	114–145	110–142
5'5"	117–149	114–146
5'6"	121–154	118–150
5'7"	125–159	122–154
5'8"	129–163	126–159
5'9"	133–167	130–164
5'10"	137–172	134–169
5'11"	141–177	—
6'0"	145–182	—
6'1"	149–187	—
6'2"	153–192	—
6'3"	157–197	—

[a](From: Dietary treatment of hypercholesterolemia: a manual for patients. American Heart Association in cooperation with The National, Heart, Lung, and Blood Institute, 6-88-1MM; 88 02 25 A, p. 9, 1988, [Adapted from the 1959 Metropolitan Desirable Weight Table].)

able body weight. It is not surprising that NHANES II reported 25.6% of Americans aged 20 to 74 years as being overweight.

Dietary Modifications to Alter Blood Cholesterol

General Guidelines

Many persons equate a cholesterol-lowering diet with limiting intake of eggs and red meat. This is an oversimplification, because saturated fats, cholesterol, and excess calories all account for the relatively high levels of cholesterol in Americans, and this situation should be addressed by altering all food intake in which fats, cholesterol, and excess calories occur rather than targeting specific foods. For some persons, the dietary changes require elimination of few obvious sources of fat and cholesterol, without a need for radical changes in dietary habits. But for most persons, dietary counseling is essential to make the necessary changes. The lack of enthusiasm for dietary

Table 10.6. Comparison of the Average American Diet with the Two Recommended Treatment Diets[a]

Dietary Component	Average American	Step 1	Step 2
		Percent Calories	
Total Fat	37%	<30%	<30%
Saturated Fat	16%	<10%	< 7%
Polyunsaturated Fat	7%	<10%	<10%
Monounsaturated Fat	14%	10–15%	10–15%
Carbohydrates	(46%)	(55%)	(55%)
	28% Complex	45% Complex	45% Complex
	18% Refined sugars	10% Refined sugars	10% Refined sugars
Protein	16%	10–20%	10–20%
Cholesterol	304 mg/day (women) 435 mg/day (men)	< 300 mg/day	< 200 mg/day
Total Calories		To achieve and maintain desirable weight	

[a] (Adapted from: National Cholesterol Education Program. Report of the expert panel on detection, evaluation, and treatment of high blood cholesterol in adults. NIH Publication No. 88-2926, November 1987.)

therapy by many physicians stems from unsuccessful personal experience with patients who have received only brief counseling, often without educational materials and without follow-up.

The dietary approach most often used is that advocated by the AHA and affiliated organizations—a two-step procedure with emphasis on a permanent change in eating behavior. The Step-1 Diet calls for the total fat intake to be reduced to below 30% of total calories, with less than 10% consisting of saturated fat, and restricting the cholesterol intake to less than 300 mg per day (*see* Table 10.6). The Step-2 Diet reduces saturated fat even more, to <7%, and dietary cholesterol to <200 mg per day. The translation of these percentages into actual grams of fat for various caloric intakes is shown in Table 10.7. Each percent decrease in calories from saturated fat is expected to produce an average decrease of up to 3 mg/dL (0.08 mmol/L) in blood cholesterol levels. In a similar way, reduction of dietary cholesterol by 100 mg per 1000 calories consumed will decrease blood cholesterol by 8 to 10 mg/dL (0.21 to 0.26 mmol/L). On average, the Step-1 Diet can be expected to decrease total cholesterol by 40 to 50 mg/dL (1.03-1.29 mmol/L), or by about 15%. Advancing to the Step-2 Diet definitely requires the assis-

Table 10.7. Translation of Percent Fat to Daily Fat Intake in Grams for Different Levels of Total Daily Caloric Intake

Total Calories	Percent Fat		
	30%	10%	7%
1000	33[a]	11	8
1500	50	17	12
2000	66	22	16
2500	84	28	19
3000	100	33	23

[a] Fat intake in grams.

tance of a nutritionist. At this level of dietary restriction, another reduction of blood cholesterol approximating 15 mg/dL (0.39 mmol/L) can be anticipated. The total absolute reduction in cholesterol is not likely, however, to be greater than 50 to 60 mg/dL (1.29 to 1.55 mmol/L) by dietary changes alone. In most patients who have total cholesterol levels approaching 300 mg/dL (7.76 mmol/L), diet alone will not be successful. The approximate expectations of lowering the total cholesterol from the various dietary components of the Step-1 and Step-2 diets are listed in Table 10.8.

Nutritional information on the content of saturated fat and cholesterol, which are responsible for the majority of the cholesterol-raising potentials of food, is available on

Table 10.8. Approximate Expectations of Lowering Total Cholesterol Using Step-1 and Step-2 Diets[a]

Dietary Component	Current American Intake	Step-1 Maximum Cholesterol Change (mg/dL)	Step-2 Maximum Cholesterol Change (mg/dL)
Saturated fatty acids (% Calories)	17%	−19	−27
Cholesterol (mg/day)	500	−10	−15
Total caloric intake	High	−10	−10
Soluble fiber	Low	−5	−5
Total Change:		−44	−57

[a](Adapted from: American Heart Association, Physicians' Cholesterol Education Program, 1988.)

about 50% of all processed foods. It is wise to avoid prepackaged goods that do not have such information. But current labeling allows major misconceptions, and the terms *low fat* and *no cholesterol* often are used regardless of saturated fat content, the dietary constituent that has the greatest effect on blood cholesterol. Nevertheless, when read properly, labels often do provide information that assist in making food choices. Several commercial practices are expected to be instituted in the future to facilitate the selection of proper foods, including the newly proposed FDA labeling rules for cholesterol.

Attention also must be given to the methods of food preparation. Cooking methods that add little or no fat, such as steaming, baking, broiling, or grilling, are preferred. The introduction of the microwave, nonstick pans, and substances that line these pans (such as vegetable cooking sprays), permit cooking without fat. Soups, stews, and other foods that contain fats should be chilled after cooking so that the fat can be skimmed off. For those persons who must eat away from home, meals should be ordered without high-fat sauces, butter, or salad dressing.

Specific Food Groups

Meats, Poultry, and Seafood. Because animal sources contribute most of the saturated fatty acids in our diet, it is necessary to select lean cuts of meat, trim all visible fat, and prepare by broiling or grilling. Lean cuts of beef or pork include round, rump, loin

(sirloin and tenderloin), and flank. Daily meat proteins should be reduced, with care taken to supplement the required amounts of iron and vitamins by other means.

Some products contain significant quantities of cholesterol, but their low content of saturated fat makes them preferable to red meat. This is true for chicken and turkey, the fat of which can be reduced mostly by removal of the skin and any underlying fat. Even though fish contain some cholesterol, they usually are also low in saturated fat and provide a good source of protein. The same considerations apply to shellfish, such as shrimp and crawfish, and it therefore is possible to substitute reasonable portions of shellfish (e.g., 5 to 6 shrimp) for meat products. The benefit of fish and shellfish can be negated, however, by frying in highly saturated fat or preparing the food with fat-rich sauces.

For all meats, it is advisable to keep the portion size at approximately three ounces after cooking, which is about the size of a deck of playing cards. With a caloric intake of 2000 kilocalories per day, the ingestion of two portions of meat, fish, or poultry per day still permits a reasonable amount (5 to 8 teaspoons) of saturated fat from dairy products, vegetable oils, or margarine. In contrast, the fat in processed meats, such as hot dogs, sausage, bacon, and bologna, often exceeds 50% by weight and should be avoided.

Eggs. Egg yolks contain on average about 213 mg of cholesterol each. Although this is lower than older data that estimated

274 mg per egg, it is more than two-thirds of the RDA cholesterol allowance. Eggs therefore should be limited to no more than three per week for the Step-1 Diet and only one per week for the Step-2 Diet. The cholesterol content of eggs is confined to the yolks, so that separated egg whites may be eaten as often as desired. It must be remembered that egg yolks often are used in commercially prepared foods and in dishes such as bakery goods, potato salad, and tuna salad, so that allowances must be made for those sources as well.

Fats, Oils, and Nuts. Most vegetable oils contain mostly unsaturated fat and need not be overly restricted in cholesterol-lowering diets unless calories are a major concern. Satisfactory oils include corn, cotton seed, olive, rapeseed (canola), safflower, soybean, peanut, and sunflower. The recent emphasis on monounsaturated fats suggests that canola and olive oil might be particularly desirable selections. Peanut butter also is a good source of monounsaturated fat; although commercial peanut butter products contain some hydrogenated fat (additives) to prevent separation of the product during storage, they still contain predominantly monounsaturated fatty acids. Nuts and seeds are predominantly monounsaturated fatty acids or polyunsaturated fatty acids, although cashews and macadamia nuts contain more saturated fat and should be used sparingly. Commercial mayonnaise and salad dressings are made from polyunsaturated fats, usually soybean oil, but as with all fats and oils, they should be used sparingly because of their high fat and caloric values.

In contrast, some fats and oils are high in saturated fat and cholesterol and should be avoided as much as possible. These include coconut oil, palm kernel, and palm oil, which often are used commercially for pastries, cereal products, and other processed foods because they are less expensive than most other vegetable oils and have a stable structure that resists oxidation. Chocolate usually is made with cocoa butter and should also be eaten in moderation. Fully hydrogenated oils should be avoided altogether and partially hydrogenated oils should be used in moderation.

Dairy Products. Whole milk contains at least 3% fat by weight, most of which is saturated; half of the calories available from whole milk come from this fat. For this reason, dairy products derived from 1% fat or skim milk are preferred. Daily use of low-fat dairy products maintains calcium intake and is part of a well-balanced diet. Acceptable cheeses are those made partly from skim milk and that contain 2 g of fat or less per ounce. Substitution of low-fat yogurt or low-fat cottage cheese for sour cream and salad dressings is desirable. Cream and butter contain too much fat and should be avoided altogether by substituting skim milk or soft margarines made from vegetable oils. Regular ice cream should be replaced with ice milk, sherbet, or sorbet. Frozen low-fat yogurt usually has only one-fourth the fat of ice cream. Although nondairy coffee creamers and other substitutes, such as whipped toppings, frequently are made from coconut or palm oil, they are advertised as containing no cholesterol; however, when used infrequently they probably are not an important source of saturated fat.

Fruits and Vegetables. Fruits and vegetables, rich in vitamins and water-soluble fiber as well as other nutrients, constitute an important part of the diet. Fruits can be substituted for pastries, candy bars, and other snacks and desserts. Olives and avocados have a high monounsaturated fat content. Coconut and coconut products should be avoided because of the high saturated fat content. Peas, beans, pintos, and other vegetables are good sources of protein, contain little fat, and are high in water-soluble fiber.

Bread, Cereal, and Whole Grain Products. Breads, cereals, pasta, and starchy vegetables prepared properly are low in fat and can be used readily. This food group should not be restricted unnecessarily unless total calories or triglyceride levels are of concern. Grain products furnish complex carbohydrates, fiber, and appropriate vitamins and minerals.

At least four to six servings daily of this food group provides B vitamins and iron.

There can be wide variation in fat among grain products. Biscuits, croissants, and some cereals contain large amounts of fats because of their preparation and they should not be used.

Other. Acceptable desserts include sherbet, angel food cake, frozen low-fat yogurt, and ice milk. As mentioned previously, fruits also can be used for desserts and snacks. Choices for snacks include popcorn (air popped), graham crackers, melba toast, some ready-to-eat cereals, vegetables, and bagels. Commercially prepared snack crackers and chips contain saturated fat and, therefore, should be avoided.

Weight Reduction

For many persons, weight loss is nearly as important as the composition of the diet. An approximation of desirable weight can be estimated by the following formula:

> Women: Desirable weight (lbs) = 100 + 5 for every inch above 5 feet.
> Men: Desirable weight (lbs) = 110 + 6 for every inch above 5 feet.

To succeed in weight loss, a sustained commitment to diet is essential. The rate of weight loss should not be so slow as to discourage patients nor so fast as to encourage fad diets or other meal plans that provide rapid weight loss but cannot be maintained. Persons who lose weight often regain it, especially women, leading to the so called *rhythm method of girth control*. Data on such cyclic weight reduction indicate that it becomes progressively more difficult to lose weight, presumably because of a gradual decrease in metabolic rate. Such fluctuations in weight actually may be detrimental and increase long-term morbidity and mortality. A caloric deficit of only 100 kilocalories per day would result in a 10-pound weight loss over a period of 1 year. It is most practical to design an eating plan that is reasonably comfortable and that is associated with a moderate calorie reduction and modest goals. Weekly weight loss goals could be set at about 1% of body weight. For those who are moderately obese (130 to 200% of

desirable weight), a loss of about six pounds per month is ideal.

Each gram of fat has more than twice the calories as the same amount of protein or carbohydrate. By reducing the amount of dietary fat and substituting foods high in complex carbohydrates, calories will be reduced. Alcohol, at seven calories per gram, may be a major source of calories for some persons, and moderation is recommended. It should be recognized that there are certain social benefits from eating that may make dieting difficult. Not infrequently, social events center around food—business lunches, family get-togethers, birthday parties, and so forth. In addition, some people live to eat and are reluctant to give up dietary pleasures. Finally, both metabolism and activity levels change with aging, complicating the opportunities for successful weight loss even more. For these reasons, the importance of life style changes cannot be overemphasized.

Alternative Methods of Weight Reduction

The short-term success of many commercial weight programs has been attributed to frequent weigh-ins and accountability. Very low calorie diets that initially supply less than 700 kcals per day were discarded because of safety issues relating to case reports of sudden death and other complications. Current very low calorie diets are more nutritionally sound and higher in calories than the original diets and may play a definite role in exceptional cases (such as morbid obesity). Yet, such diets require careful medical supervision. The problem of long-term maintenance of weight loss with these diets has not yet been solved, and emphasis on life-long behavioral changes is crucial. For these reasons, the management plan for weight control should be easily adaptable to dietary preferences and, in most cases, provide no less than 1200 calories. Appetite suppressant drugs may facilitate weight loss, but weight is regained more often after drug therapy than with other methods.

Desirable body weight ranges are shown in Table 10.5. To reduce weight, a low-calorie diet should be compatible with either the Step-1 or Step-2 dietary recommendations. A sample of diet of 2000 calories for use with the Step-1 Diet is given in Appendix D.

As a typical routine, persons should lose weight at a steady rate of 1 to 2 pounds per week until the ultimate goal is reached. Typical caloric levels for men are 2000 (sedentary) to 2500 (physically active) calories for weight maintenance, and 1600 to 2000 calories for weight loss. Corresponding values for women are 1600 to 2000 calories for weight maintenance and 1200 to 1600 calories for weight loss. To produce the desired weight loss, calculate the total amount of calories needed daily; this allows estimation of the amount of caloric expenditure required to lose 1 pound per week. A simple technique for estimating daily caloric requirements for four different physical activity levels is given in the table below. Because the caloric equivalent of one pound of fat is 3500 kcal, a deficit of 500 kcal/day is required to achieve a weight loss of one pound per week (500 kcal × 7 days). For example, to reduce at the rate of one pound per week a very sedentary 50 year old man who weighs 90 kg (198 lbs) and is 175 cm tall (69 inches) require 1834 basal kcal and 458 activity kcal. Daily caloric requirements to lose

one pound per week would be 1792 kcal [(1834 + 458) − 500].

Because basal energy requirements vary by as much as 100% in persons of similar sex, height, and weight, these values are averages and serve only as guides. Consequently, the effectiveness of a diet must be evaluated individually. Failure to lose weight on a specific diet does not necessarily indicate noncompliance and may require only adjustment of caloric intake or expenditure.

Exercise is a valuable adjunct to weight control programs. At a constant caloric intake, augmented physical activity will lead to weight loss. Also, exercise confers additional health benefits gained from improved physical fitness and alteration of risk factors. The increased caloric expenditure provides a greater allowance of calories, permitting a more varied diet with adequate nutrients. Studies are inconsistent in terms of whether exercise in combination with calorie restriction alters lipoprotein patterns more favorably than caloric restriction alone. Most studies indicate that such a combination results in a preferential loss of fat rather than lean body mass, as well as better maintenance of weight loss. Exercise also provides an alternative activity to eating.

Daily Caloric Requirements

A. Basal Calories (Harris Benedict Equation)

BMR (kcal) Males = $66 + (13.7 \times W) + (5 \times H) - (6.8 \times A)$ = _____

BMR (kcal) Females = $655 + (9.6 \times W) + (1.7 \times H) - (4.7 \times A)$ = _____

W = weight in kg; H = height in cm; and A = age in years

B. Activity Calories (choose one)

Very sedentary	25% × _____	(BMR)	= _____
Sedentary	30% × _____	(BMR)	= _____
Moderately active	40% × _____	(BMR)	= _____
Very active	50% × _____	(BMR)	= _____

C. Total Calories for weight loss of 1 lb per week

(A. Basal Calories + B. Activity Calories) − 500 kcal = _____

Diet Reference Manual, ed. 2, Massachusetts General Hospital, Department of Dietetics, 1984.

Although desirable for general health, it is not crucial to exercise at levels that will induce cardiovascular fitness. The intensity of exercise is of little importance, because the main consideration is the duration of activity, which determines the total number of calories expended. An exercise routine that can be followed on a regular basis should be established. The exercise should be maintained for at least 20 to 30 minutes, four times or more per week. As long as the activities are acceptable and pleasurable to the patient, there is no need to tailor them to any specific program. In addition to the structured routine, exercise should be incorporated into daily activities, such as walking and climbing stairs.

Suggestions for losing weight have been formulated by the AHA as follows:

1. Look at your present eating habits. Write down everything you eat or drink in one day. Be sure to note the time of day you ate and the circumstances that surrounded the eating (such as watching TV, after an argument, and so forth).
2. Include a variety of nutritious foods in your meals—lean meat, poultry, and fish; fruit, vegetables, and grain products; and low-fat dairy products.
3. Eat regular, well-balanced meals each day. If you really get hungry between meals, save a small piece of fruit, some raw vegetables, or a glass of low-fat milk from one of your regular meals. If you feel an urge to eat when you really are not hungry, try exercising instead. Drink a large glass of water or tea, refresh yourself by taking a shower or brushing your teeth, or get busy with a favorite hobby.
4. Calorie-proof your house. Get rid of brimming candy dishes, sweet soft drinks, tubs of ice cream, giant bargain bags of snack foods, and that bowl of nuts by the TV.
5. Gradually build more physical activity into your day. Take a walk at lunch time, use the stairs instead of the elevator, and spend at least 30 minutes at least three times per week enjoying an active sport,

such as jogging, brisk walking, cycling, or swimming.

Counseling

The rationale for dietary counseling is based on the need to lower blood cholesterol and triglyceride levels maximally with minimal use of drugs. Dietary control may obviate the need for drugs or at least reduce drug dosage with fewer side effects at less cost. The role of diet in the overall management of the lipid disorder must be clearly outlined.

Before instituting treatment, just as in other spheres of medicine, the physician should take an adequate dietary history. The initial history should cover mealtime practices, including eating out and previous experiences with special diets. A dietary habits worksheet can prove useful (Appendix B). Key points include the following:

1. At what times of the day does the patient usually eat?
2. Are some meals routinely skipped?
3. At what time does the patient eat the largest meal?
4. Where are meals typically prepared (e.g., in a restaurant, fast food chain, or at home)?
5. Are meals eaten at home prepared from packaged convenience foods or are they prepared at home from foods purchased in the market?
6. What are the patient's favorite foods and what foods are disliked?
7. Which foods will be most difficult to increase or decrease?
8. Who shops and who prepares the food?

A food frequency checklist (Appendix C) can be helpful in summarizing a person's usual dietary habits. Attention to the patient's social habits, cooking facilities, and knowledge of food preparation and nutrition is necessary before realistic dietary goals can be established.

The nutritional assessment also should include a series of weights: the patient's current weight, weight as a young adult (high school, college graduation, or military service), and maximum and minimum weights ever. Generally, the weight as a young adult represents an ideal weight for many patients and provides a useful guideline.

Characterization of physical activity levels and patterns should be documented. These data should include the type, intensity, frequency, and duration of exercise performed (e.g., jogs 3 miles in 60 minutes, four times per week). This information allows an estimation of caloric expenditure, as well as likely benefits and safety of the routine. Until such information has been acquired, the patient cannot be counseled effectively regarding recommendations and goals.

Most patients prefer an adequate dietary trial before institution of life-long drug therapy. As a prologue to any dietary therapy, it is best to review with the patient the common terms used, such as the types of fatty acids and the recommended percentage of fats, proteins, and carbohydrates to be consumed. Patients must also be made aware of the nutrient content of various foods and how preparation alters the contribution of foods toward blood cholesterol levels. Patients must be able to interpret food labels, ask the appropriate questions when dining out, and have access to reputable cookbooks. Much of the available literature and data on dietary recommendations contain erroneous information, and the patient must be able to recognize this. Once patients are aware that recommended diets include a wide choice of acceptable foods and modes of preparation, long-term maintenance of acceptable diets becomes easier.

The response to dietary change should be evaluated continuously; adjustment to new eating patterns should be discussed with the patient; and lipids should be remeasured at approximately 1 and 3 months after implementation of the Step-1 Diet (Fig. 10.7). If the goal has not been achieved with these dietary recommendations, then more instruction on the Step-1 Diet or advancement to a Step-2 Diet is indicated.

The shift in cholesterol will occur mainly in LDL-C, although weight loss, smoking cessation, and exercise may provide benefits in HDL-C as well. It usually takes 3 to 6 months for patients to make the necessary behavioral changes to implement these dietary modifications, and the physiologic response to such a diet takes another 3 to 8 weeks. Behavioral strategies and written instructions will help most patients. If progress is unsatisfactory, consideration should be given to using alternative methods of motivation or contemplating the earlier use of drugs to emphasize to the patient how important it is that the lipid goal be reached. When patients are faced with the use of drugs, they sometimes reconsider a lackadaisical attitude toward diet.

The Physician's Role. The physician should given an overall appraisal of the patient's diet and give each patient a realistic expectation of the potential lipid changes that could occur as a result of successful dietary modifications. For the diet to succeed, however, the physician must be an enthusiastic advocate, even if the details of the dieting are delegated to a nutritionist. Success of dietary therapy depends largely on the physician's attitude, knowledge, and skills in motivating the patient to follow the appropriate diet. A good understanding of the patient's dietary habits will facilitate evaluation of necessary modifications in eating behaviors. These activities take considerable time, however, and may be delegated more effectively to a nutritionist or other trained staff member. The value of a nutritionist in implementing cholesterol-lowering diets cannot be overstated.

Educating patients, or taking even a brief dietary history, may require more time than most physicians are willing to take. A number of aids and alternatives can be used. A plethora of booklets and audio-visual materials are available from local affiliates of the AHA, the NIH, and the USDA. In addition, excellent video tapes on many of these topics now are available. (Many of these materials are described in Chapter 13 on Educational Resources.) Physicians should consider developing a patient education area in which patients actually can review these materials and watch the video tapes in the physician's office. Because some patients

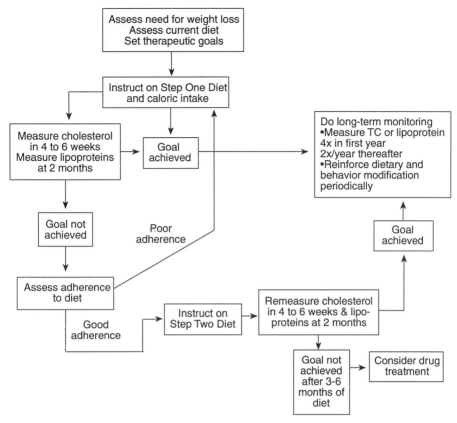

Figure 10.7. General outline of dietary management for dyslipidemic persons. After initial assessment of the baseline diet and need for weight loss, a Step 1 diet is initiated. Persons who do not achieve their therapeutic goals after several months on this diet should be put on the more stringent Step 2 diet. Pharmacologic therapy generally should be reserved for those persons who have not benefited from a dietary therapy of at least 3-6 months. If the goal is achieved by diet alone, however, long-term monitoring is necessary. (Adapted from: Report of the National Cholesterol Education Program expert panel on detection, evaluation, and treatment of high blood cholesterol in adults. Arch Intern Med 1988;148:36–69.)

may wish to take these materials home, a system for lending or renting such materials would be an important consideration.

The Dietitian's Role. The availability of a registered dietitian familiar with lipid reduction and behavioral modification aids the physician in the successful implementation of these dietary plans. The physician may wish to implement the Step-1 Diet without assistance, although few physicians have the knowledge to counsel patients adequately on specific details. The Step-2 Diet requires even more details on use and preparation of foods.

Referral to community sources for nutritional counseling can be very useful. Nutri-

tionists or registered dietitians usually are available in private practice and in official health agencies such as county health departments. Many of the large metropolitan areas have a *Dial A Nutritionist* program in which qualified dietitians provide advice and materials to interested parties over the telephone. The registered dietitian should help individualize general dietary guidelines. Although most diets are useful for lowering cholesterol and triglycerides, there are differences that must be clearly emphasized. Perhaps the most important dictum to follow in prescribing diets is to tailor the diet to the person's likes and dislikes. The plan must stress permanent modification of eating and life-style habits, while maintaining individual food preferences. Fad and crash diets, or even excellent diets not meeting individual food preferences, are doomed to long-term failure. Follow-up counseling is the key to successful intervention; this can be accomplished by periodic group sessions or individual sessions to modify behavior.

Behavioral Strategies. The following points may prove to be useful behavioral strategies for the nutritionist:

1. Convince the patient of your enthusiasm for implementing dietary change. The patient must share your enthusiasm for the dietary management scheme and feel comfortable enough to ask questions.
2. Speak in terms of dietary changes and healthy eating patterns rather than *diet*. Because the goal of dietary intervention for hypercholesterolemia is long-term change, it is important to speak in terms of diet modification to avoid the implication that the diet will be terminated at some point.
3. Consider counseling to small groups. Group interaction may lend additional support and motivation to some persons, who, when in the group, can take active roles in assisting and influence their own behavior.
4. Avoid speaking of failures. When goals are not met, focus on what can be done to achieve the goals, making it *a learning experience* rather than a failure. Most patients are likely to be unsuccessful in achieving at least some goals. Break the goal down into smaller segments if the original goal is too large. Lack of success in achieving one goal over a short time will have little bearing on the overall objective.
5. Focus on the patient's behavior rather than changes in cholesterol values or changes in weight. Such changes do not always reflect the patient's effort or actual compliance.
6. Praise the patient for achieving desired behavior changes and encourage self-

rewards for making positive changes. Praise should be used for every favorable change, even if the changes were not sufficient to achieve the objectives.

7. Set only a few, short-term, achievable objectives. Patients are more likely to be successful in making long-term changes if successive short-term goals are encouraged. A few changes at a time are easier for the patient who is trying to change life-long habits.
8. Set flexible objectives. Allow the patient to reduce or modify certain desirable foods rather than eliminate them altogether.
9. Set realistic objectives. Objectives that have a high probability of achievement should be set. A useful adage to keep in mind is to set goals that are small enough to achieve but large enough to count.
10. Assess possible barriers to achieving the goals and engage the patient in problem solving for solutions. Such barriers arise from family, social, financial, or logistical sources, such as family members who do not like the new foods or friends who push high-fat or high-cholesterol foods onto patients; social situations in which food choices may be limited to high-fat or high-cholesterol foods; and the patient's concerns over the high cost of foods necessary to make dietary changes. Try to get the patient involved in generating solutions.
11. Make sure that the patient understands the reasons for making changes, the objectives that are being set, and the methods for achieving these objectives. Many patients sit and nod their heads when they do not really understand what is being said to them. The best method of determining whether the patient understands what is being said is to say: "Now that I have explained this to you, I would like to hear your interpretation of what I said just to make sure that you understand me." Use written material to reinforce verbal instructions.

12. Participation of spouses and family often is important for needed social support or to enable the food preparer to understand the recommended changes. The ability of the family and others to provide encouragement and reinforce good dietary behaviors cannot be emphasized enough.

13. Counseling is an ongoing and evolving process. At first, the patient will not know what to ask or how best to achieve goals. Dietary modification is a continual process that requires perio·lic assessment and reinforcement. Patients will be tempted to return to former habits because of urges or cravings from various sources and situations. Teaching strategies to cope with these impulses and using alternative behaviors require repetitive and habitual learning processes.

Sequence of Dietary Therapy

Unless there is a registered dietician on hand, initial instruction on the Step-1 Diet should be done by the physician or a staff member. If the goals of dietary therapy have not been met at follow-up, and especially if a Step-2 Diet is contemplated, the patient should have access to additional dietary information from a dietitian through the physician's office or through community resources. Often the patient will not know what questions to ask and how to comply with the diet until actually experiencing the changes. Also, periodic reinforcement by concerned professionals at subsequent contacts, either visits or telephone calls, should facilitate adherence to the new diet routine.

According to guidelines from the NCEP Report, blood cholesterol should be measured at the end of one month and again at three months. At this time, the physician should be able to assess whether the patient has been able to adhere to the diet and whether the response, based on the blood total cholesterol level, is satisfactory (see Chapter 12 for more details on therapeutic goals). If the response is satisfactory, long-term monitoring with periodic reevaluation

and reinforcement should commence (see Fig. 10.7). The patient should be counseled by a physician, a staff member, or a registered dietitian at least quarterly for the first year of long-term monitoring and every 6 months thereafter. As with weight control programs, relapse to less stringent diets and, consequently, elevated cholesterol concentrations once again remain a distinct possibility. In such situations, individual discretion must be used in additional management. If possible, reinstitution of dietary therapy usually is successful.

If lipid values are not at goal levels, then the patient should be referred to a registered dietitian for more instruction on the Step-1 Diet or for progression to the Step-2 Diet. After additional treatment for a period of another two months, the response to the therapeutic diet should be reassessed. If at this time the therapeutic goal still has not been met, serious consideration should be given to drug therapy unless there is reason to believe that prolongation of dietary therapy would be useful.

Not all persons who fail to achieve target goals for LDL-C should be discouraged. Patients who have severe elevations of blood cholesterol will not be able to achieve goals by diet alone, no matter how strict the diet. Some persons appear to be inherently resistant to LDL-lowering by diet modification alone. Others will be unwilling or unable to adhere to dietary recommendations despite the best efforts of the physician and staff. A prolonged period of dietary therapy may be necessary for some patients. These persons may show a late response because of additional training or additional time required to adjust to the new dietary regimen or to evoke an appropriate metabolic response.

Pediatric Considerations

The guidelines for nutritional modifications in adolescents and children above 2 years of age, as recommended by the NCEP Report of the Expert Panel on Blood Cholesterol Levels in Children and Adolescents, parallel those of the adult population. The average dietary intake of fat and cholesterol

Table 10.9. Current Compared to Recommended Nutrient Intake in Children and Adolescents[a]

	Current	Recommended
Saturated fatty acids	14% calories	<10% calories
Total fat	35–36%	Avg. no more than 30%
Polyunsaturated	6%	Up to 10%
Monounsaturated	13–14%	10–15%
Cholesterol	193–296 mg/day	<300 mg/day

[a](From National Cholesterol Education Program. Report of the expert panel on blood cholesterol levels in children and adolescents. Pediatrics 1992 [In press].)

in these age groups is shown in Table 10.9. The goal of the recommended modifications is to reduce the dietary content of saturated fat and cholesterol to levels similar to those of the Step-1 Diet in adults, with emphasis on eating a wide variety of foods. Because this age group is still actively growing and developing, the total intake of calories should be adequate to support growth and development and to reach and maintain desirable body weight. The required amount of energy will depend on the height, weight, rate of growth, and level of physical activity. A fat intake that corresponds to 30% of total calories usually is adequate. A lower fat intake usually is not necessary and, for some children and adolescents, may make it difficult to provide enough calories and minerals for growth and development.

Because children eat a large portion of their meals in school and learn dietary habits there, the school becomes an important place to initiate nutrition intervention efforts. Current school food programs provide 25% to 40% of total calories, with 15% to 17% of calories derived from saturated fatty acids. The NCEP Panel's recommendations for food choices for school meals are given in Appendix E. In addition, children today eat more meals and snacks in fast food restaurants, which generally provide foods high in fat, cholesterol, and calories (Table 10.10). The approach to changing eating patterns in these age groups should include education in appropriate food choices in places outside the home.

The toddler age group (2 to 3 years old) represents a transition period when children gradually assume the eating patterns of the

rest of the family. This requires more flexibility in applying the recommended nutritional guidelines. Because the RDAs for zinc, iron, and calcium are relatively high at these ages, the diet generally should contain a higher proportion of lean meat (iron and zinc), low-fat dairy products (calcium), and fortified or enriched bread and cereal products, peas, and beans (iron).

Below the age of two years, there is general agreement that no restriction of fat and cholesterol should be applied. Breastmilk and infant formula provide about 50% of calories from fat, which meet the high caloric requirements of infants. Because the mother's diet has no major effect on the fat content of breastmilk, there is no contraindication for the lactating mother to adopt a low-fat, low-cholesterol diet. Skim milk is not recommended for infants, because it provides a higher than desirable solute load from calories laden with mineral salts and protein.

Smoking Cessation

Cigarette smoking is related directly to the total cholesterol and inversely to HDL-C. Among patients who have lipid abnormalities, cessation of cigarette smoking can modify the risk of CHD substantially through its effect on improvement of blood lipid values as well as through other physiologic mechanisms. Physician counseling alone can result in as much as a 10% long-term cessation rate, which can be heightened by use of appropriate educational material and nicotine gum or patches. Even greater success is possible through various group programs that

Table 10.10. Most Popular Restaurant Foods for Children and Adolescents[a]

	Food Item Percent of Eater Occasions[b]	
	Children Under 6 Years	Children/Adolescents Between 6 and 17 Years
Soft drinks	35	43
Hamburgers/cheeseburgers	24	24
French fries	33	30
Pizza	18	21
Fried chicken	12	10
Ice cream	9	9
Milk	8	NA
Other desserts	8	8
Mexican food	4	4
Juice	3	NA
Breads[c]	NA	7
Side-dish salads	NA	6

[a](From National Cholesterol Education Program. Report of the expert panel on blood cholesterol levels in children and adolescents. Pediatrics 1992 [In press].)
[b]Expressed as percent of eater occasions on which food item was ordered. All foods items are not represented in each age category; missing values indicated by NA.
[c]Includes all breads, such as biscuits, toast, and other side orders of bread.

use behavioral principles and training in conjunction with physician counseling. Such efforts can result in a 30% quit rate at 1 year. Repetitive efforts can bring about larger reductions in cigarette smoking. The AHA, the American Cancer Society, and the American Lung Association provide detailed instructional booklets and self-help aids (*see* Chapter 13).

Physical Activity

Habitual physical activity offers a nondrug approach to improving blood lipids. There are two basic kinds of physical activities: (1) dynamic, also known as endurance or isotonic activity, in which muscle tension results in movement through a range of motion such as jogging, bicycling, or swimming; and (2) static (isometric) exercise, used in strength training and characterized by muscle tension without appreciable movement. Actually, most dynamic activity contains static elements, but static exercises do not always have dynamic components. Cardiorespiratory fitness and lipid modification result primarily from dynamic exercise at levels adequate to induce training adaptations.

Although it is clear that the magnitude of the lipoprotein changes depends on the type and amount of exercise, the exercise dose required for optimal lipid modification has not yet been established. Although training on alternate days is recommended for minimizing musculoskeletal stress, this may not be the best method for influencing lipids, because some data suggest it is better to exercise every day. The greater the energy expenditure per session and the longer the time period, the more likely a substantial change will be experienced. The benefits of mild exercise have been recommended recently for bed-rested patients or very sedentary persons. Infrequent physical activity or exercise of brief duration with less strenuous work has not evoked lipid changes consistently, especially among untrained persons. Generally, data indicate a less favorable lipoprotein response to exercise in women compared to men. The response to exercise may be impaired by estrogen as more favorable changes occur in estrogen deficient post menopausal women.

A reasonable exercise regimen of moderate intensity should increase the heart rate to approximately 70% to 85% of its maximal value (equivalent to 60% to 80% of max

VO$_2$) for 30 to 40 minutes at least three times per week. As much physical activity as possible between exercise sessions, such as stair climbing and walking, is advisable. Longer durations and increased frequency of activity, even at the lower range of recommended intensity, may permit more beneficial lipid changes than high-intensity sports activity. Until more specific recommendations for lipid reduction are available, exercise recommendations should follow available routines outlined by the American College of Sports Medicine or the AHA.

Because of the variability in response, the extent of lipid changes with exercise cannot be estimated with any precision; however, average triglyceride decreases on the order of 20% and HDL increases of about 10% can be anticipated. Alteration of body habitus, with decreased body fat and increased lean body mass, may be partially responsible for any favorable lipid changes. Still, the only well-documented physical activity for altering lipid profiles is dynamic exercise—brisk walking, jogging, swimming, or active sports.

Among the nondrug approaches available to increase HDL mass, the most useful is exercise, especially when associated with weight loss and cigarette smoking cessation. Weight loss augments the exercise effect, whereas continued cigarette smoking tends to negate the HDL-C benefits of exercise. Weight loss, combined with endurance training, offers the best potential for reducing triglycerides. Exercise does little to decrease the blood LDL cholesterol concentrations. Whether or not exercise favorably alters the distribution of lipids or proteins among the various LDL subfractions is not yet known.

Suggested Readings

AHA Special Report: Recommendations for the treatment of hyperlipidemia in adults: a joint statement of the Nutrition Committee and the Council on Arteriosclerosis of the American Heart Association. Circulation 1984;69:443A–68A.

Atkinson RL: Low and very low calorie diets. Med Clin North Am 1989;73(1):203–215.

Connor WE, Connor SL. Dietary treatment of familial hypercholesterolemia. Arteriosclerosis 1989;9 (suppl I):91–105.

Dietary guidelines for healthy American adults: a statement for physicians and health professionals by the nutrition committee of the American Heart Association. Circulation 1988;77:721A–724A.

Grundy SM, Barrett-Conner E, Rudel LL, Miettinen T, Spector AA. Workshop on the impact of dietary cholesterol on plasma lipoproteins and atherogenesis. Arteriosclerosis 1988;8:95–101.

Harvey A, Wesler CA, Krahn JL, West ME (eds): Cooking Light '90. Birmingham: Oxmoor House, 1990.

Mattson FH, Grundy SM. Comparison of effects of dietary saturated, monounsaturated, and polyunsaturated fatty acids on plasma lipids and lipoproteins in man. J Lipid Res 1985;26:194–202.

Meichenbaum D, Turk DC. Facilitating treatment adherence: a practitioner's guidebook. New York: Plenum, 1987.

National Cholesterol Education Program. Report of the expert panel on blood cholesterol levels in children and adolescents. Pediatrics 1992 (In Press).

National Research Council (U.S.). Committee on Diet and Health. Diet and health: implications for reducing chronic disease risk. Committee on Diet and Health, Food and Nutrition Board, Commission on Life Sciences. National Academy Press, Washington, DC, 1989.

Report of the Expert Panel on Detection, Evaluation, and Treatment of High Blood Cholesterol in Adults: National Cholesterol Education Program. Arch Intern Med 1988; 148:36–69.

United States Department of Agriculture. Human Nutrition Information Service. Nutritive Value of Foods. Home and Garden Bulletin No. 72, 1986.

United States Department of Health and Human Services, Coronary Heart Disease. In: The Surgeon General's Report on Nutrition and Health, Pub. No. 88–50210:83–137, 1988.

CHAPTER 11

Pharmacologic Therapy of Lipid Disorders

General Approach to the Management of Dyslipidemia

Therapy for dyslipidemia always should begin with dietary and other nonpharmacologic modifications. The approach to nonpharmacologic management should be methodical and planned; simply handing patients printed diets and advising them to follow the plans is not generally effective. Whenever possible, supervision and direction of the diet by a trained nutritionist is desirable. In our experience, many patients referred for management of dyslipidemia claim to have been following a diet. On referral of such patients to trained dietitians, additional reductions occur in their blood lipids, so that as many as 25% to 30% no longer qualify for pharmacologic therapy. Because dyslipidemia ordinarily does not constitute a medical emergency, except in patients who have the hyperchylomicronemia syndrome or in whom recurrent pancreatitis is a problem, pharmacologic therapy usually can be delayed until the effects of nonpharmacologic interventions have been maximized.

The therapeutic goals for dyslipidemia are discussed in Chapter 12. Should dietary therapy prove inadequate to reach these goals, then pharmacologic therapy should be added. The major drugs used for treating dyslipidemia, their mechanisms of action, and their adverse effects are given in Table 11.1. The decision to proceed to pharmacologic therapy must not be taken lightly, and consideration should be given to the consequences of the cost, as well as possible side effects, of many drugs. Nevertheless, persons at high risk for CHD should not be deprived of appropriate pharmacologic therapy solely for these reasons. The decision for using drugs should be based on the severity of the lipid abnormality, the family history, other associated risk factors for CVD, and the presence or absence of CHD (*see* Chapter 2). It should be reemphasized that adequate control of associated risk factors such as hypertension, cigarette smoking, obesity, alcohol consumption, and diabetes mellitus must be given high priority.

Attempts also should be made to correct any secondary causes of dyslipidemia before initiating definitive lipid-lowering drug therapy. Where lipid abnormalities can be reasonably attributed to a medication that the patient is taking, an effort should be made to discontinue that medication, reduce its dosage or route of administration, or switch to another drug that is more lipid neutral.

It often is possible to predict, by the magnitude of the lipid disorder, that dietary therapy alone will not suffice to control the lipid abnormality. In fact, for those patients who have a total cholesterol level at or above 300 mg/dL (7.76 mmol/L), it is likely that both dietary and pharmacologic therapy will be necessary. Because hypercholesterolemia presents no urgency, diet can be pre-

Table 11.1. Drugs Used To Treat Dyslipidemia

Drug	Daily Dose	Major Side Effects	Suggested Mechanisms of Action
Nicotinic acid (Niacin)	1–6g	Flushing Pruritus/dry skin Gastrointestinal symptoms[a] Aggravation of duodenal ulcer Hyperglycemia Hyperuricemia Hepatotoxicity	↓ Lipolysis in adipocytes ↓ Hepatic triglyceride production ↓ Synthesis of VLDL ↓ Clearance of HDL
Cholestyramine (Questran) Colestipol (Colestid)	8–32g 10–30g	Gastrointestinal symptoms[a] Hypertriglyceridemia Decreased absorption of fat-soluble vitamins and certain medicines	Binds bile acids in intestine, interrupting enterohepatic circulation of bile acids ↑ LDL clearance through ↑ LDL receptor activity
Gemfibrozil (Lopid)	1200mg	Gastrointestinal symptoms[a] Hepatotoxicity Myopathy Bone marrow depression Interaction with anticoagulants	↑ Activity of lipoprotein lipase ↑ nonsplanchnic catabolism of VLDL and possibly ↑ Synthesis of HDL
Lovastatin (Mevacor)	20–80mg	Gastrointestinal symptoms[a] Headache Insomnia Hepatotoxicity Myopathy	Competitively inhibits the early stage of cholesterol biosythesis ↑ LDL clearance through increased LDL receptor activity
Probucol (Lorelco)	500–1000mg	Gastrointestinal symptoms[a] Prolongation of QT interval	↑ LDL clearance by nonreceptor pathways ↓ Synthesis of HDL May inhibit LDL oxidation and foam cell production

[a] Abdominal pain, constipation, flatulence, diarrhea, bloating.

scribed for 3 to 6 months before initiating pharmacologic therapy. This may not be the situation with severe hypertriglyceridemia, where diet alone may not produce the desired degree of control promptly enough, and drug therapy must be added almost from the outset. Once hypertriglyceridemia is sufficiently well controlled, however, it may be prudent to discontinue pharmacologic therapy, particularly if contributing factors such as uncontrolled diabetes mellitus, obesity, alcoholism, and adverse drug effects have been corrected adequately. If the triglyceride levels remain within a satisfactory range, pharmacologic therapy may not have to be resumed.

A major consideration in the treatment of dyslipidemia is the cost of medications. Average prices for these medications are shown in Table 11.2. Unfortunately, the high cost of these drugs limits their use for some patients who otherwise might benefit from them. Nicotinic acid is the least expensive form of therapy and can be prescribed in generic forms available over the counter in most drug and nutrition stores; however, it is advisable to select preparations produced by manufacturers that are regulated adequately, such as well known pharmaceuticals. Bile acid resins in bulk form are considerably cheaper than those in packets or in flavored Cholybars. Because of the expense of several drugs, particularly the HMG CoA reductase inhibitors, the use of several drugs at lower doses may reduce cost and improve the effectiveness of the regimen because of complementary actions. The possibility of an adverse drug-to-drug interaction must always be kept in mind, particularly because experience with combination therapy is not extensive. Beside the direct cost of these agents, it also must be realized that there will be additional sizeable costs related to screening for chemical toxicity (CBC, chemistry profiles, and so forth).

Regardless of cause, selection of an appropriate drug depends on whether the patient has hypercholesterolemia, hypertriglyceridemia, or combined dyslipidemia. Although many hypolipidemic agents affect more than one lipoprotein, the degree of modification of each lipoprotein differs greatly among the agents; the optimal drugs should be tailored to each lipoprotein profile, as well as to each individual patient. The lipid-lowering drugs have other metabolic effects that may contribute to their efficacy in reducing atherosclerosis. Fibric acid derivatives reduce platelet aggregation and decrease the effect of platelet derived growth factor on smooth muscle. Niacin inhibits thromboxane synthesis and platelet

Table 11.2. Cost of Representative Lipid-Lowering Drugs[a]

Generic Name	Trade Name	Form	Daily Dosage[b]	Cost/Month[b]
Cholestyramine	Questran®	Dose packets	4-8g bid	$64-$127
Cholestyramine	Questran®	Bulk powder	4-8g bid	$40-$80
Cholestyramine	Cholybar®	Bars	4-8g bid	$65-$130
Colestipol	Colestid®	Dose packets	5-10g bid	$50-$100
Colestipol	Colestid®	Bulk powder	5-10g bid	$40-$80
Gemfibrozil	Lopid®	Tablets	600mg bid	$50
Lovastatin	Mevacor	Tablets	20-40mg[c]	$52-$104
Pravastatin	Pravachol®	Tablets	10-20mg	$38-$77
Simvastatin	Zocor®	Tablets	10-20mg	$50-$90
Niacin (Regular)	generic	Tablets	1g bid-1g tid	$3-$4
Niacin (Sustained Release)	Nicobid®	Tempules	1g bid-1g tid	$80-$119
Probucol	Lorelco®	Tablets	500mg bid	$54

[a] Prices are based on average wholesale price (AWP), as listed in the 1991 Drug Topics Red Book. Oradell, New Jersey. Medical Economics. Retail prices vary markedly by quantity purchased, locale, and type of pharmacy and may even be less than AWP.
[b] Based on usual dosage level
[c] 10mg dose is available but cost is only slightly less than 20mg dose; developed primarily for transplant patients who are being treated with cyclosporin

aggregation and stimulates prostaglandin formation, whereas probucol has an antioxidant effect and inhibits interleukin-1 release. The rationale for the choice of the drug should be explained carefully to the patient who should share in the decision-making process. The considerations of costs, adverse effects, and target lipid values to be obtained may all influence the patient's willingness to adhere to treatment. There also is considerable interindividual variability in the response to these drugs, so that it often is necessary to have therapeutic trials and comparisons. Although the effectiveness of most drugs can be determined within several weeks after initiation of therapy, doses often need to be adjusted.

Hypercholesterolemia

The pharmacologic agents of first choice for the treatment of hypercholesterolemia are the bile acid resins and nicotinic acid (Table 11.3). This conclusion is based on the long-term experience with these drugs, their minimal serious toxicity when adequately supervised, and their effectiveness for those patients that can tolerate them. Both of these drugs have been shown to provide protection against the development of CHD. Although they are associated with a high incidence of side effects, these usually are

Table 11.3. Estimated Efficacy of Commonly Used Lipid-Lowering Agents in Treating Dyslipidemias

	HC[a]	HC and HTG	HTG[b]
Bile acid resins	+++	±	0
Nicotinic acid (NA)	+++	+++	+++
Reductase inhibitors (RI)	+++	++	+
Gemfibrozil	+	++	+++
Probucol	+	+	0
BAS and NA	+++	++	0
BAS and RI	+++	++	0
BAS and gemfibrozil	na	+++	0
RI and NA	+++	+++	0
NA and gemfibrozil	na[c]	+++	+++
Fish Oil	0	0	+

[a] HC = hypercholesterolemia
[b] HTG = hypertriglyceridemia
[c] na = nonapplicable

tolerable, easy to monitor, and reversible. The experience with these two classes of drugs suggests that approximately 50% of patients will be intolerant or unwilling to take one of them because of adverse effects. Improvements have been and may continue to be made in palatability of the bile acid resins, however, appropriate use and precautions can reduce the incidence of side effects with niacin. Although a listing of past medical problems and questions regarding tolerance to the most likely adverse effects will help in the implementation of therapy, a therapeutic trial using one of these agents must be undertaken to assess tolerability. A bile acid resin may be preferable with mildly elevated LDL-C levels, because it is nonsystemic and has over 20 years of careful evaluation. If the patient is intolerant of one of these first-line agents, then the other should be tried. A combination of the two drugs at low doses may potentiate the effect at reduced adverse effects.

Should patients be intolerant of both medications, or if the maximal response attainable is insufficient, then the use of reductase inhibitors are indicated. By way of contrast, they are as effective as either of these drugs in reducing the LDL concentration and are well tolerated with little immediate toxicity. Only approximately 2% of patients on lovastatin must have the drug discontinued because of adverse biochemical effects. The major problem with these drugs, however, is uncertainty about potential long-term toxicity. There have been only six years of clinical experience with lovastatin but less with pravastatin and simvastatin and there is concern that long-term adverse effects will appear that are not apparent currently. For this reason, and because the need for medication is lifelong, it is difficult to embark on a therapeutic regimen using reductase inhibitors in young or middle-aged persons unless there is evidence that the risk of untreated hypercholesterolemia is greater than the potential risk of future adverse effects. In other words, trivial hypercholesterolemia may be best left untreated. The bile acid resins, nicotinic acid, and reductase inhibitors can be used in combination, because many of their actions are complementary. There is some concern about myotoxicity with niacin-lovastatin combinations, although our experience with the combination has been good.

For those patients who have hypercholesterolemia and established CHD, a more aggressive approach may be indicated. Recent studies suggest that a target total cholesterol of 150 mg/dL (3.88 mmol/L) and an LDL-C of around 100 mg/dL (2.59 mmol/L) may be worth the attempt to retard progression and perhaps initiate regression of atherosclerotic plaques in patients who have established CHD.

Hypertriglyceridemia

For patients who have hypertriglyceridemia, there are only two medications that are likely to be effective; these are nicotinic acid and gemfibrozil. It should be emphasized that obesity, noninsulin-dependent diabetes mellitus, drugs, and the liberal use of alcohol are frequent contributors to the hypertriglyceridemic state, and the elimination or treatment of these disorders should be undertaken before (or in addition to) implementation of pharmacologic therapy. There appears to be no controversy about the importance of treatment of hypertriglyceridemia to prevent eruptive xanthoma, acute pancreatitis, or the chylomicronemia syndrome. There is still controversy, however, over whether hypertriglyceridemia represents an important risk factor for the development of cardiovascular disease (*see* Chapter 2). Because patients who have only moderate hypertriglyceridemia (250 to 500 mg/dL) frequently have coexistent low levels of HDL-C, as well as smaller and denser LDL particles, there is ample reason to believe that hypertriglyceridemia is directly related to, or is associated with, metabolic abnormalities that predispose to the development of CVD. Also, low HDL-C may relate to other metabolic derangements that

increase VLDL and chylomicron remnants (*see* Chapter 2).

Although triglyceride levels of 250 to 500 mg/dL (2.82 to 5.65 mmol/L) may be considered to be borderline elevations by some and not in the range that threatens development of acute pancreatitis or xanthomas, it is our opinion that these levels indicate the need for treatment when there is a strong history of cardiovascular disease, such as in familial combined hyperlipidemia. Although there are suggestions that high apoprotein B levels make it easy to distinguish a hypertriglyceridemic patient who is at risk of developing CHD from one who is not, a more pragmatic approach would be to base treatment decisions on the family history.

Either nicotinic acid or gemfibrozil can be used effectively in such patients. Because gemfibrozil usually is better tolerated, it may be the drug of choice. It can reduce the triglyceride concentration substantially and may be associated with a 10% to 25% increase in HDL-C, particularly when the baseline HDL-C is low. Curiously, in such patients, the LDL-C concentration often increases as the triglyceride concentration decreases. This is usually not important unless the LDL-C concentration increases to above 130 to 160 mg/dL. Overall, gemfibrozil's effects on LDL-C depend on the lipid phenotype (Table 11.4): in Type IIA, gemfibrozil will reduce LDL-C by approximately 10%; in Type IIB, by approximately 5%, whereas in Type IV it increases LDL-C by approximately 10%. Nicotinic acid is an ideal drug for treating hypertriglyceridemia because it usually decreases both triglyceride and

Table 11.4. Effect of Gemfibrozil on Plasma Lipids and Lipoproteins by Phenotype

Pattern	LDL-C	HDL-C	TG
Type IIA	−10%	+10%	—
Type IIB	−4%	+10%	—
Type III	−45%	+65%	−65%
Type IV	+10%	+12%	−23%
Type V	−20%	+48%	−75%

LDL-C concentrations and increases HDL-C. In our experience, however, the decrease in LDL-C is less in hypertriglyceridemics than in normotriglyceridemics. Usually a minimum of 1500 mg of nicotinic acid per day is required to produce favorable changes in the atherogenic particles (LDL and VLDL), whereas lower doses (1000 mg per day) may be associated with a significant increase in the HDL-C concentration. The relations between changes in VLDL-C, triglyceride, LDL-C and HDL-C produced by nicotinic acid are not well understood.

Combined Hyperlipidemia

When patients have simultaneous hypertriglyceridemia and hypercholesterolemia (Type IIB), nicotinic acid or gemfibrozil may be used effectively. Nicotinic acid probably is a better drug for the treatment of patients who have mixed dyslipidemia, because it produces a greater reduction in LDL-C levels. The use of these drugs in combination with other drugs to potentiate the cholesterol-lowering effect or to prevent the compensatory increase in triglyceride concentration that commonly occurs with bile acid resins should be considered.

As a rule, bile acid resins and nicotinic acid should be started at very low doses and advanced slowly, so that patients can become acclimated to them without incurring severe effects that would necessitate stopping the medication. No complete agreement exists about this approach, and some authorities prefer to advance doses very rapidly to distinguish quickly those patients who can tolerate the medication from those who cannot. Another strategy that sometimes is used is the administration of a large dose of nicotinic acid (500 mg) with a cup of coffee as a one-time provocative test in the office; this exaggerates the flushing and demonstrates to the patients the type of skin reaction that they are likely to encounter when they take this agent. If this approach is used, the patient should be told that proper precautions will reduce the extent of the flushing substantially.

Hypoalphalipoproteinemia

The issue of using drugs to increase HDL-C in patients who have isolated low HDL-C is debatable but gaining in favor. The independent effect of increasing isolated low HDL-C on CHD risk has not yet been studied, although there are indirect data from large clinical trials that suggest benefit from such action (*see* Chapter 2). Although few physicians would treat persons who have low HDL and who are not otherwise at risk for CHD, a drug trial may be considered appropriate for those patients who have CHD or who are at high risk for developing the disease.

Initial therapy should be directed at those factors known to reduce HDL-C—obesity, cigarette smoking, physical inactivity, a very high carbohydrate diet, and certain drugs (Table 11.5). Estrogen replacement therapy may be of some benefit in postmenopausal women and probably is beneficial to every postmenopausal woman who does not have a contraindication to its use. If the HDL-C is still below the therapeutic goals (*see* Chapter 12), then drug therapy should be initiated

Table 11.5. Effects of Drugs and Other Factors on HDL

Increase	Decrease
Female sex	Male sex
Exercise	Type I and II diabetes
Ethanol	Smoking
Chlorinated	Obesity
hydrocarbons	Chronic uremia
Weight reduction	Diet high in polyunsatu-
Familial hyperalpha-	rated fat
lipoproteinemia	High-carbohydrate diet
Drugs	Hypertriglyceridemia
Estrogens	Primary HDL deficiency
Nicotinic acid	syndromes (e.g., Tan-
Gemfibrozil	gier disease)
Reductase inhibitors	
Corticosteroids	**Drugs**
Phenytoin	Androgens
Carbamazepine	Progestins (oral
Beta-adrenergic	contraceptives)
agonists	Beta-adrenergic
Alpha-adrenergic	blockers
blockers	Probucol
	Retinoids

with nicotinic acid at low doses or gemfibrozil.

Pediatric Considerations

The decision to use drugs for control of dyslipidemia in children and adolescents is complex. Atherosclerosis begins in childhood and already is established in many young adults, suggesting that early prevention of CHD is important. However, lipid-lowering therapy requires life-long intervention, which magnifies the risk and cost of drug therapy that may not be as obvious and frequent in adults. Because children are still growing and developing, care must be taken to ensure that the medication does not interfere with these processes. Finally, the stigma and discomfort involved in the daily administration of a drug to a child, especially if associated with side effects, should not be ignored. Drug therapy therefore should be reserved for children who have severe dyslipidemia that is insufficiently responsive to dietary therapy, especially when the family history suggests a high risk for CHD.

The NCEP Expert Panel on management of cholesterol in children and adolescents recommends considering drug therapy in children ages 10 years and older if, after an adequate trial of diet therapy (6 months to 1 year) LDL-C remains at or above 190 mg/dL (4.91 mmol/L) or LDL-C remains at or above 160 mg/dL (4.14 mmol/L) and there is a positive family history of premature CHD, or two or more other risk factors are present despite rigorous attempts to control them. These cut-points, although arbitrary, are based on data that suggest that LDL-C of 164 mg/dL (4.24 mmol/L) best differentiates between children and adolescents who have familial hypercholesterolemia and those who have other types of hypercholesterolemia.

The only drugs recommended for treatment of hypercholesterolemia in this age group are the bile acid resins. These drugs are not absorbed systemically and have been shown to be effective and apparently safe in children. The starting dose of the drug is not

Table 11.6. Initial Dosage Schedule for Treatment of Familial Hypercholesterolemic Children and Adolescents with a Bile Acid Sequestrant[a]

Daily doses of bile acid sequestrant[b]	Total cholesterol (TC) and low density lipoprotein cholesterol (LDL-C) levels (mg/dL) after diet	
	TC	LDL-C
1	<245	<195
2	245–300	195–235
3	301–345	236–280
4	>345	

[a]These are generally recommended doses and may require adjustment based on the patient's response.
[b]One dose is the equivalent of a 9-g packet of cholestyramine (containing 4g cholestryramine and 5g filler), or 5g of colestipol.
(From: National Cholesterol Education Program, Report of the Expert Panel on Blood Cholesterol Levels in Children and Adolescents. Pediatrics 1992 [In Press].)

related to the child's body weight but rather to the postdietary LDL-C levels (Table 11.6). The doses should be increased slowly to achieve the required effect. In addition to lipid levels, follow-up should include monitoring of height and weight, as well as appropriate analyses to ensure the absence of specific vitamin deficiencies (fat soluble vitamins and folic acid).

Experience with nicotinic acid in growing children is limited, and its use requires caution; this drug should be prescribed only after referral to a lipid specialist and if lipid-lowering therapy by diet and resins has not reached its specific goals. The HMG CoA reductase inhibitors, probucol, gemfibrozil, and clofibrate, are not recommended as routine drugs for use in children and adolescents.

Lipid-Lowering Drugs (Fig. 11.1)

Bile Acid Sequestrants (Resins)

Resins have been used successfully for several decades to reduce blood cholesterol levels. Two available resins, cholestyramine and colestipol, appear to be equally effective at equivalent doses although they differ somewhat in their physical characteristics (size of particle) and patient acceptability. These drugs should be used primarily to

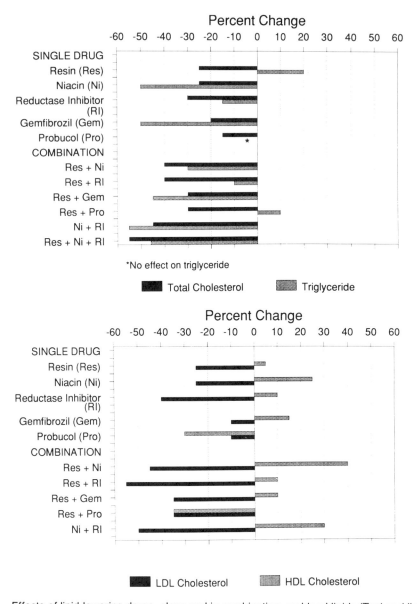

Figure 11.1. Effects of lipid-lowering drugs, alone and in combination, on blood lipids (Top) and lipoproteins, HDL-C and LDL-C (Bottom). The percent changes represent the average changes at maximal doses. (*Note:* The effects of gemfibrozil on LDL-C depend on the initial lipoprotein profile; LDL-C decreases most in Type IIa, may decrease or increase in Type IIb, and usually increases in Type IV.)

lower LDL-C levels and should not be used alone when the plasma triglyceride exceeds 300 mg/dL (3.39 mmol/L), because they increase the synthesis of the triglyceride-rich VLDL and unmask existing defects in VLDL clearance. These resins can be considered, however, if the triglyceride concentration is only slightly increased or if used in combination with other drugs that reduce triglyceride levels (e.g., niacin or gemfibrozil).

The clinical efficacy of cholestyramine has been demonstrated by two large clinical trials in the United States. In the Lipid Research Clinics-Coronary Primary Prevention Trial (LRC-CPPT), cholestyramine lowered the concentrations of the total and LDL cholesterol, as well as the incidence of CHD, in asymptomatic, hypercholesterolemic men (*see* Chapter 1). The reduction in serum cholesterol for the entire group was 9%, whereas the reduction in the relative risk of developing CHD was 18%. Among men who had a 20% lowering of blood cholesterol, the reduction in CHD relative risk was 40%. Both the number of packets of cholestyramine taken and the concomitant reduction in blood cholesterol were correlated to the reduction in CHD risk. In an angiographic study of men who had established CHD and cholesterol levels at or below the 95th percentile, cholestyramine and dietary therapy were compared with diet alone. Combined therapy stabilized coronary artery lesions and reduced the rate of progression, even though the degree of cholesterol lowering achieved in this study was modest.

Cholestyramine has been used continuously for over two decades in adults and children, and no unanticipated complications have been associated with its use. Its palatability and frequent gastrointestinal side effects do remain a problem. Newer formulations may improve patient compliance and adherence to medication schedules. The resins also are effective in children, and abnormalities of growth or development have not been reported with their use.

Mechanism of Action. The resins bind bile acids in the gut, thereby blocking the enterohepatic recirculation of bile acids and decreasing the size of the bile acid pool (Figs. 11.2 and 11.3). In doing so, the conversion of cholesterol to bile acids in the liver is accelerated, resulting in reduced content of cholesterol within the liver cell. As the intracellular content of cholesterol in the hepatocyte decreases, mechanisms are activated to restore the intracellular cholesterol concentration into an optimum range. This is accomplished in two ways: (1) by increasing the synthesis of the cell-surface LDL receptor responsible for binding and internalization of circulating LDL-C, and (2) by increasing intracellular synthesis of cholesterol through activation of the rate-limiting enzyme of cholesterol synthesis, HMG CoA

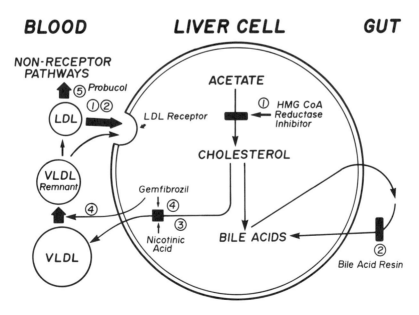

Figure 11.2. Site of action of commonly used drugs in treating dyslipidemia. Dotted arrows represent inhibited steps, broad arrows represent accelerated steps. As shown, some medications probably affect several steps.

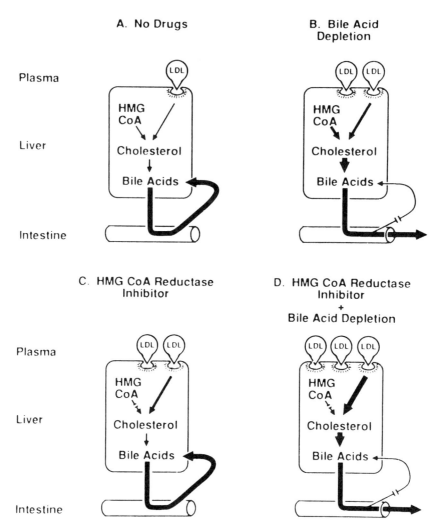

Figure 11.3. Mechanisms of action for bile acid sequestrants and HMG CoA reductase inhibitors. The conversion of HMG CoA to mevalonate is the rate-limiting step in cholesterol synthesis. The LDL receptors are also increased by treatment with resins and HMG CoA reductase inhibitors. (From: Bilheimer DW, East C. Lipid-lowering agents. In: Williams RL, Brater DC, Mordenti J, eds. Rational therapeutics: A clinical pharmacological guide for the health professional. New York: Marcel Dekker, 1990:689–755. "Reprinted by courtesy of Marcel Dekker, Inc."

reductase. These changes result in accelerated removal of circulating LDL-C and a decrease of its blood concentration. Unfortunately, the accompanying increase in intracellular cholesterol synthesis prevents the maximum expression of cell-surface LDL receptors, thus limiting the cholesterol-lowering effects of resins. Although one might imagine that the resins bind cholesterol in the intestine, making it unavailable

for absorption, this is not the case, because intestinal cholesterol is a neutral sterol and is not in the anionic state that can be bound by resins.

Resins exert their primary effect on LDL-C and, when used in maximum effective doses, can be expected to lower the LDL-C level by as much as 30%. Any effect on HDL-C concentrations is minimal, and the marked improvement that occurs in the

LDL:HDL cholesterol ratio is caused primarily by the reduction that occurs in the LDL-C concentration.

Interestingly, and for reasons not completely understood, the resins increase the synthesis of VLDL particles by the liver. As a result, there usually is a modest increase in the serum triglyceride concentration, although this generally has no clinical consequence, particularly when initial triglyceride levels are normal. Occasionally, patients who have mild hypertriglyceridemia and subtle defects in the control of VLDL synthesis or removal are extremely sensitive to this effect and will develop significant hypertriglyceridemia when treated with resins. In patients who have combined dyslipidemia (hypertriglyceridemia associated with hypercholesterolemia), severe hypertriglyceridemia can occur as a consequence of the use of resins alone. Although the resins decrease LDL-C, the total cholesterol concentration increases paradoxically as the triglyceride concentration rises. This occurs because the triglyceride: cholesterol ratio in the VLDL particle approximates 5:1, and a marked augmentation in the synthesis and release of VLDL particles automatically increases the VLDL-C concentration.

Adverse Effects.   The resins do not act systemically and tend to cause few systemic complications. The major adverse side effects related to their use are gastrointestinal, with symptoms of bloating, abdominal pain, distention, constipation, and exacerbation of preexisting hemorrhoids. Mild transient increments in alanine aminotransferase and alkaline phosphatase can be encountered, although the mechanisms are not understood. Whether these changes are related to the drugs or simply reflect the normal spontaneous variability in these parameters in otherwise healthy persons is not clear. Routine biochemical monitoring is not required in patients who use resins.

The bile acid resins are anionic exchange resins and, consequently, may also bind other anionic compounds. In this regard, the resins can interfere with the absorption of various medications used by the patient. Because it is not possible to study the interaction of all drugs with these resins, patients should be instructed to take other medications 1 hour before or 3 to 4 hours after the resin. In addition, when used in very large quantities, the resins produce a bile acid deficiency, induce fat malabsorption, and, consequently, create deficiencies of fat-soluble vitamins. These complications are unlikely at the doses of resins used by most patients, even over a long-term period.

The effect of resins on triglyceride metabolism has been mentioned previously. Although in most cases this is mild and negligible, the occasional patient may develop severe hypertriglyceridemia; thus, resins are contraindicated in patients who have uncontrolled triglyceride elevations.

Clinical Use.   Most investigators recommend that cholestyramine and colestipol be started at low doses and increased slowly, as tolerated by the patient. Quick acceleration of the dose may lead to the development of adverse effects that convince patients that they will never be able to tolerate the medication. To some extent, this problem is accentuated by failure of the physician to appreciate the potential discomfort to the patient. Because therapy is life-long and there are no benefits to acute reduction of the cholesterol concentration, it probably is unimportant whether it takes two, four, or even six months to arrive finally at the appropriate dose.

Each packet, or scoop, contains 4 g of cholestyramine or 5 g of colestipol. Treatment usually starts with 1 or 2 packets (or scoops) of the resin taken once daily, preferably 30 minutes before the nighttime meal. There is evidence that cholesterol-lowering drugs are more effective when given at night than at other times of the day, probably because cholesterol synthesis is greater at night. Because the evening meal usually is the largest meal for Americans, more bile acids will be secreted and the resins will be more effective in depleting the bile acid pool. Taking the resin before rather than after the meal reduces the incidence of bloating and other gastrointestinal complaints and may also reduce the amount of food consumed because of earlier satiety.

If tolerated, the dose can be increased to 2 or 3 packets (or scoops) with the nighttime meal. Thereafter, additional increases in dose should be distributed throughout the day or administered just before breakfast. A twice-daily dosing schedule generally is effective with resins.

The greatest relative cholesterol reductions occur with the initial drug doses (Fig. 11.4). At a dose of two scoops per day, resins should lower the LDL level by 10% to 20%. Doubling the dose reduces LDL-C another 4% to 8%. Little additional effect occurs when resin doses exceed 4 scoops per day (i.e., 16 g per day of cholestyramine or 20 g per day of colestipol). At these doses, any potential benefit in lipid reduction is likely to be offset by adverse effects and lessened patient compliance. It may be preferable to add a second drug before reaching maximum doses of the resins (32 g cholestyramine, 30 g colestipol). This allows one to take advantage of the relatively greater response to the initial doses of the second drug and fewer adverse effects may be expected than with the use of a single drug at maximum doses. This is true whether a resin is added to another drug or vice versa.

It is impossible to determine clinically which patients will be able to tolerate the drug and which patients will not. Some patients can take the maximum recommended doses without any significant symptoms, whereas others develop significant and unacceptable symptoms at very low doses. Consequently, trial and error remains the only way to determine acceptability and tolerance of the drug. Discussion of side effects and ways to minimize any associated discomfort will facilitate adherence. The problem of constipation can be offset effectively by the use of a stool bulking agent, especially during initial stages of therapy. In this regard, psyllium (metamucil) or oat bran is effective in counteracting the constipating effects of the resins and may have the addi-

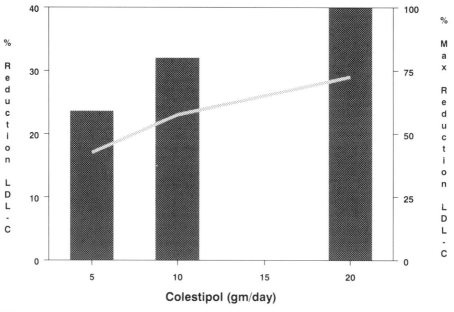

Figure 11.4. Dose-response curve of LDL-C to treatment with colestipol. Most of the effect of treatment (20+% LDL-C reduction) is seen at a dose of 10 g/day (2 scoops/day), with relatively little additional benefit occurring after additional dosage increases. The bars represent the percent of maximal LDL-C reduction achievable; about 80% reduction with 10 g/day and 100% at 20 g/day.

tional benefit of potentiating the cholesterol-lowering effect.

Overall, the lack of palatability of the resins has been one of their major drawbacks. Patients simply do not like the sandy or gritty quality of the medication and the oral sensations that these agents produce. As a consequence, it is important to disguise the medication as effectively as possible. Advanced preparation of the resin and storage in the refrigerator will enhance palatability; however, no more than a three-day supply should be made, and the mixture should be mixed carefully each time before ingestion. The medication can be suspended in fruit juice, mixed with applesauce or cereal, or incorporated into bran muffins. Another formulation, Questran® Light has a finer texture and contains a sugar substitute to reduce the caloric content. This preparation requires less liquid for mixing. Recently, cholestyramine has been formulated as a candy-like bar (Cholybar™); these bars have been shown to be as effective as regular cholestyramine in terms of cholesterol-lowering effects but retain the high incidence of side effects. The bars contain 60 calories each and, if required in large doses, might be a problem for overweight patients. In addition, it is yet unclear whether most patients prefer the bars over the original powder, although the selective use of a Cholybar™ at otherwise inconvenient times might increase compliance and may be an excellent way to administer the resin to a child.

Because of its anion-binding capacity, suspension of the resin in artificially sweetened drinks will remove saccharin, cyclamate, and aspartate and change the perceived sweetness. Patients should be encouraged to use their own creativity and ingenuity in disguising the medication. It can be incorporated into virtually any food as long as it is taken with adequate liquid and is acceptable to the patient. Although the two resins are equally effective, patients generally prefer cholestyramine to colestipol because of its finer particle size and, therefore, greater palatability. On the other hand, colestipol is somewhat less expensive than cholestyramine (*see* Table 11.2).

Cholestyramine and colestipol can be used effectively in combination with other drugs, such as nicotinic acid, lovastatin, fibric acid derivatives, and probucol. These combinations will be discussed at the end of this chapter.

Nicotinic Acid

Nicotinic acid (niacin) is vitamin B-3, which is required in low doses (less than 50 mg/day) for the normal function of cellular metabolic pathways. When used in pharmacologic doses, nicotinic acid has a favorable effect on lipoprotein metabolism and has been used for the treatment of hypercholesterolemia and hypertriglyceridemia since the early 1950s. The drug lowers the concentrations of both LDL-C and VLDL-C and raises HDL-C. A dose response relation exists between the amount of nicotinic acid taken and the changes that occur in triglyceride, LDL-C, and HDL-C concentrations. At doses of 2 to 6 g/day, a 30% to 40% reduction in triglyceride, a 30% to 35% reduction in LDL-C, and a 30% to 35% increment in the HDL-C concentration can be expected. The effects of nicotinic acid on lipids and lipoproteins are completely independent of its action as a vitamin. Nicotinamide (Niacinamide), which is also effective as a vitamin, has no effect on dyslipidemia.

Despite its association with a high incidence of adverse side effects, nicotinic acid has not been associated with any permanent or unexpected long-term toxicity and is now one of the time-honored forms of therapy. In the opinion of some experts, nicotinic acid is the drug of choice for the treatment of most forms of dyslipidemia. Among the hypolipidemic agents it is the only drug known to be significantly associated with a reduced number of coronary events and a long-term reduction in total mortality. In the Coronary Drug Project, daily treatment with 3 g nicotinic acid reduced the 5-year incidence of nonfatal myocardial infarction by approximately 25%, although total mortality was not altered. Follow-up of surviving patients for an additional 5 to 9 years after discontinuing therapy revealed a significant 11% long-term reduction in all-cause mortality, compared to those patients who had received placebo. Nicotinic acid was the only drug of those patients evaluated in the Coronary Drug Project that reduced long-term mortality.

Mechanism of Action. The exact mechanisms by which nicotinic acid affects lipoprotein metabolism are not clearly understood. Available information suggests that the drug is antilipolytic and inhibits the mobilization of free fatty acids from adipose tissue, thus decreasing the substrate for hepatic triglyceride synthesis. Nicotinic acid also directly inhibits the hepatic synthesis of triglyceride. The result of these effects is a reduction of VLDL secretion by the liver (see Fig. 11.2). Because the LDL particle is a byproduct of the lipolytic cascade and the metabolism of VLDL, it is not surprising that LDL-C concentrations usually also decrease. We have no explanation for the occasional patient in whom LDL-C increases during niacin treatment of hypertriglyceridemia, although this effect may be more common at the lower doses of nicotinic acid.

Nicotinic acid also can increase the concentration of HDL-C, especially HDL_2-C, substantially. The precise relation of the changes in HDL-C to the changes in the metabolism of VLDL and LDL are unknown. Because HDL-C concentrations may be increased at low doses of nicotinic acid that appear to have no effect on the concentrations of VLDL and LDL, it is possible that the mechanisms that underlie such changes are unrelated. It should be noted, however, that significant changes may occur in the secretion and catabolic disposal of both VLDL and LDL in the absence of any changes in their concentrations. Although the effects of nicotinic acid have not been studied in great detail, the increase in HDL concentration occurs most likely as a consequence of a reduction in the rate of HDL-C catabolism rather than an increase in the rate of HDL

synthesis. Because the HDL particle is metabolized by hepatic triglyceride lipase, the effect of nicotinic acid to raise the HDL-C levels could be explained by its known antilipolytic effect, but this is conjectural.

Adverse Effects. Nicotinic acid often is associated with substantial and limiting side effects. Consequently, its use should be undertaken in a deliberate way, and therapy should be initiated at very low doses. There is evidence to suggest that patients become tolerant to some of the adverse effects of nicotinic acid after 14 days of therapy at doses that exceed 500 mg.

One of the most common and distressing adverse effects of nicotinic acid is cutaneous flushing, which is caused by a prostaglandin-mediated vasodilation and increase in skin temperature. In this regard it is said to be similar to the hot flash that menopausal women experience. The majority of patients (80%) experience flushing when the drug is started and when the dose is increased. Some patients have more extreme reactions, and, for this reason, it may be advisable to give a small challenge dose in the office to allow the patient to experience the drug effects under careful supervision. Flushing episodes usually resolve or at least improve substantially, within 2 weeks of changing the dose. To counteract the flushing that nicotinic acid produces

1. Initiate therapy at a low dose and gradually increase it to the desired level. Any increases in dose also should be done in the same way.

2. Always prescribe the medication to be taken with a meal, because this appears to diminish flushing.

3. Suggest that the drug be taken with a cold liquid; hot liquids and alcoholic beverages are to be avoided, because they increase the vasodilation.

4. Advise that one-half to one adult aspirin be used 30 minutes before ingestion of the nicotinic acid dose during initial therapy and temporarily during dose increases. Aspirin and other prosta-

glandin synthetase inhibitors inhibit the prostaglandin-mediated flushing effectively.

If the above guidelines are adhered to rigorously, many patients will be able to tolerate nicotinic acid despite the side effects. Once the dose has been stabilized, most patients become tolerant to the flushing and report either no or only mild symptoms. Flushing episodes can recur, however, if the patient misses several doses.

Other significant dermatologic side effects that may be encountered in patients who use nicotinic acid are dry skin or pruritus (50%), generalized rash (20%), and, rarely, acanthosis nigricans. In patients who take large doses of nicotinic acid, gastric irritation, gastrointestinal complaints, and abnormalities of liver function also can be encountered. Gastrointestinal symptoms, which include abdominal pain, heartburn, nausea, anorexia, and diarrhea, may occur alone or in combination in 25% of patients who take unmodified forms of nicotinic acid. Peptic ulcer is a relative contraindication to the administration of nicotinic acid, although the data to support such a recommendation are not impressive.

Asymptomatic increases in liver enzymes are not uncommon at large doses of niacin (> 3 g/day). Gradually increasing the dose of nicotinic acid appears to minimize elevations of liver enzymes at larger doses. Modest changes in SGOT and alkaline phosphatase levels also may occur, but, if they are not greater than three times the upper limit of normal, nicotinic acid need not be discontinued. Stopping or reducing the nicotinic acid usually will reverse any abnormal liver function tests. Clinical hepatitis is much rarer, but both hepatocellular and cholestatic jaundice have been described in poorly monitored patients, including some reports of fulminant hepatitis. The incidence and severity of hepatitis appear to be significantly greater with sustained-release nicotinic acid preparations, even at relatively low doses. Some evidence suggests that this phenomenon is especially prominent in patients who switch from immediate-release to sustained-release preparations. It should be noted that jaundice in users of nicotinic acid also can be caused by hemolysis, as well as by an increase in indirect hyperbilirubinemia in patients who have Gilbert's syndrome.

Niacin also can interfere with glucose metabolism and worsens glycemic control in approximately 50% of established diabetic patients. Occasionally, nondiabetics can develop impaired glucose tolerance and even fasting hyperglycemia. Patients on nicotinic acid are several times more likely to have fasting glucose concentrations above 110 to 120 mg/dL and 1-hour plasma glucose concentrations above 220 to 240 mg/dL than persons who take a placebo. In many such cases, the modest increase in fasting glucose concentration is of no apparent clinical consequence, whereas the increase in the severity of hyperglycemia in established diabetic patients usually can be managed simply by increasing the dose of insulin or oral hypoglycemic agent. The mechanism of the diabetogenic action of nicotinic acid is not known but appears to be caused by insulin resistance. Diabetes is considered to be a relative contraindication to the use of nicotinic acid, but this need not be invariant. Because insulin may be an independent risk factor for the development of atherosclerosis, increasing the dose of insulin to control the hyperglycemia may not be in the best interest of the patient. Alternatively, other appropriate lipid-active agents should be selected for diabetic patients.

An increase in the serum uric acid concentration may occur in as many as 50% or more of the patients placed on nicotinic acid. In many of them, however, the increment remains within or only slightly above the normal range. In 10% to 15% of the patients, severe hyperuricemia (≥10 mg/dL) develops and may be associated with the development of gout. Preexisting symptomatic gout also is a contraindication to the use of nicotinic acid.

Miscellaneous adverse effects include an increase in body hair, brownish skin discoloration, an offensive body odor, worsening of psoriasis, fatigue, lack of energy, and occasionally, hypotension. Rare adverse effects include toxic amblyopia, atrial arrhythmias, and panic attacks. Modest elevations of creatine kinase (CK) to 150 to 200 IU/L occur commonly (50%) in patients on nicotinic acid, and nicotinic acid-associated myopathy also has been reported. This effect could contribute to skeletal muscle toxicity when used in combination with lovastatin. Such patients should be followed carefully.

On the basis of this long list of possible adverse effects, niacin may sound like an extremely toxic drug that should be avoided whenever possible. In fact, when used in carefully selected patients and when appropriately regulated, nicotinic acid has not been associated with any permanent or unexpected long-term toxicity, and most of the side effects are reversible on discontinuation of therapy. The long-term safety of niacin has been documented in a number of large, long-term studies as well as in worldwide treatment of numerous patients. It would appear prudent to exclude the use of nicotinic acid in patients who have liver disease, active peptic ulcer, and gout; the presence of diabetes mellitus remains a relative contraindication.

Clinical Use. To minimize potential adverse reactions, gradual escalation of the dose is desirable (*see* Table 13.2). There are two alternative methods:

1. Using 100-mg tablets—therapy can be initiated with 50 to 100 mg of nicotinic acid three times daily, taken with meals. The dose of nicotinic acid can be increased progressively by 100 mg at each meal at 1-week intervals. Once the patient has worked up to four 100-mg tablets with each meal (usually over a period of four weeks), additional dose increases should be attempted by using 500-mg tablets.

2. An alternative schedule is to start with a single 250-mg dose (one-half of a 500-mg tablet) and increase the dose each week by 250 mg to a maximal dose of 3 to 6 g.

The target dose is 1 to 2 g per day taken with meals, depending on the patient's tolerance to the medication as well as the pharmacologic effect achieved. Beneficial effects of nicotinic acid are seldom achieved at doses under 1.5 g/d, although 1 g/d has been demonstrated to increase the HDL-C by approximately 30% with relatively inconsequential changes in the total cholesterol and triglyceride concentrations. Although the drug is traditionally taken three to four times per day, some of our patients find a twice-daily dosage to be easier to tolerate.

There has been and continues to be significant interest in sustained or slow-release forms of nicotinic acid because they seem to be better tolerated and are as effective as the unmodified forms. Sustained-release forms of nicotinic acid are recommended for patients in whom the cutaneous flush is particularly bothersome. The use of these forms of nicotinic acid to minimize flushing is supported in the literature, but it is by no means a panacea for this problem. In a recent study, flushing occurred in 100% of patients who took the unmodified form of nicotinic acid and in 82% of those who took a sustained-release form. Several studies document that other side effects of nicotinic acid are accentuated in patients who take the sustained-release form of nicotinic acid— nausea, vomiting, diarrhea, and fatigue occur with increased frequency. Furthermore, heartburn and indigestion, indicative of gastritis, and abnormalities in liver function also occur more frequently and at lower doses in patients who take the sustained-release form of nicotinic acid. In some patients, the elevated liver enzymes observed with sustained-release forms disappear when regular nicotinic acid tablets are substituted. There also may be a difference in pharmacologic effectiveness of nicotinic acid in the tablet and sustained-release forms. One study suggested that the hypocholesterolemic effects of the two forms of nicotinic acid were similar, but that the sustained-release forms were less effective than the regular nicotinic acid tablets in reducing the triglyceride and raising the HDL_2 concentration. These results emphasize that there also may be important pharmacologic differences between the sustained-release and unmodified forms of nicotinic acid.

A plethora of slow-release forms of nicotinic acid have appeared on the market and can be obtained from pharmacies and health food stores. Little if any clinical information is available on individual preparations. Many appear to be effective and are well tolerated; however, it is impossible to know whether this will be true for all preparations. Currently, regular nicotinic acid is preferred to the sustained-or slow-release forms for the reasons noted above.

Almost 60% to 70% of patients will tolerate nicotinic acid if started appropriately and taken faithfully. Many physicians and patients become discouraged early in therapy, however, and the overall compliance with nicotinic acid regimens is only approximately 50%, even in those patients who may be taking as little as 1000 mg per day.

Nicotinic acid is a useful drug in combination with resins and may be particularly effective in the treatment of combined dyslipidemia. The combined use of nicotinic acid and resins minimizes their adverse effects and permits use of lower doses of each to achieve the same effect.

HMG CoA Reductase Inhibitors

The HMG CoA reductase inhibitors comprise a new and exciting class of drugs for the treatment of hypercholesterolemia. Three drugs, lovastatin (Mevacor®), pravastatin (Pravachol®) and simvastatin (Zocor®) are currently on the market in the United States, but several additional drugs in this class currently are under development.

The original compound, compactin, is a byproduct isolated from a *Penicillium* fungus, whereas Mevinolin is a fermentation product derived from an *Aspergillus* species fungus. Pravastatin and simvastatin, both currently under clinical testing, are derived

from *Nocardia* and *Aspergillus* species. These substances are potent competitive inhibitors of HMG CoA reductase, the enzyme that catalyzes the rate-limiting step of cholesterol synthesis: the conversion of hydroxymethylglutaryl CoA to mevalonate (*see* Fig. 11.2). All of the drugs that inhibit HMG CoA reductase competitively resemble HMG CoA. Several are pro-drugs that are inactive in their administered form but are converted to the active form in the body. Minor structural differences between the drugs may have implications with regard to potency, efficacy, and toxicity.

For the most part, HMG CoA reductase inhibitors are well tolerated and have relatively few adverse effects. It is likely that long-term reduction in cholesterol with these agents also will be associated with protection against the development of CHD. There is no reason to think that this would not be the case in view of the fact that diet, nicotinic acid, resins, and gemfibrozil all have reduced the risk of developing CHD. However, it is still too soon to know whether there will be any unanticipated long-term serious adverse effects as a result of the use of these drugs. This is not inconsequential, because treatment of hypercholesterolemia, by its nature, requires life-long administration of medication. Consequently, the HMG CoA reductase inhibitors should not be considered first-line drugs at this time, even though they appear to be the most efficacious for lowering LDL-C. Particular caution should be used in younger patients because of the need for life-long therapy.

Mechanism of Action. This class of drugs works by inhibiting cellular cholesterol synthesis, thereby reducing the intracellular content of cholesterol (*see* Fig. 11.2). As the intrahepatic concentration of cholesterol decreases, cellular mechanisms are activated to restore cellular cholesterol to optimum levels. Ordinarily, this is accomplished in two ways: (1) by increasing the synthesis of cell-surface LDL receptors that bind and internalize circulating LDL, and (2) by increasing the rate of cholesterol synthesis. Because the second response is blocked by the drug, the cellular cholesterol level can be restored toward normal only through the mechanism of increased synthesis and expression of LDL receptors. As a consequence of these changes, the fractional catabolic rate for LDL and the cholesterol that it carries is enhanced, and the blood LDL-C concentration decreases. Although the drug inhibits cholesterol synthesis, the major mechanism by which it lowers the cholesterol concentration is by enhancing removal of LDL-C from the blood. It has been suggested that 75% of its cholesterol-lowering effect is mediated in this way.

Like many other drugs, the effectiveness of this class of drugs depends on the ability of liver cells to synthesize and express the LDL receptor. This drug is effective then in patients who have polygenic hypercholesterolemia and in patients who have the heterozygotic form of familial hypercholesterolemia. It is not effective, however, in patients who have homozygous familial hypercholesterolemia, in which receptors are totally absent or defective. Because polygenic hypercholesterolemia is a heterogeneous disorder and may include patients who have abnormal apoprotein B as well as patients who have subtle point mutations in the binding portion of the LDL receptor, not all patients will respond in a dramatic or gratifying fashion.

Adverse Effects. Because cholesterol is necessary for the synthesis of cellular and subcellular membranes, bile acids, and adrenal and gonadal steroid hormones, there was initial concern that the HMG CoA reductase inhibitors would be dangerous. Recent studies demonstrate that total body cholesterol is unchanged in patients who take lovastatin; thus, the effect on cholesterol appears to be confined primarily to circulating LDL-C. This indicates that cellular cholesterol content remains within normal limits and the chance for serious toxicity is minimal. To date, no serious adverse effects involving electron transport or steroid hormone synthesis have been reported. Furthermore, none of the HMG CoA reductase inhibitors have produced lithogenic bile or predisposed to the development of cholelithiasis.

The reductase inhibitors usually are well tolerated and only infrequently has it been necessary to discontinue their use. Patients may develop mild gastrointestinal complaints such as flatulence and diarrhea. In some studies, gastrointestinal complaints were no more frequent than those experienced by patients who took placebo. Liver enzyme changes may occur but generally are minor; increased transaminase levels occur in less than 5% of patients, and only about 2% of patients treated with lovastatin will have an increase in the alanine aminotransferase to more than three times the baseline value. The abnormalities in liver enzyme tests appear to be dose-related.

A complication of greater concern is skeletal myopathy, as reflected by the elevation in creatine kinase levels in some patients. The mean increase in creatine kinase level is statistically significant only for those patients who receive the highest dose, 80 mg per day. It has been suggested that the myopathic syndrome reflects susceptibility to the drug at the level of the myocyte, because plasma mevalonate (precursor at the HMG CoA reductase step) concentrations are similar in those patients who do and do not have the myopathy. Lovastatin-associated rhabdomyolysis has been reported with increased frequency in patients who have renal failure and in patients who have organ transplantation, particularly those who receive cyclosporine A. Myopathy or rhabdomyolysis has also been reported in lovastatin-treated patients who concomitantly received gemfibrozil; rarely in patients who also received nicotinic acid; and more recently, in patients who were treated with erythromycin. In the latter two settings, blood levels of lovastatin have been elevated markedly, suggesting that an interaction occurs that slows the rate of metabolism of lovastatin and leads to toxicity. We suspect that postmarketing surveys will disclose adverse drug-to-drug interactions with a greater frequency than observed in the early trials that used rigid exclusion criteria. It is estimated that about 30% of patients who take lovastatin and cyclosporine A, 5% of patients who take lovastatin and gemfibrozil, and 0.2% of patients who take lovastatin alone will develop myopathy. The myositis resolves within weeks of discontinuing lovastatin. In one large series of patients who had received lovastatin on a long-term basis (average 2.6 years), 2.2% were withdrawn from the drug because of adverse effects, only one of whom had myopathy. The mechanism by which lovastatin causes myopathy is unknown, but clinical experiences suggest that the drug may sensitize the myocyte to exercise-induced cellular damage. Myopathy also occurs with pravastatin and simvastatin.

Because of prior experience with triparanol (MER-29), there has been great concern about the possibility of development of lens opacities in patients treated with lovastatin. A comparison of lens opacities in patients who received lovastatin with those who received cholestyramine demonstrated no significant differences with regard to baseline opacities, new opacities, lost opacities, or opacities present at completion of the study. Similar observations have been made with pravastatin (Pravachol®) and simvastatin (Zocor®). The FDA has permitted the pharmaceutical companies marketing these drugs to remove the warning about lens opacities and the need for ophthalmologic examination from the drug inserts.

A relatively inconsequential complication of lovastatin has been an alteration in sleep pattern. The total duration of sleep may be shortened by one to three hours. This problem has been reported by 15% to 20% of patients in one medical center. There is little pharmacokinetic information available on the use of this class of drugs in patients who have impaired renal or hepatic function, and, in patients who have these problems, the drugs should be used cautiously. Patients who have nephrotic dyslipidemia appear to tolerate lovastatin without much difficulty, but the number of patients studied has been small.

Because of tissue selectivity of certain of the new HMG CoA reductase inhibitors (e.g., pravastatin), some adverse effects, such as the potential for muscle abnormalities may be reduced. However, selective concentration of the drug in the liver theoretically could increase hepatic dysfunction or adverse drug-to-drug interaction. It really is entirely too soon to know whether any of the drugs under development will have greater efficacy or different toxicities.

Clinical Use. Numerous studies demonstrate that lovastatin, pravastatin, and simvastatin reduce the concentration of total and LDL-C by approximately 30% to 45% at doses of the drug that range from 40 to 80 mg daily (lovastatin) or 20-40 mg (pravastatin and simvastatin). They are as effective in patients who have polygenic (or nonfamilial) hypercholesterolemia as in those who have heterozygotic form of familial hypercholesterolemia. These drugs can be used effectively alone, but their cholesterol-lowering effects can be potentiated by combining it with various cholesterol-lowering drugs (resins, nicotinic acid, probucol), as well as with ileal bypass. Although their primary effects are to reduce the concentration of LDL-C, they frequently are associated with a modest increase, up to 10%, in HDL-C, and a 15% decrease in the triglyceride concentration. The mechanism by which the HDL-C concentration increases is unknown; the reduction in the triglyceride concentration is caused by decreased VLDL synthesis and accelerated removal of small VLDL particles (IDL) that normally bind to and are removed by the LDL receptor.

Lovastatin therapy should be initiated at a dose of 20 mg daily and pravastatin and simvastatin at doses of 10 mg daily, administered with the nighttime meal, when cholesterol synthesis is at its maximum. However, the recently released 10 mg dose may be preferable for some persons. After four to six weeks, the dose can be increased to 20 or 40 mg as a single dose, depending on the drug selected. Eighty to ninety percent of the reduction in the LDL-C is achieved at half the maximum recommended doses (Fig. 11.5). Consequently, it usually is prudent and cost-effective to add a second drug rather than to increase the dose to 60 or 80 mg daily.

Lovastatin is effective in patients who have noninsulin-dependent diabetes and produces changes in blood levels of LDL-C, HDL-C, LDL-apoprotein B, VLDL-C, and triglyceride that are similar to those seen in nondiabetics. No unusual side effects or toxicities have been reported during short-term lovastatin therapy in patients who have noninsulin-dependent diabetes. It is also effective and well tolerated in the elderly.

Lovastatin and simvastatin also have been used effectively in a small number of patients who have nephrotic dyslipidemia. In such patients, lovastatin decreases LDL-C concentrations by reducing LDL production. This appears to be a result of enhanced removal of IDL by the LDL receptor, thereby decreasing the availability of substrate for LDL synthesis. Furthermore, lovastatin also reduced VLDL triglyceride levels by enhancing VLDL catabolism.

Fibric Acid Derivatives

Fibric acid derivatives are indicated primarily for the treatment of patients who have hypertriglyceridemia. Of the two approved drugs in this class in the United States, gemfibrozil is the agent of choice. Clofibrate, a first-generation fibric acid derivative, was shown to have a high incidence of adverse effects in several large clinical trials and now is seldom used. Additional drugs of this class (fenofibrate, bezafibrate) are available in Europe and may be approved by the FDA for future use in the United States.

Gemfibrozil acts primarily to decrease triglyceride levels, and this often is associated with increases in HDL-C. Its effect on LDL-C is less marked and varies according to the initial LDL-C level. For patients who

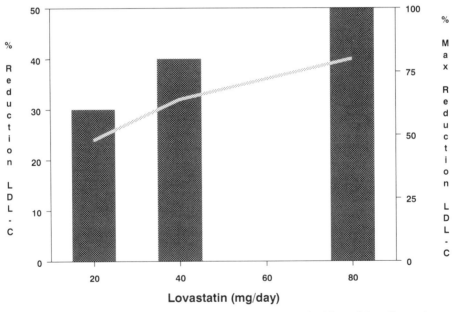

Figure 11.5. Dose-response curve of LDL-C to treatment with lovastatin. Most of the effects of treatment are seen at a dose of 40 mg /day (30+% LDL-C reduction), with relatively little additional benefit occurring after additional dosage increases. The bars represent the percent of maximal LDL-C reduction achievable; about 80% reduction with 40 mg per day and 100% at 80 mg per day.

have increased triglyceride and relatively low LDL-C, the LDL-C levels actually increase as triglyceride levels fall. In persons who have normal triglyceride levels, LDL-C often decreases; however, the reduction of LDL-C generally is in the range of 10% to 15%, and only patients who have mild or borderline hypercholesterolemia are likely to attain their therapeutic goals. Gemfibrozil is not approved for use in patients who have hypercholesterolemia alone.

Clofibrate was the first of the fibric acid derivatives to be used. It is less potent than gemfibrozil in reducing triglyceride and increasing HDL-C concentrations. Although the incidence of nonfatal MIs was reduced by 25%, there was no difference in the rate of fatal first heart attacks. In addition, there was a 36% higher mortality rate because of noncardiovascular causes in the clofibrate-treated group than in the placebo group, half of which were caused by malignancy. Because all-cause mortality increased in the clofibrate-treated group, great concern was expressed. In the Coronary Drug Project Study, noncardiovascular mortality rates were not significantly greater in the clofibrate-treated group than in the placebo group, but clofibrate provided no protection from cardiovascular events. As a consequence of these studies, clofibrate is no longer recommended in the treatment of patients who have dyslipidemia.

Gemfibrozil is a second-generation fibric acid derivative. It is used commonly and is generally well tolerated. Its efficacy in preventing heart disease recently has been demonstrated in the Helsinki Heart Study in asymptomatic dyslipidemic men 40 to 55 years of age. Entry into this trial depended on a non-HDL (total cholesterol minus HDL-C) cholesterol greater than 200 mg/dL (5.17 mmol/L). The use of gemfibrozil, at a dose of 1200 mg/d, reduced fatal and nonfatal myocardial infarctions, as well as sudden death caused by coronary artery disease, by 34% (from 41/1000 to 27/1000). This effect was more pronounced than in any previous large-scale clinical trial.

Mechanism of Action.   The basic mechanism by which the fibric acid derivatives reduce the triglyceride and cholesterol concentrations is not known with certainty. The drugs appear to affect several sites in the lipoprotein pathway (*see* Fig. 11.2), inhibiting VLDL production as well as enhancing VLDL clearance. Reduction of hepatic triglyceride synthesis occurs in two ways: (1) inhibition of lipase activity in adipose tissue, thereby reducing the mobilization of FFA to the liver; and (2) direct inhibition of FFA incorporation into triglyceride in the liver. Enhanced peripheral disposal of VLDL triglyceride occurs by activation of lipoprotein lipase. In contrast to gemfibrozil, virtually all of the triglyceride-lowering effects of clofibrate result from increasing the rate of peripheral clearance. These differences may explain partly the greater efficacy of gemfibrozil in reducing the VLDL and triglyceride concentrations. Fenofibrate, a third-generation fibric acid derivative, has effects similar to those of gemfibrozil.

The mechanisms by which gemfibrozil and fenofibrate reduce the LDL-C concentration are not clear but may be related to inhibition of HMG CoA reductase; exactly how these drugs might do this is not understood. It is curious that the overall LDL-C response ranges from a reduction of 10% in Type IIa (isolated hypercholesterolemia) to an increase of 10% in Type IV (combined hypercholesterolemia and hypertriglyceridemia); however, the greatest LDL-C lowering occurs among those persons who have Type III (45%) and Type V (20%) (*see* Table 11.4). The response tends to be greater in those persons who have higher LDL values. Fenofibrate appears to have the same effect on the LDL-C concentration in these lipoprotein disorders. Because gemfibrozil decreases VLDL-C, increases HDL-C, and has a variable effect on LDL-C, the total cholesterol cannot be used to assess the effectiveness of this drug. Interestingly, gemfibrozil has no effect on the Lp (a) concentration.

Adverse Effects.   Side effects associated with the use of fibric acid derivatives are relatively mild. Although clofibrate is associated with an increased frequency of cholelithiasis and biliary tract disease (caused by increased bile lithogenicity) as well as gastrointestinal malignancy, these problems seem less frequent with gemfibrozil. Gastrointestinal symptoms are more frequent in patients who take gemfibrozil (4%) than placebo (2.5%), but the overall rate is small and symptoms usually are mild. Gemfibrozil is contraindicated in patients who have active peptic ulcer disease. Gemfibrozil potentiates the action of coumadin and also can induce a myopathy, particularly in combination with lovastatin or when used in patients who have renal failure.

Clinical Use. It is difficult to generalize about the effects of gemfibrozil on plasma lipid and lipoprotein levels, because the response varies, depending on the initial lipoprotein profile. In general, gemfibrozil decreases the plasma triglyceride concentration by as little as 20% to as much as 75%, depending on the underlying lipoprotein phenotype. In patients who have Types II-B and IV (mixed dyslipidemia), the triglyceride-lowering effect is around 25%, whereas in patients who have Types III and V, the reduction may reach 70%. Gemfibrozil may be most effective in those patients who have the highest triglyceride levels.

In addition to reducing the triglyceride concentration, gemfibrozil increases the concentration of HDL-C. In most dyslipidemic patients the increase in HDL-C ranges from 10% to 25%, being greatest in those patients whose initial HDL-C levels are the lowest. In patients who have Types III and V hyperlipoproteinemia, the increase in HDL-C is approx. 65% and 50%, respectively. The HDL increases because of augmented HDL synthesis, rather than diminished HDL catabolism. Both HDL_2 and HDL_3 fractions increase concurrently with enhanced synthesis of apolipoproteins A-I and A-II. Although the direction of the changes in triglyceride and HDL-C levels are predictable, the magnitude of the changes will vary.

Gemfibrozil is well tolerated by patients who have diabetes mellitus, and, curiously, some patients show an improvement in glycemic control as their hypertriglyceridemia is corrected while glycemic control deteriorates in others. Improvement is compatible with the concept that increased metabolism of FFA leads to insulin resistance and that fibrates decrease lipolysis and release of FFA from adipose tissue.

Patients who have the nephrotic syndrome, by virtue of their defects in lipoprotein metabolism, are excellent candidates for gemfibrozil, but because gemfibrozil is excreted by the kidney, the dose must be modified (50% reduction in dose). Such patients may develop myopathy as a consequence of treatment with gemfibrozil.

Probucol

Probucol has been available for the treatment of hypercholesterolemia for almost 20 years. It has never become a first-line drug, however, and has not been demonstrated to reduce the relative risk of CHD, although it has been reported to bring about sizeable reductions in skin and tendon xanthomas in homozygous FH patients. A major problem of this drug has been that the lipoprotein responses to it vary greatly, and many patients have little, if any, reduction in LDL-C. In addition, probucol consistently lowers HDL-C levels in most patients.

Mechanism of Action. The ability of probucol to reduce the total cholesterol and LDL-C concentrations is caused by an enhanced clearance of LDL from the circulation (*see* Fig. 11.2). The precise mechanism by which probucol increases the fractional catabolic rate for LDL is not known, but it is not mediated by the LDL receptor. Probucol does not increase the excretion of cholesterol or bile acids, does not alter the composition of bile, and cannot inhibit the absorption of cholesterol or influence the activities of lipoprotein lipase and hepatic lipase. The fact that it is able to reduce the cholesterol concentration in patients who have homozygous familial hypercholesterolemia suggests that it does not depend on the LDL receptor for its hypocholesterolemic effect and that it is one of the few forms of treatment that may be useful in this disorder.

Although the use of probucol has resulted in the regression of tendon xanthomas, there has always been some concern about its use as a primary form of therapy for hypercholesterolemia, because it reduces HDL-C as well as LDL-C concentrations. In fact, the relative reduction in HDL-C concentration is greater than the reduction that occurs in LDL-C concentration, so that the LDL-C-to-HDL-C ratio actually increases, a change that theoretically affects adversely the risk of developing CHD. Probucol reduces the plasma total cholesterol concentration by 10% to 20%, but the magnitude of reduction in LDL-C is relatively modest.

The reduction in HDL-C and apoprotein A-I concentrations could reflect enhanced reverse cholesterol transport, but the reduction in apoprotein A-I is actually caused by inhibition of apoprotein A-I synthesis. Probucol enhances the efflux of cholesterol from cells and the release of cholesterol from macrophages in the presence of HDL (reverse cholesterol transport). The reduction that occurs in HDL-C may be caused in part by enhanced uptake of HDL by the liver and the presence of smaller, denser HDL reputed to be more effective in reverse cholesterol transport.

Although there is a great deal of uncertainty about the mechanism of action of probucol, there is new and exciting information concerning its potential role in the prevention of atherosclerosis by a mechanism that appears to be separate from its lipid-lowering effects (*see* Chapter 1). Evidence has accumulated that oxidative modification of LDL enhances its atherogenicity. Once modified, LDL particles are removed more efficiently by the scavenger receptor present on the surface of endothelial cells, smooth muscle cells, and macrophages in the vessel wall. Probucol is incorporated into the LDL particle and protects it against oxidative modification. The LDL particles isolated from the blood of hypercholesterolemic patients who have been treated with probucol are highly resistant to peroxidation, suggesting that probucol may inhibit atherogenesis by limiting oxidative modification of LDL and subsequent foam cell formation. The effectiveness of probucol in preventing atherosclerotic lesions has been demonstrated in an experimental animal model that is highly susceptible to atherosclerosis. In Watanabe rabbits, probucol was more effective in inhibiting the formation of atherosclerotic aortic lesions than was lovastatin, despite the fact that lovastatin was more effective in lowering the concentrations of LDL-C. Although probucol may lower the LDL-C concentration, such changes are modest at best, and its ability to inhibit atherogenesis may be more a result of protection of the LDL particle. Regression of xanthomatous lesions occurs despite a reduction in the HDL-C concentration. Probucol may prove to be as effective or more effective than other agents in causing regression of atheromatous lesions despite the fact that it is not particularly impressive as a cholesterol-lowering agent.

Adverse Effects. Probucol usually is well tolerated, and side effects, when present, are relatively mild and gastrointestinal in origin. Loose stools may be the only significant symptom, and these usually do not require that probucol be discontinued. When probucol is used in combination with bile acid resins, this side effect may be advantageous in offsetting the constipation that is characteristic of this class of drugs. Although probucol has been associated with prolongation of the QT interval in experimental animals, this has not been a problem with its clinical use.

Clinical Use. The effect of probucol to lower the blood cholesterol usually is additive to that of the diet. Probucol will reduce the cholesterol level by an average of about 15% and perhaps by as much as 25% in patients who also are following a low-fat diet (*see* Fig. 11.1). Because of its different mode of action, probucol should potentiate the hypocholesterolemic effects of other drugs.

Probucol can be used successfully in combination with resins or with nicotinic acid. The combination of probucol with bile acid resins will reduce the total cholesterol concentration by approximately 25% to 30% and the LDL-C concentration by 30%. The effect of these two drugs in combination appears to be additive. Interestingly, the effect of probucol on HDL-C concentrations predominates in the combination, so that the HDL-C level is reduced by about 30%. Probucol does not add to the cholesterol-lowering effect of lovastatin. Presently, probucol should be considered only as a secondary drug for treatment of hyper-cholesterolemia.

Estrogens

Although estrogens are not usually thought of as hypolipidemic agents, they may be considerably beneficial in postmenopausal women who have mild-moderate hypercholesterolemia. Estrogen deficiency may be associated with a 10% to 20% increase in LDL-C and only modest reductions in HDL-C, approx. 5 mg/dL (.13 mmol/L). Estrogen replacement therapy in such a woman will decrease the LDL-C substantially and may make the need for other hypolipidemic therapy unnecessary. These effects are expressed optimally when conjugated estrogens are administered at a dose of 0.625 mg/d. This is also the minimum dose proven to protect the skeleton against osteoporosis. It has been estimated that approx. 50% of the cardiovascular protection from estrogen is expressed through changes in lipoproteins. There are benefits, however, that are mediated by improved coronary vasodilation and reductions in systolic blood pressure, diastolic blood pressure, fasting glucose level, and adiposity. Because of the increased risk of endometrial and, possibly, breast cancer, estrogen replacement for menopausal women must be considered on a patient-by-patient basis.

Neither hypertension nor venous thrombosis is increased by postmenopausal estrogen replacement.

Recently, transdermal estrogen has been used for correction of estrogen deficiency. It is well tolerated and probably will be as efficacious as oral and parenteral estrogen, but long-term studies are not available. Hypertriglyceridemia is less likely to occur with transdermal estrogen, and this may be the preferred preparation in women in whom this is a consideration. Little or no change occurs in LDL-C and HDL-C with transdermal estrogen.

Combination Drug Therapy

Combined drug therapy is being used increasingly to treat hypercholesterolemia and the combined dyslipidemias. Combinations of lipid-lowering drugs may be particularly useful in patients who have resistant hypercholesterolemia and often are necessary in patients who have total cholesterol levels above 300 mg/dL (7.76 mmol/L). In general, the effects of drugs tend to be additive for those lipid disorders on which they have similar effects and offsetting for those lipids in which their effects are opposite. Consequently, the use of two drugs that both lower LDL and raise HDL would tend to be additive for these two lipoproteins. In this regard, the use of nicotinic acid with lovastatin and probably other reductase inhibitors multiplies the effectiveness of these two drugs on the LDL-C concentration. By way of contrast, the combination of bile acid resins with probucol accentuates the reduction in the LDL-C concentration but converts a mild to modest positive effect of the resin on HDL-C to a negative one. The use of bile acid resins alone for hypercholesterolemia may be associated with an increase in the triglyceride concentration, but when used in combination with a drug that lowers the triglyceride concentration, such as gemfibrozil or nicotinic acid, the triglyceride concentration is reduced.

Lovastatin can be used effectively and safely in combination with resins and nicotinic acid but probably not gemfibrozil. It also can be used with probucol, but the additional effects are variable and not of great magnitude. It is particularly effective in patients who receive resins, because it accentuates LDL receptor synthesis and expression by inhibiting the compensatory increase in cholesterol synthesis that occurs when resins are used alone. Consequently, the combination can reduce the LDL-C concentration by 50% to 60%.

When nicotinic acid is used in combination with lovastatin, the beneficial effects of both drugs on the LDL-C and HDL-C concentrations are preserved, but, in addition, there is a substantial lowering of the triglyceride concentration that would not be seen otherwise. With combinations of appropriately selected drugs, the total cholesterol can be reduced as much as 55%; the LDL-C, 60%; and the HDL-C can be increased by as much as 45%.

In patients whose LDL-C is unusually resistant to therapy, resins, nicotinic acid, and HMG CoA reductase inhibitors can be used simultaneously; under these circumstances, the LDL-C level may be reduced by 65% to 70%, and the HDL-C level may be increased by approximately 40%. Most of the increase in the HDL-C is attributed to the associated use of nicotinic acid. These drugs work in a complementary fashion. In patients in whom combinations of drugs are used to treat isolated hypercholesterolemia or hypercholesterolemia with mild hypertriglyceridemia, the triglyceride concentration may be reduced by as much as 50%.

In situations in which hypertriglyceridemia is the predominant problem, the use of two drugs that lower triglyceride concentration by presumably different mechanism, such as nicotinic acid which inhibits the production of VLDL and gemfibrozil which primarily accelerates the disposal of VLDL, allows marked reduction in the serum triglyceride concentration with additive or greater effects on the HDL-C concentration.

Although fish oil has been used to treat hypertriglyceridemia, little is known about its efficacy and safety in combination with other drugs.

Nicotinic acid appears to be the drug of choice in patients who have combined hyperlipidemia, because it reduces synthesis of both VLDL and LDL. Gemfibrozil also can be used in combination with cholestyramine or colestipol.

The results of selected drug combinations are demonstrated in Figure 11.1.

Suggested Readings

A symposium: cardiovascular disease in the elderly. Am J Cardiol 1989;63:1H–19H.

Blum CB, Levy RI. Current therapy for hypercholesterolemia. JAMA 1989; 261:3582–3587.

Bradford RH, Shear CL, Chremos AN. Expanded Clinical Evaluation of Lovastatin (EXCEL) study results. Arch Intern Med 1991;151:43–49.

Brown MS, Goldstein JL. Drugs used in the treatment of hyperlipoproteinemias. In: Gilman AG, Rall TW, Nies AS, Taylor P, eds. The pharmacologic basis of therapeutics. New York: Pergamon Press, 1990: 874–896.

Havel RJ. Lowering cholesterol, 1988. Rationale, mechanisms, and means. J Clin Invest 1988;81:1653–1660.

Henkin Y, Oberman A, Hurst DC, Segrest JP. Niacin revisited: clinical observations on an important but underutilized drug. Am J Med 1991;91:239–246.

Illingworth DR. Drug therapy of hypercholesterolemia. Clin Chem 1988; 34 (suppl B): B-123–B-132.

Kane JP, Malloy MJ. When to treat hyperlipidemia. Ann Intern Med 1988; 33:143–164.

National Cholesterol Education Program, Report on the Expert Panel on Blood Cholesterol Levels in Children and Adolescents. Pediatrics 1992 (In Press).

Pierce LR, Wysowski DK, Gross TP. Myopathy and rhabdomyolysis associated with lovastatin-gemfibrozil combination therapy. JAMA 1990;264:71–75.

Rifkind BM, ed. Drug treatment of hyperlipidemia. New York: Marcel Dekker, Inc., 1991.

Saku K, Gartside PS, Hynd BA, Kashyap ML. Mechanism of action of gemfibrozil on lipoprotein metabolism. J Clin Invest 1985;75:1702–1712.

Schaefer EJ. When and how to treat the dyslipidemias. Hospital Practice 1988;23:69–80, 83–84.

Witztum JL. Current approaches to drug therapy for the hypercholesterolemic patient. Circulation 1989;80:1101–1114.

CHAPTER 12

A Practical Approach to Patient Management

In this chapter we incorporate the theoretical principles discussed previously into an overall approach to patient management. A systematic approach with documented goals is necessary to achieve the lipid levels required to prevent CHD. The strategy depicted in this chapter is based on a modification of the National Cholesterol Education Program (NCEP) guidelines as practiced by the staff of the Atherosclerosis Detection and Prevention Clinic at the University of Alabama at Birmingham. This regimen should only serve as a guideline for managing patients, because specific patients often require an individualized approach.

Initial Evaluation

Screening

The recommendations for cholesterol screening in adults have been described in Chapter 9 (*see also* Fig. 9.2). Briefly, a non-fasting blood cholesterol should be measured in all adults 20 years of age and over at least once every five years. Values above 200 mg/dL (5.17 mmol/L) should be confirmed by repeat testing. Four groups of persons are required to undergo lipoprotein analyses (Fig. 12.1):

1. All persons who have an elevated blood cholesterol ($\geq$ 240 mg/dL, 6.21 mmol/L);
2. Persons who have borderline cholesterol levels (200 to 239 mg/dL, 5.17 to 6.18 mmol/L) and two or more additional risk factors;
3. Persons who have borderline cholesterol levels who also are likely to have high HDL-C levels (premenopausal women, estrogen-treated postmenopausal women, athletes, and persons with familial high HDL-C); and

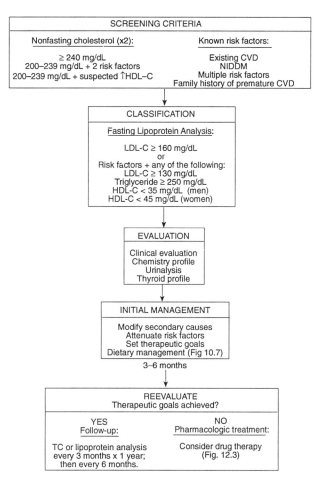

Figure 12.1. General outline of management of lipid disorders. Lipoprotein analysis is done in patients found to have elevated cholesterol levels at screening or borderline elevation of cholesterol likely caused by elevated HDL and in every patient who shows evidence of atherosclerotic cardiovascular disease, noninsulin-dependent diabetes mellitus, a strong family history of coronary heart disease, or other coronary risk factors. Classification of those requiring additional evaluation as based on LDL-C, triglyceride, and HDL-C level in conjunction with other risk factors. After evaluation and modification of secondary causes and other risk factors, the therapeutic goals are set on the basis of lipoprotein values and overall risk profiles. Therapy is always begun with dietary counseling. Patients who do not achieve their therapeutic goals after 3 to 6 months of nondrug therapy should be considered for pharmacologic management. (CVD = cardiovascular disease; NIDDM = noninsulin-dependent diabetes mellitus; TG = triglyceride.)

4. Persons with CHD or considered to be at special risk for atherosclerosis (*see* Fig. 12.1), should proceed directly to the fasting lipoprotein profile without a nonfasting blood cholesterol meaurement.

The NCEP guidelines suggest an individualized approach for those persons in the borderline range who have only one additional risk factor or who are younger (aged 20–39 years).

Assessment of Cardiovascular Risk Profile

Additional assessment of the overall cardiovascular risk is based on the LDL-C level, the presence of preexisting CHD, and the number of risk factors (Table 12.1) (*see* Chapter 2), as given below:

Low risk (Desirable)—LDL-C below 130 mg/dL (3.36 mmol/L)

Table 12.1. Major Risk Factors for Coronary Heart Disease Other than Total and LDL Cholesterol Abnormalities

Factor	Criteria for Classification
Sex	Male or female after menopause[a]
Hypertension	SBP ≥ 140 mmHg or DBP ≥ 90 mmHg
Cigarette smoking	> 10 cigarettes per day
Obesity	≥ 30% overweight
Sedentary activity level[a]	No regular strenuous physical activity, an energy expenditure < 2000 kcal/week
Diabetes mellitus	Fasting plasma glucose ≥ 140 mg/dL or clinical diagnosis
Family history of premature CHD	Parent or sibling who manifests CHD or sudden death before age 55
Low HDL-C level	Below 35 mg/dL (0.91 mmol/L) in men
	Below 45 mg/dL (1.16 mmol/L) in women[a]
Other CVDs	History of cerebrovascular or peripheral vascular disease

[a] Not included as risk factor by National Cholesterol Education Program Expert Panel on Detection, Evaluation, and Treatment of High Blood Cholesterol in Adults

Moderate risk (Borderline high risk)— LDL-C 130 to 159 mg/dL (3.36–4.11 mmol/L) without CHD or two or more risk factors

High risk—LDL-C 130 to 159 mg/dL (3.36–4.11 mmol/L) with CHD or two or more risk factors

LDL-C 160 mg/dL (4.14 mmol/L) or greater

In addition to the risk factors outlined in the NCEP guidelines, we consider the post-menopausal state and sedentary physical activity to constitute independent risk factors. A more quantitative assessment can be achieved by using the formula adopted by the American Heart Association (*see* Fig. 2.19); this formula estimates the 10-year probability of CHD based on each of six risk factors, including HDL-C. Although it is probably unnecessary for establishing the need to initiate therapy, it can be used for educating the patient as well as for evaluating the patient's success in modifying the risk profile.

The NCEP guidelines do not provide any clear recommendations for CHD risk assessment of persons who have isolated hypertriglyceridemia. The National Institutes of Health Consensus Development Conference on Treatment of Hypertriglyceridemia in 1984 defined triglyceride levels as follows:

Desirable—below 250 mg/dL (2.82 mmol/L)

Borderline—250 to 500 mg/dL (2.82–5.64 mmol/L)

Elevated—above 500 mg/dL (5.64 mmol/L)

The European Consensus Conference classified persons who have plasma cholesterol levels below 200 mg/dL (5.17 mmol/L) and triglyceride levels of between 200 and 500 mg/dL (2.26–5.64 mmol/L) as having isolated hypertriglyceridemia, and those persons who have triglyceride levels above 500 mg/dL (5.64 mmol/L) as having severe hypertriglyceridemia. In such cases, the risk for CHD should be judged from the presence or absence of other risk factors, the presumed origin of the hypertriglyceridemia (familial hypertriglyceridemia, familial combined hyperlipidemia, and so forth) and the presence of premature cardiovascular disease in the family history.

The classification of isolated low HDL-C is even less well defined and varies according to the patient's gender. Until other guidelines are established, HDL-C below 35 mg/dL (0.91 mmol/dL) in men and below 45 mg/dL (1.16 mmol/L) in women should probably be considered low. Risk stratification in such patients is unclear and should take into account the family history and presence of other risk factors.

Therapeutic Goals

The next step involves making a decision on what the final outcome of therapy should be. Establishing therapeutic goals is a prerequisite for achieving changes that will be optimal for reducing CHD. Such a goal should be established for each individual risk factor, because it provides the patient and the physician with a target value. Once this goal is reached, it also can be used as a basis for additional follow-up and maintenance. It should be realized, however, that this therapeutic goal should serve only as a guideline and need not always be achieved at any cost. The cost-benefit ratio of attaining the goals should be evaluated and reevaluated whenever complications arise or when modifications of the therapy (increase of drug dose, addition of a new drug) are contemplated.

Our guidelines for therapeutic goals are shown in Figure 12.2. For primary preven-

tion of CHD, the recommended goals depend on the estimated risk. For persons who have few other risk factors and who are at low risk, the LDL-C should be reduced to less than 160 mg/dL (4.14 mmol/L). In persons at higher risk, a more stringent goal of 130 mg/dL (3.36 mmol/L) or lower should be targeted. Despite the absence of national recommendations, we believe that persons at high risk for CHD should have triglyceride levels less than 250 mg/dL (2.82 mmol/L).

In patients who have existing CHD *(secondary prevention)*, consideration should be given to a more aggressive approach aimed at reaching lipoprotein levels at which regression of existing plaques is likely to occur. Angiographic studies suggest that this occurs most frequently at total cholesterol levels at or below 150 mg/dL (3.88 mmol/L) and at LDL-C levels at or below 100 mg/dL (2.59 mmol/L) (*see* Chapter 1). Because increases in HDL-C levels were correlated in-

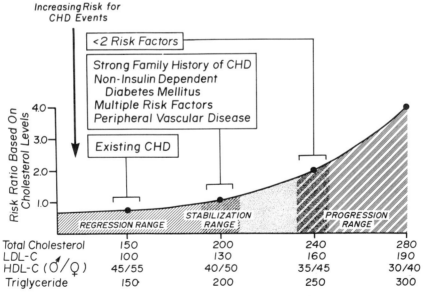

Figure 12.2. Therapeutic goals for treatment of dyslipidemia. The goals of therapy depend on the overall cardiovascular risk for the individual patient. Persons who have existing CHD should receive more aggressive therapy aimed at the regression of existing atherosclerotic plaques *(bracket on left)*. Persons who have a high-risk profile but show no evidence of underlying CHD should aim at stabilization, an intermediate goal *(middle line)*, whereas persons who have few risk factors and no CHD should be treated least aggressively *(bracket on right)* to prevent progression. The lipid and lipoprotein values shown below the graph approximate the target values based on the assessment of need.

dependently to regression, we usually aim toward HDL-C levels at or above 45 mg/dL (1.16 mmol/L) in males and at or above 55 mg/dL (1.42 mmol/L) in females. In patients who have type IIb or type IV dyslipidemia, and especially in those who have familial combined hyperlipidemia, we attempt to reduce the triglyceride levels to below 150 mg/dL (3.88 mmol/L) because of the atherogenic potential of higher levels.

Another group of persons at very high risk for CHD but without overt disease, the so-called *intermediate group*, can also be identified; this includes patients who have NIDDM or peripheral vascular disease, or those who have a strong family history of CHD or those with multiple risk factors. In our opinion, the goal for such persons should be somewhere between the primary prevention and the secondary prevention groups. The allocation of patient to this group is often arbitrary and depends on the physician's clinical judgement.

Finally, special considerations should be used for patients who have isolated hypertriglyceridemia. In those patients considered to be at increased risk for CHD (mainly those who have familial combined hyperlipidemia), the above guidelines should be applied. Some persons who have hypertriglyceridemia do not have evidence of increased CHD risk, however, and the only consideration should be prevention of pancreatitis. Because pancreatitis often occurs at triglyceride levels above 1000 mg/dL (11.29 mmol/L), the goal of therapy should aim to maintain triglyceride levels well below this value. In patients who have a past or present history of pancreatitis, lower levels—below 500 mg/dL (5.64 mmol/L)—should be achieved to ensure prevention of a recurrent episode of pancreatitis. An individualized approach should be applied to persons who have symptoms suggestive of the chylomicronemia syndrome (lethargy, memory deterioration, confusion, and the like), a strong family history of CHD, or secondary hypertriglyceridemia.

Nonpharmacologic Management

Therapy of dyslipidemia should be implemented in stages (*see* Fig. 12.1), beginning with modification of nonlipid cardiovascular risk factors (smoking, blood pressure, obesity, blood sugar, and physical inactivity) and should attempt to eliminate secondary causes of dyslipidemia (*see* Chapter 8). This is followed, if necessary, by dietary modifications.

Although a prudent diet is recommended for the adult population in general, a stricter approach should be initiated in dyslipidemics. After assessment of the patient's current dietary intake and the need for weight loss (based on the patient's current weight, height, and fat distribution), therapy begins with a Step-1 Diet, as recommended by the American Heart Association (*see* Fig. 10.7). Although some physicians may have the time and knowledge to educate the patient themselves, we believe that in most cases a dietitian should be used for such purposes. An initial evaluation of the patient's baseline diet should be followed by specific guidance in making dietary choices, changing eating habits, and reading nutritional labels on food products.

Once the diagnosis of dyslipidemia is clearly established and therapy has been recommended, the patient must be reevaluated at regular intervals. Questions often arise concerning the frequency with which patients should be seen by their physician or dietitian and how often the lipids should be remeasured. Most authorities recommend that diet therapy is given a three-to-six-month trial before pharmacologic therapy is considered and initiated. Because of variability in lipid measurement, it is not wise to rely on values obtained at great intervals. Furthermore, because reinforcement of dietary principles is important for success in this form of therapy, we recommend lipid measurements at two-month intervals during the induction phase of therapy. An interim cholesterol measurement after four weeks of dietary intervention may increase patient motivation and compliance.

Adherence to these dietary recommendations should be reevaluated by measurement of the total cholesterol (or lipoprotein profile), as well as dietary reassessment. Patients who achieve the therapeutic goals at these stages should be commended, provided support for long-term maintenance to these dietary and life-style habits, and scheduled for follow-up evaluations every three months during the next year. In many patients, continuous dietary support and supervision are necessary to ensure long-term success.

The principles of dietary management of hypertriglyceridemia are similar to those in hypercholesterolemia. More emphasis should be placed on weight reduction, however, and restriction of concentrated sweets and simple sugars. Alcohol should be avoided completely, because many hypertriglyceridemics are sensitive to the extent that even small amounts of alcohol can raise triglyceride levels. In patients with low HDL-C, a greater proportion of the calories should come from monounsaturated fats rather than polyunsaturated fatty acids and carbohydrates.

Patients who have not achieved the target goals despite appropriate dietary changes and patients who are already on a Step-1 (or equivalent) diet at the initial evaluation should receive instructions for the Step-2 diet. These additional restrictions require a greater degree of dietary modification and follow-up and necessitate the use of a trained dietitian in almost all cases. Like the Step-1 diet, the Step-2 diet needs careful follow-up of adherence and response.

Patients who have not made the initial appropriate dietary changes should receive supplementary guidance on the Step-1 diet and should be followed for three additional months. Occasionally the physician may be impressed that the patient cannot (or will not) make such modifications in life-style and dietary habits. In this situation, deci-sions about advancement to pharmacologic therapy should be based on the likelihood of success.

Pharmacologic Management

Pharmacologic therapy should be reserved for those persons who have not achieved sufficient risk reduction by the above modifications (*see* Chapter 11). According to the NCEP guidelines, pharmacologic therapy should begin after 3 to 6 months of intensive nonpharmacologic interventions in all persons who have LDL-C at or above 190 mg/dL (4.91 mmol/L), as well as in persons who have LDL-C at or above 160 mg/dL (4.14 mmol/L) and have two or more risk factors. Lower LDL-C levels may be sufficient for initiating drug therapy in patients who have existing CHD, patients in the intermediate risk group, in Fig. 12.2 and persons who have familial combined dyslipidemia.

In selected circumstances, it may be appropriate to proceed to pharmacologic therapy at an accelerated pace. These circumstances include

1. LDL cholesterol above 225 mg/dL (5.82 mmol/L) or instances in which persons at very high risk, such as those who have CHD, urgently need to reduce cholesterol;
2. Evidence of chylomicronemia syndrome with triglycerides at or above 750 mg/dL (8.47 mmol/L) or a history of abdominal pain or pancreatitis;
3. A patient who already is maintained on a Step-2 Diet for a period of 3 months; and
4. Any other situation in which the judgement of the physician is that drug therapy should be initiated.

The initial choice of drugs depends on the specific lipoprotein abnormality, anticipated side effects, contraindications, and long-term safety and efficacy in reducing CHD morbidity and mortality. The hypolipidemic drugs can be categorized as those used primarily to lower LDL-C levels and those used primarily to lower triglyceride levels, although some overlap exists. Consideration also should be given to the HDL-modifying effect of each drug. Specific details about the mode of use of each drug, potential side effects, follow-up scheme, and cost are discussed in Chapter 11.

The bile acid resins and nicotinic acid were advocated as the drugs of first choice by the NCEP, because both have been shown to reduce cardiovascular morbidity and mortality, and they have a long, established record of safety. Resins are used for reducing LDL-C levels in hypercholesterolemia, but their use in combined dyslipidemia is complicated by the prospect of exacerbating the hypertriglyceridemia. Nicotinic acid (niacin), the drug of choice in combined dyslipidemia, also is effective in pure hypercholesterolemia and in isolated hypertriglyceridemia. It has the most potent HDL-C-elevating effect of any of the drugs currently used.

The fibrates and the HMG CoA reductase inhibitors still constitute the second-line of treatment for dyslipidemia. Although gemfibrozil (Lopid®) has only been approved by the FDA for the treatment of hypertriglyceridemia and combined hyperlipidemia, results of the Helsinki Heart Study provide evidence of its safety and efficacy for persons who have hypercholesterolemia. Clofibrate (Atromid-S®) is rarely used, however, as a result of the worrisome adverse effects previously reported in clinical trials. Lovastatin (Mevacor®) is gaining popularity as additional studies and clinical experience support its efficacy and short-term safety in treatment of hypercholesterolemia as well as mixed dyslipidemia. Its transition to a first-line drug awaits further proof of its long-term safety and efficacy in preventing CHD. Such considerations apply also to simvastatin, pravastatin, and other HMG CoA reductase inhibitors.

Controversy still surrounds the clinical application of probucol (Lorelco®). The drug's modest and variable efficacy in LDL-C reduction, coupled with the frequently observed reductions in HDL-C, has hampered its widespread clinical application. More enthusiasm for its use may ensue if the preliminary observations, suggesting that the drug impedes atherosclerosis by preventing the oxidation of LDL, are substantiated. At this stage we have not included it in our treatment protocol. The value of oral estrogens in preventing atherosclerosis is better established, and their beneficial lipoprotein effects can be useful in dyslipidemic women. Estrogen can raise triglyceride values abruptly to very high levels, and observations on the lipid profile within a month of therapy are desirable. The long-term effects of estrogen-progestin combinations on CHD progression are still unclear.

Figure 12.3 delineates our approach to drug therapy according to the patient's lipoprotein phenotype. For practical purposes, the triglyceride levels used to define mixed dyslipidemia (type IIb) are set at 250 mg/dL (2.82 mmol/L), because this is the level around which the use of resins and estrogens becomes inappropriate. The maximum lipid-lowering effect of each dose of any hypolipidemic agent is seen within 4 to 6 weeks, and, consequently, evaluation of the efficacy of a fixed-dose drug, such as gemfibrozil or probucol, can be assessed quickly. For those drugs in which the dose is titrated, such as nicotinic acid, the effect of each dose also can be assessed at 4 to 8 week intervals. At reevaluation, a lipoprotein analysis and an evaluation of any adverse drug effects (in addition to reenforcement of exercise and diet and modification of risk factors) are needed. A chemistry profile also is required at this time for monitoring side effects of niacin (glucose, uric acid, and liver function tests), gemfibrozil (liver function tests), and lovastatin (CPK, liver function tests) (Table 12.2).

Treatment of dyslipidemia should be instituted in stages, similar to the individualized step-care approach to hypertension. In general, treatment should be initiated with one drug at less than the full dosage. Patients who achieve their therapeutic goals and can tolerate the drug are encouraged to

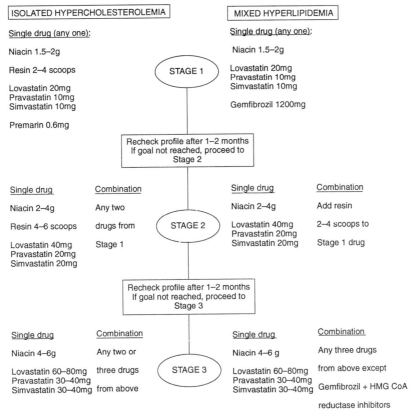

Figure 12.2. Therapeutic goals for treatment of dyslipidemia. The goals of therapy depend on the overall cardiovascular risk for the individual patient. Persons who have existing CHD should receive more aggressive therapy aimed at the regression of existing atherosclerotic plaques *(bracket on left)*. Persons who have a high-risk profile but show no evidence of underlying CHD should aim at stabilization, an intermediate goal *(middle line)*, whereas persons who have few risk factors and no CHD should be treated least aggressively *(bracket on right)* to prevent progression. The lipid and lipoprotein values shown below the graph approximate the target values based on the assessment of need.

Table 12.2. Follow-up Scheme for Patients on Drug Treatment for Dyslipidemia

	Evaluation for diet & exercise, risk factors, adverse drug effects, lipid profile, and dosage adjustments (each visit)	Liver enzymes, muscle enzymes, if symptomatic (each visit)	Glucose/uric acid (each visit)	Complete blood count (at 6 & 12 mos for first yr)
Bile acid resins	X			
Cholestyramine				
Colestipol				
Nicotinic acid	X	X	X	
HMG CoA reductase inhibitor	X	X		
Lovastatin				
Pravastatin				
Simvastatin				
Fibric acid derivative	X	X		X
Gemfibrozil				

continue that treatment. In patients who have intolerable adverse effects, an alternative drug should be substituted. If the drug is well tolerated but the goal is not achieved, the dose of the drug can be increased or an additional drug can be added. This approach should result in the use of the least number of medications at the lowest effective doses for the patient.

Isolated Hypercholesterolemia (Type IIa)

Any of the hypolipidemic drugs potentially can be used to lower LDL-C levels in isolated hypercholesterolemia. The initial choice depends on the degree of LDL-C elevation, the perceived risk of the lipid disorder, the therapeutic goal, and the age of the patient.

Stage one initiates therapy with single-drug therapy, including any one of the following:

1. Cholestyramine or colestipol—one or two scoops or packs twice daily.
2. Niacin—choose one of the schedules shown in Table 13.2 for initiating treatment with regular niacin. Increase the daily dose by 250 to 300 mg every week, up to a total daily dose of 1.5 to 2.0 grams; this usually takes 6 to 8 weeks. (*See* Chapter 11 regarding ways to manage the accompanying flushing.)
3. Lovastatin (20 mg), pravastatin (10 mg) or simvastatin (10 mg) with the evening meal. (Until long-term safety is established, this alternative should be used infrequently in young persons and in low-risk patients.)
4. Estrogen—in postmenopausal women who have no contraindications to estrogen, therapy can be initiated with 0.625 mg of conjugated estrogen tablets (Premarin®) or an equivalent dose of another estrogen. Addition of medroxyprogesterone is indicated in women who still have their uterus.

Stage-two therapy offers the following options:

1. Single-drug therapy
   Cholestyramine or colestipol—increase to 3 scoops or packs twice daily.

Niacin—increase the daily dose by 250 to 300 mg every week, up to a total daily dose of 3.0 grams; this usually takes 4 weeks.
   Reductase inhibitors—increase to 40 mg of lovastatin with the evening meal or 10 mg bid of pravastatin or simvastatin.
2. Combination therapy
   Combine any two of the drugs at the doses shown in stage one.

Stage-three therapy offers the following options:

1. Single-drug therapy
   Niacin—increase the daily dose by 250 to 300 mg every week, up to a total daily dose of 4 to 6 grams.
   Reductase inhibitors—increase the dose at intervals of 6–8 weeks.
2. Combination therapy
   Any combination of resins, niacin, or reductase inhibitors at the above doses; use the lowest possible dose of the most tolerable combination for the patient to achieve the target goal.

Note: As the dose of medications increases, a reevaluation of the therapeutic goals and the cost-benefit ratio should be made, because the incidence of side effects and the financial cost of the drugs markedly increase. Combination drug therapy is preferred to single drug therapy in stage three. It is prudent to consult a lipid specialist in difficult cases. Although gemfibrozil has a modest LDL-C lowering effect, it is rarely used in isolated hypercholesterolemia unless there are contraindications to the use of other medications. These recommendations do not apply to persons who have homozygous familial hypercholesterolemia; however, such patients are generally nonresponsive to medications and require a special approach in a clinic that specializes in lipid disorders.

Mixed Dyslipidemia

Type IIb. The drug of choice in Type IIb dyslipidemia in nicotinic acid, because it reduces both LDL-C and VLDL-C while increasing HDL-C. In patients who have markedly elevated VLDL-C, however, the levels of LDL-C may rise during initial therapy, and higher doses of nicotinic acid then are required to reduce it.

Gemfibrozil also is useful in decreasing elevated VLDL-C and in decreasing HDL-C levels, but the associated increase in LDL-C in these cases less than with nicotinic acid and the LDL-C may rise. Despite potential elevations in LDL-C, the Helsinki Heart Study showed a decrease in CHD mortality in type IIb patients treated with gemfibrozil.

Reductase inhibitors can be used in patients who have mild to moderate elevations in VLDL-C; although they decrease VLDL-C less efficiently than gemfibrozil, the effect on LDL-C is more favorable. In contrast, cholestyramine and estrogens are not recommended for treatment of this entity, because the risk for severe hypertriglyceridemia is substantial.

Stage one generally involves single-drug therapy, with the following drugs:

1. Niacin—initiate therapy as previously described to an initial dose of 1.5 to 2.0 grams. If there are contraindications to the use of nicotinic acid or if the patient cannot tolerate the associated side effects, use one of the alternative drugs.
2. Reductase inhibitors—start with the lowest dose with the evening meal. (Until long-term safety is established, this alternative should be used sparingly in young persons and in low-risk patients.)
3. Gemfibrozil—600 mg twice daily.

Stage two offers the following options:

1. Single-drug therapy
   Niacin—increase the daily dose by 250 to 300 mg every week, up to a total daily dose of 3.0 grams.
   Reductase inhibitors—increase to 40 mg of lovastatin in the evening or 10 mg twice daily of pravastatin or simvastatin.
2. Combination therapy
   If the triglyceride levels are under control, add cholestyramine or colestipol, 1 to 2 scoops or packs twice daily, to any of the drugs in stage one. If the triglyceride levels are still above 250 mg/dL (2.82 mmol/L), proceed to stage 3.

In Stage three therapy, if the therapeutic goals have still not been reached, use one of the following schemes:

1. Single-drug therapy
   If niacin is well tolerated, increase its dose gradually up to 4 to 6 grams per day given in 3 to 4 divided doses.
   Reductase inhibitors—increase to 60 to 80 mg of lovastatin or 30 to 40 mg of pravastatin or simvastatin in two divided doses.
2. Combination therapy
   Any combination of resins with niacin, gemfibrozil, or reductase inhibitors can be used, as well as regular niacin with reductase inhibitors at Stage-1 doses (see note below) or niacin with gemfibrozil at Stage-1 doses.

(Note: As in isolated hypercholesterolemia, a reevaluation of the therapeutic goals and the cost-benefit ratio should be made when high-dose medications become necessary. Combination-drug therapy is preferred to single-drug therapy in stage three. The use of reductase inhibitors with gemfibrozil carries an increased risk of rhabdomyolysis and should be avoided. An increased awareness of this complication also maybe required for reductase inhibitors-niacin combinations and possibly for gemfibrozil-niacin combinations.)

Familial Dysbetalipoproteinemia (Type III). This rare disorder is characterized by the accumulation of chylomicron remnants, and β-VLDL with elevations of both cholesterol and triglyceride levels (*see* Chapter 7). It is associated with premature atherosclerosis.

The drug of choice for this disorder is gemfibrozil, 600 mg twice daily. Niacin and reductase inhibitors also are effective and can be used as second-line drugs as described previously. Although clofibrate is very effective in this disorder, we recommend its use only if the patient cannot tolerate any of the above drugs.

Isolated Hypertriglyceridemia (Types IV and V)

As already explained, the primary purpose of reducing elevated triglyceride levels is the prevention of pancreatitis and the chylomicronemia syndrome, both of which usually occur only at triglyceride levels above 1000 mg/dL (11.29 mmol/L). In patients who have an increased risk for CHD, and especially in those patients who have familial combined hyperlipidemia, more rigid goals may be appropriate.

Only two classes of drugs currently are recommended for treatment of hypertriglyceridemia: the fibrates (gemfibrozil and clofibrate) and nicotinic acid. The fibrates may be more efficacious in severe hypertriglyceridemia, whereas nicotinic acid may have an advantage in familial combined dyslipidemia because of its greater LDL-C-lowering effect. Their mode of use in this disorder is similar to that in type IIb dyslipidemia.

In cases that are refractory to therapy with fibrates and nicotinic acid or when side effects preclude the use of adequate drug doses, pharmacologic doses of fish oil (Promega® or Max EPA®) (5 to 20 g/day) may be used as primary or adjunctive therapy. Fish oil contains the omega-3 polyunsaturated fatty acids—eicosapentanoic acid and docosahexanoic acid, both of which have a triglyceride-lowering effect (see Chapter 10). The optimal mode of treatment and the long-term effects of fish oil remain unclear.

Isolated Low HDL-C

The value of treating isolated HDL-C, in the absence of elevated LDL-C or VLDL-C, for prevention of CHD has not yet been established by appropriate clinical studies. The indications for such treatment are unclear at this time. Cautious treatment may be worthwhile in patients who have established CHD or a strong family history of CHD. Nicotinic acid is the most effective drug for this purpose, and even doses as low at 1.0 to 1.5 grams per day may elevate HDL-C by 10% to 15%. Other drugs (gemfibrozil, lovastatin, and the resins) have a more modest HDL-elevating effect but may be useful in reducing the atherosclerotic risk by further lowering LDL-C. Recently, chromium (200 mg tid) has been shown to increase HDL-C.

Special Subgroups

Diabetes Mellitus

Non-insulin dependent diabetes mellitus (NIDDM) frequently is associated with type IV hyperlipidemia, decreased HDL-C levels, and structural abnormalities in LDL (see Chapter 8). Although male sex generally is considered to be a risk factor for CHD, diabetic women do not have the same protection as nondiabetic women, and their age-adjusted risk for development of CHD is greater than for diabetic men. Because many of the other risk factors are more common in NIDDM than in the general population, and because diabetes itself is an independent risk factor for atherosclerosis, an aggressive approach to the detection and treatment of dyslipidemia is indicated. In our opinion, a complete fasting lipoprotein profile should be obtained in every diabetic patient.

Although improved glucose control does not always normalize the lipoprotein abnormalities, it can be associated with marked improvement in the lipoprotein profile and should be the first therapeutic goal. Every effort should be made to achieve euglycemia with appropriate use of diet and, when necessary, hypoglycemic medications. Great emphasis on weight reduction, physical activity, and control of other risk factors (hypertension, smoking) is mandatory.

The low-fat diets recommended for nondiabetic patients who have dyslipidemia (Step 1 and 2 diets) generally are appropriate for diabetic patients as well. A few points are worthy of emphasis:

1. A low-fat, low-cholesterol diet requires that the carbohydrate content be increased to maintain isocaloric intake. This diet has been associated with aggravation of hypertriglyceridemia in some patients who have NIDDM, and this is a source of some controversy. If a high-carbohydrate diet is used, triglyceride

levels should be monitored. The NCEP offers diabetics an alternative diet that consists of a lower carbohydrate intake (40% to 45% of total calories). Monounsaturated fats (present in relatively high concentrations in olive and other selected oils) (*see* Fig. 10.2) may have some advantage over polyunsaturated fats, because they have been shown to reduce LDL-C without decreasing HDL-C. Consequently, diets lower in carbohydrates and rich in monounsaturated fatty acids may result in better glycemic control, reduced triglyceride and VLDL-C levels, and higher HDL-C levels. Although the advisability and practicality of such diets are still debated, they may prove useful in NIDDM patients who have high-serum triglyceride or low HDL-C levels.

2. Increased fiber intake may improve glycemic control, but caution should be exercised in advocating high-fiber diets in the presence of autonomic gastropathy.
3. Omega-3 fatty acids worsen glycemic control in NIDDM and probably should not be used, except in severe hypertriglyceridemia that is unresponsive to all other forms of therapy.

The principles of pharmacologic therapy in diabetes also are different than in the general population. Two of the first-line drugs, niacin and the bile acid resins, have an unusually high incidence of side effects in diabetics. Large doses of nicotinic acid (3000 mg/d) cause or worsen hyperglycemia, which may require additional insulin. Because insulin levels are thought to be associated with atherosclerosis, this may be unwise, at least theoretically. It is not clear whether lower doses of nicotinic acid can be used in NIDDM without these adverse changes. The resins may increase triglyceride levels even more and their gastrointestinal side effects may be accentuated in diabetic patients who have autonomic visceral neuropathy. These drugs should be used in diabetics with great caution.

Both lovastatin and gemfibrozil are effective and probably safe in diabetics; lovastatin has a greater LDL-lowering effect and gemfibrozil a greater TG-lowering effect in these persons, although both drugs seem to be equally effective in controlling mild hypertriglyceridemia.

Renal Disease

Chronic renal insufficiency and the nephrotic syndrome both are associated with an increase incidence of dyslipidemia and CHD (*see* Chapter 8). Experience in treatment of lipoprotein disorders in these diseases is limited, and dietary treatment of dyslipidemia often has to be modified in the presence of renal disease because of constraints on protein intake. A registered dietitian is the best source of advice for these patients.

The pharmacologic management of dyslipidemia in renal disease has not been evaluated sufficiently. Cholestyramine may increase the hypertriglyceridemia that is present in many patients. Fibrates have been used to treat the lipid abnormalities associated with renal disease and have been shown to reduce triglyceride concentrations and increase HDL-C in patients on hemodialysis, but they do not correct the dyslipidemia. They also improve the activity of lipoprotein lipase in patients on hemodialysis but have no effect on undialyzed patients who have chronic renal failure. The clearance of fibric acid derivatives is reduced in patients who have chronic renal failure and, consequently, their doses should be reduced by about 50% and patients should be followed carefully for the development of myositis.

Lovastatin and simvastatin have been used successfully in patients who have the nephrotic syndrome and hypercholesterolemia. These drugs reduce LDL-C by 30% to 40%. In a small number of patients, lovastatin has been shown to reduce concentrations of total cholesterol, triglyceride, VLDL-C, and LDL-C and to raise the concentration of HDL-C. Lovastatin is excreted primarily by the liver. These drugs should be used cautiously in patients who have chronic renal failure, however, because relatively little experience with their use is documented under these circumstances.

Pregnancy

Because pregnancy alone often is associated with elevations in all of the lipoprotein fractions, it may aggravate any preexisting dyslipidemia. Nonpharmacologic management of such lipoprotein abnormalities should remain the primary treatment in pregnant women, because the safety of hypolipidemic drugs in this situation is unknown. Women who have severe dyslipidemia during pregnancy probably should be referred to a clinic that specializes in lipid disorders. Postpartum reductions in some of the lipoproteins often require several months, and patience is required before resorting to aggressive treatment. Long-term follow-up on such patients should be considered because they may be at special risk of developing lipid problems.

The Elderly

Although controversy still exists about the advisability of treatment of dyslipidemia in the elderly (*see* Chapter 2), many authorities believe that such treatment is appropriate in persons who are otherwise in good health and have no other debilitating or life-threatening disorders. The life expectancy for a 75-year-old person is estimated to be around 8 to 12 years, and for an 85-year-old person, 5 to 7 years. In addition to any possible increases in life-expectancy, appropriate lipid modifications in such persons may increase the symptom-free period and, therefore, the quality of life.

The principles of dietary management among the elderly are similar to those in younger persons, but some aspects require special attention. Life-style habits acquired over many years may be difficult to change; the capacity to master new skills needed to prepare low-fat foods may be impaired; and weight loss and exercise often are hard to achieve in the elderly. In addition, the diets of elderly persons often lack important nutrients, and care should be taken not to aggravate such deficiencies.

Pharmacologic therapy requires special considerations. Elderly persons often have associated illnesses and organ impairment and may already be taking several medications. They also may be more susceptible to adverse drug effects such as those related to nicotinic acid and the resins. Older persons tend to be more reliable in taking medication, however, and the risk of long-term adverse effects may not be as worrisome as in younger patients. Lovastatin and gemfibrozil, with their low incidence of side effects, appear to be preferable for the elderly.

Children and Adolescents

Pediatric considerations of evaluation (*see* Chapter 9), dietary management (*see* Chapter 10) and pharmacologic management (*see* Chapter 11) have been outlined previously. It is generally accepted that prevention of atherosclerosis should begin early in life, and this is reflected by extension of the NCEP population approach to the pediatric age groups. This approach aims at lowering the average blood cholesterol levels among all Americans through population-wide changes in nutrient intake and eating patterns. Debate still exists, however, about the optimal methods of detecting and managing persons at high risk for developing CHD.

The NCEP Expert Panel on Blood Cholesterol Levels in Children and Adolescents has adopted a selective screening scheme aimed at detecting those persons at high risk for CHD based on family history. In this scheme, all children with a family history of high blood cholesterol ($\geq$ 240 mg/dL, 6.21 mmol/L) should have their nonfasting cholesterol checked, whereas those children with a family history of premature CHD should undergo lipoprotein analysis. Other high-risk persons may be tested according to the physician's clinical judgement. Further risk stratification is based on the LDL-C (*see* Chapter 9). The dietary management of children and adolescents is basically similar to that for adults, except that more care is required to ensure that adequate calories are ingested to support growth and development and to maintain desirable body weight. The goals of therapy for LDL-C have been designated by the panel as follows:

1. For borderline LDL-C—lower to < 110 mg/dL (2.84 mmol/L)
2. For high LDL-C—lower to < 130 mg/dL (3.36 mmol/L) as minimum goal and then lower to < 110 mg/dL (2.84 mmol/L) as ideal goal

Pharmacologic therapy should be considered for children 10 years of age or older only if after 6 to 12 months of an adequate dietary trial the LDL-C still is above 190 mg/dL (4.91 mmol/L), or 160 mg/dL (4.14 mmol/L) or greater in those patients who have a family history of premature CHD and more than one other risk factor for CHD. The drugs of choice are the resins, which are not absorbed systemically and have been shown to be safe in children and adolescents. Little experience exists with the other lipid-lowering agents in these age groups, and their use should be reserved exclusively to lipid experts.

Drug-Induced Dyslipidemia

The importance of various medications as secondary causes of dyslipidemia is reviewed in Chapter 8. Knowledge of such effects obviously is important when interpreting results of lipoprotein profiles. Because it is unclear whether the long-term consequences of drug-induced lipoprotein changes are similar to those of "natural" variations within these parameters, caution should be exercised when considering corrections of these drug-induced abnormalities. The clinical implications of a drug's effect on lipid metabolism also depend on mode of use or route of administration. Little consideration should be given to lipoprotein changes produce by short-term drug therapy (e.g., antibiotics given for acute infections), whereas even small changes in LDL-C or HDL-C assume importance in chronic treatment. The lipid effects of drugs such as estrogens can be negated by transdermal administration and by avoiding first pass through the liver.

When a drug is considered to affect lipoprotein metabolism adversely in a patient who has a high cardiovascular risk, an attempt should be made to determine the drug's contribution to the patient's dyslipidemia. In some cases (e.g., a patient who has mild hypertension treated with a diuretic), it may be possible to discontinue the drug temporarily and to check the lipoprotein profile after 3 to 4 weeks. In cases in which a dose-effect relation exists, a lower drug dose may be tried first. In other cases, an alternative treatment that does not confer the same adverse effects may be tried; clinical efficacy, cost, and other possible side effects should be considered when contemplating the use of alternative drugs in specific clinical situations. Finally, if the lipoprotein abnormalities cannot be resolved by manipulation of the drug regimen and the patient still is considered to be at high risk for CHD, thought should be given to supplement treatment with specific hypolipidemic agents and other therapeutic measures.

Illustrative Case Histories

The cases that follow will be used to demonstrate some of the principles of patient management and treatment. They all are based on patients treated at the Atherosclerosis Detection and Prevention Clinic at the University of Alabama at Birmingham.

For the sake of clarity, the lipoprotein values in the tables are presented in mg/dL only. The conversion factor for cholesterol is 1 mmol/L = 38.67 mg/dL; for triglyceride, 1 mmol/L = 88.57 mg/dL.

Case 1: A Health-Conscious Young Woman

A 36-year-old woman was found on routine screening to have a total cholesterol of 266 mg/dL (6.88 mmol/L). She had been in good health all her life, did aerobic exercises every day, and did not take any medications on a regular basis. She did not smoke cigarettes, drank alcohol on rare occasions only, and consumed a low-fat, high-fiber diet. Both her parents are still alive and healthy, and there is no history of cardiovas-

cular disease in the rest of the family. Physical examination was unremarkable.

Comment: This is a healthy, health-conscious young woman who has no apparent risk factors for dyslipidemia or cardiovascular disease. Her serum cholesterol is elevated, however, and she is very worried about it. Because this may be a laboratory error, you probably should repeat the test:

A repeat fasting total cholesterol is 257 mg/dL (6.65 mmol/L).
Lipoprotein analysis:

Total cholesterol	= 254 mg/dL
Triglyceride	= 130 mg/dL
HDL-C	= 76 mg/dL
LDL-C (calculated)	= 152 mg/dL

A chemistry profile and thyroid function tests are within normal limits.

Comment: The patient has an elevated HDL-C (not unusual in young, physically active women) and borderline elevated LDL-C. Because her overall cardiovascular risk is very low, specific therapy is not indicated. The lipoprotein analysis should probably be checked annually to detect possible increases in LDL-C.

****

Case 2: A Diet-Responsive Man

A 41-year-old man was found on routine screening to have a nonfasting cholesterol of 287 mg/dL (7.42 mmol/L). A second test in the fasting state showed total cholesterol of 280 mg/dL (7.24 mmol/L) and triglyceride of 176 mg/dL (1.99 mmol/L). He was in good general health, with no history of any chronic disease or regular ingestion of medications. He had been smoking one pack of cigarettes per day for four years but quit 20 years ago. He had been drinking four bottles of beer per week (often while watching football) and exercised heavily for 2 to 3 hours once weekly. There was no family history of dyslipidemia or cardiovascular disease. Physical examination, chemistry profile, and thyroid function tests were within normal limits. (weight 214 lbs, height 5'11").

Lipoprotein profile:

Total cholesterol	= 284 mg/dL
Triglyceride	= 110 mg/dL
HDL-C	= 41 mg/dL
LDL-C (calculated)	= 218 mg/dL

Comment: Compared to the first case, this patient has a higher cardiovascular risk. He is male, overweight, and has higher LDL-C levels. According to the NCEP guidelines, dietary therapy is indicated. Better distribution of his exercise (at least three times per week) may be warranted.

The patient volunteered to participate in an ongoing nutritional study at the University, in which he was assigned a Step-1 American Heart Association diet. The study involved intensive dietary education during the first 4 weeks, with close dietary supervision and follow-up during the following 8 weeks. No other intervention was provided during the study. The patient adhered well to the diet and lost 15 pounds.
Lipoprotein analyses during the study showed:

Week		4	12
Total cholesterol	=	189	186 mg/dL
Triglyceride	=	61	52 mg/dL
HDL-C	=	44	47 mg/dL
LDL-C (calculated)	=	130	130 mg/dL

At a follow-up visit six months after termination of the study, the patient had regained weight (217 lbs) and admitted that he was not adhering to the diet; his major difficulty was abstaining from fast foods at work and at the football games.
His lipoprotein profile is now:

Total cholesterol	= 296 mg/dL
Triglyceride	= 165 mg/dL
HDL-C	= 44 mg/dL
LDL-C (calculated)	= 219 mg/dL

Comment: This patient appears to be very diet-sensitive. Adherence to the low-fat, low-cholesterol diet resulted in significant reductions in LDL-C and increases in HDL-C to the required range. Unfortunately, his long-term dietary compliance was poor, and his lipoproteins returned to their baseline levels after termination of the study. This emphasizes the need for continuous dietary support and supervision at frequent intervals. If compliance to diet cannot be achieved, it may be necessary to resort to drug therapy. LDL-C > 190 mg/dL (4.91 mmol/L) is the level required for initiation of drug therapy in persons with less than two risk factors. Occasionally, the threat of drug therapy will persuade patients to make a greater effort to adhere to a diet.

****

Case 3: Secondary Prevention with Niacin

A 42-year-old man has had a myocardial infarction at the age of 39, following which he had angioplasty of the right coronary artery. Past medical history is negative for diabetes mellitus, hypertension, liver disease, kidney disease, or hypothyroidism. He had smoked 10 to 15 cigarettes per day for 20 years but quit smoking seven years before his myocardial infarction. He has 3 to 4 drinks of liquor per week and exercises heavily for one hour three times per week. He has been on a low-fat, low-cholesterol diet for the past year. Current medications include diltiazem, dipyridamole, and low-dose aspirin.

Family history reveals that both parents died from CHD at ages 77 and 79, and that his maternal grandmother died of a heart attack at age 60 (all smoked cigarettes). He has four healthy sisters (lipoprotein profiles unknown) and two healthy sons who have normal cholesterol levels but low HDL-C levels.

Physical examination, chemistry profile, and thyroid function tests were within normal limits. Lipoprotein profile (average of two sets):

Total cholesterol	=	214 mg/dL
Triglyceride	=	210 mg/dL
HDL-C	=	33 mg/dL
LDL-C (calculated)	=	139 mg/dL

Comment: This patient evidently has premature coronary atherosclerosis. Although his LDL-C is mildly elevated, the major lipoprotein abnormality is a low HDL-C. His only other obvious coronary risk factors include male sex and a remote history of smoking.

According to the NCEP Adult Treatment Guidelines, drug therapy is such cases is indicated only when the LDL-C is above 160 mg/dL (4.14 mmol/L), with the goal of treatment being an LDL-C of 130 mg/dL (3.36 mmol/L). The physician in this case, however, considered the patient to be at high risk for progression of atherosclerosis and chose to aim at reducing his LDL-C to below 100 mg/dL (2.59 mmol/L) and increasing his HDL-C to above 45 mg/dL (1.16 mmol/L) ("regression goal"). In addition to a low-fat diet, the physician may choose from a number of drugs. The drug of choice for this purpose probably is niacin, because is can elevate HDL-C potentially by 15% to 40%, in addition to lowering LDL-C and triglyceride. Gemfibrozil can have similar effects, although its HDL-C elevations are somewhat lower and it might not achieve a substantial LDL-C reduction in hypertriglyceridemic persons. Cholestyramine can reduce LDL-C but its HDL elevations are modest and it might aggravate the hypertriglyceridemia. Lovastatin is very efficient in lowering LDL-C but its modest HDL-C elevations and its unknown long-term safety make it less attractive as a first-choice drug in this relatively young person.

The patient was evaluated by a dietitian who confirmed that his current diet was compatible with a Step-2 AHA diet. Regular niacin was started at a dose of 250 mg/day (half a 500 mg tablet) at breakfast, and the daily dose was increased by 250 mg each week. Flushing during the first few weeks was relieved by the ingestion of aspirin and by the avoidance of hot drinks and alcohol; thereafter, tolerance to the flushing developed and the patient no longer required aspirin for this purpose. Serum chemistry profiles (glucose, uric acid, and liver tests) remained within normal limits. Lipoprotein analysis after 2, 4 and 6 months of therapy revealed:

Daily niacin dose (g)	2.0	3.0	4.0	
Total cholesterol =	183	184	159	mg/dL
Triglyceride =	50	66	85	mg/dL
HDL-C =	40	42	49	mg/dL
LDL-C (calculated) =	133	129	93	mg/dL

Comment: This patient managed to tolerate niacin well and achieved good results with minimal side effects. The key to success with this drug lies in strong motivation and support by the physician, who must warn the patient about the anticipated side effects as well as ways to prevent them. For patients who cannot tolerate high-dose niacin, lower doses may be used alone or combined with cholestyramine.

The need for pharmacologic intervention for this patient's sons is unresolved at this stage, because there are no clinical trials of treatment for persons who have isolated low HDL-C. A better evaluation of their overall risk profile, with dietary and life-style modifications, is warranted.

****

Case 4: Chylomicronemia

A 55-year-old woman was found on routine testing to have an extremely elevated serum cholesterol (around 1000 mg/dL) (25.86 mmol/L) and triglyceride (around 4000 mg/dL) (45.16 mmol/L). Her past medical history is unremarkable. She does not have hypertension or diabetes mellitus and denies previous episodes of acute pancreatitis, arthritis, abdominal pain, or cardiovascular disease. She denies recent onset of depression or memory loss but confesses to feeling tired lately. Her only medication includes oral estrogens (Premarin®), which were began five months ago because of severe postmenopausal flushing and anxiety. She does not smoke cigarettes and drinks alcohol only rarely.

Family history reveals premature CHD. Her mother had coronary bypass surgery at age 55, and four of her mother's brothers had CHD in their early sixties. Her father developed CHD at age 60 and died suddenly at age 66. She has two healthy younger brothers and two healthy sons. The patient did not know any details about lipid abnormalities in her family members.

Physical examination reveals eruptive xanthomas on the buttocks and elbows. Her weight was 177 pounds and height was 5'8". No other abnormalities were noted.

Lipoprotein analysis revealed:

Total cholesterol = 1067 mg/dL
Triglyceride = 4000 mg/dL
HDL-C = 21 mg/dL

LDL-C cannot be estimated
A creamy band was noted at the top of a plasma sample left overnight in the refrigerator. A chem-

istry profile and thyroid function tests were normal.

Comment: This woman obviously has severe hypertriglyceridemia with chylomicronemia. Her family history of premature CHD suggests the possibility of an inherited dyslipidemia, perhaps familial combined hyperlipidemia (*see* Chapter 7). The institution of oral estrogens may have exacerbated her baseline hypertriglyceridemia.

The presence of chylomicronemia poses a danger of acute pancreatitis and requires prompt action to lower the triglyceride levels. Although a strict diet and discontinuation of the estrogens may prove sufficient to lower the triglyceride to safe levels, drug therapy may be justified initially to enhance such effect. The drug of choice in this case is gemfibrozil.

The patient was instructed by a dietitian to follow a low-fat diet and was encouraged to lose some weight. The estrogens were discontinued, and gemfibrozil was started at a dose of 600 mg twice daily.

At follow-up two months later, the patient weighed 170 lbs, and claimed to feel more energetic; however, the postmenopausal symptoms were bothering her again. The eruptive xanthomata had disappeared. A repeat lipoprotein analysis revealed:

Total cholesterol = 227 mg/dL
Triglyceride = 213 mg/dL
HDL-C = 43 mg/dL
LDL-C (calculated) = 142 mg/dL

Lipoprotein analyses of the family revealed elevated cholesterol and triglyceride in one brother and mildly elevated cholesterol levels in one of her two sons.

Comment: This patient had a remarkable response to the above interventions, and the initial goal of preventing pancreatitis had been achieved.

The second goal is prevention of CHD. Because the patient's family history is suggestive of familial combined hyperlipidemia, she may still be at increased risk for CHD. The patient therefore was encouraged to continue on a low-fat, low-cholesterol diet, lose some more weight, increase her exercise activity, and continue gemfibrozil therapy. The postmenopausal symptoms were controlled with transdermal estrogens, which usually have fewer effects on lipoprotein metabolism than oral estrogens, without evidence of recurrence of hypertriglyceridemia.

****

Case 5: Familial Hypercholesterolemia

A 45-year-old woman developed a myocardial infarction and had a coronary artery bypass graft. Five years ago she was found to have hypertension and elevated cholesterol levels and has been on a low-fat, low-sodium diet since. She is not known to have diabetes mellitus, liver disease, kidney disease, or hypothyroidism. The patient does not drink alcohol and exercises heavily for one hour three times per week. She continues to smoke 20 cigarettes per day. Current medications include diltiazem, dipyridamole, and low-dose aspirin.

Family history reveals that her mother suffered from elevated cholesterol levels and developed CHD at age 50. Her father died at age 64 of an unknown cause. A younger brother suffered a myocardial infarction at age 43 and a younger sister is healthy but has cholesterol levels about 500 mg/dL (12.93 mmol/L). She has two healthy sons whose cholesterol levels she did not know.

Physical examination revealed corneal arcus, xanthalesmas, and thickened, irregular Achilles tendons, but no other tendinous xanthomas. (Weight 132 lbs, height 5'5", blood pressure 122/68.)

Lipoprotein analysis revealed:

Total cholesterol	=	456 mg/dL
Triglyceride	=	145 mg/dL
HDL-C	=	54 mg/dL
LDL-C (calculated)	=	373 mg/dL

A chemistry profile and thyroid function tests were normal.

Comment: The patient has heterozygous familial hypercholesterolemia, as manifested by her elevated LDL-C, premature CHD, family history of premature CHD, and tendon xanthomas. Aggressive cholesterol-lowering therapy and risk-factor modification obviously are required. Although both nicotinic acid and the resins can be used for initial therapy, the physician decided to initiate therapy with lovastatin.

The patient was evaluated by a dietitian, and her diet was found to comply with the AHA Step-2 diet. She refused to quit smoking, despite expressing understanding about the dangers associated with that habit. Lovastatin was initiated at a single daily dose of 20 mg in the evening and then was increased to the maximal dose of 40 mg twice daily.

The following lipoprotein profiles were obtained at successive follow-up visits:

Daily lovastatin dose (mg)	20	40	80	
Total cholesterol =	353	315	292	mg/dL
Triglyceride =	121	119	115	mg/dL
HDL-C =	59	55	60	mg/dL
LDL-C (calculated) =	270	236	209	mg/dL

The patient tolerated the drug well, and chemistry tests remained normal.

Comment: The patient's response to the lovastatin was gratifying (around 44% reduction in LDL-C). Her LDL-C levels are still elevated, however, and she still smokes. This is the kind of situation in which combination therapy should be considered. As in the previous case, particular attention must be given to screening other family members and institution of early therapy where necessary.

Cholestyramine was added gradually, beginning at a dose of 1 scoop twice daily (8 grams/day) and increased gradually to 3 scoops twice daily (24 grams/day). The dose of lovastatin was kept constant at 40 mg twice daily. Although mild constipation was a problem during the early stages of therapy, it was relieved by metamucil. The patient also lost 6 pounds and continued to exercise regularly.

Her most recent lipoprotein profile shows:

Total cholesterol	=	198 mg/dL
Triglyceride	=	84 mg/dL
HDL-C	=	54 mg/dL
LDL-C (calculated)	=	127 mg/dL

Comment: This patient has made an appreciable progress, and her overall CHD risk has decreased substantially. Unfortunately, all efforts to persuade her to discontinue smoking have failed.

****

Case 6: Familial Dysbetalipoproteinemia (Type III)

A 56-year-old woman was found to have elevated cholesterol and triglyceride on two occasions during routine screening. Her medical history was unremarkable, with no evidence for diabetes mellitus, hypertension, liver disease, kidney disease, or cardiovascular symptoms. She was not using any medication on a chronic basis, did not smoke cigarettes, and did not drink alcohol. Her weight had increased by 15 pounds over the past year, and she rarely exercised.

Family history revealed one brother who had elevated cholesterol and triglyceride and suffered a heart attack at age 59; there was no other history of dyslipidemia or cardiovascular disease in the family.

Physical examination revealed an overweight woman who had multiple nodular lesions on both elbows and knees and xanthelasma on both eyelids. There was no evidence of tendon xanthomas or palmar crease xanthomas. The rest of the examination was normal, as were the serum chemistry and thyroid function tests.

Lipoprotein profile:

Total cholesterol	*= 364 mg/dL*
Triglyceride	*= 218 mg/dL*
HDL-C	*= 44 mg/dL*
LDL-C (calculated)	*= 276 mg/dL*

Comment: The presence of unusual skin lesions in a dyslipidemic person should arouse suspicion of xanthomas. In this case, the xanthomas are not located around tendons and, thus, are not suggestive of familial hypercholesterolemia. Rather, they resemble tuberous xanthomas, and the association of such xanthomas with elevated serum cholesterol and triglyceride should arouse the suspicion of familial dysbetalipoproteinemia (Type III dyslipidemia, *see* Chapter 7).

The physical hallmarks of familial dysbetalipoproteinemia, which is manifested by abnormal accumulations of cholesterol-enriched remnants of VLDL and chylomicrons, include tuberous and tuberoeruptive xanthomas of variable sizes (most often on the elbows and knees) and orange/yellowish discolorations in the creases of the palms (xanthomas striata palmaris).

Because this dyslipidemia often is associated with premature atherosclerosis, dietary therapy and drug therapy are indicated. The drug of choice in this disorder is gemfibrozil or clofibrate.

The presence of Type III dyslipidemia was confirmed by the findings of an elevated level of IDL-C (128 mg/dL, 3.31 mmol/L), with a relatively low true LDL-C (86 mg/dL) (2.22 mmol/L) on ultracentrifugation and the presence of an apo E_2/E_2 isotype.

Treatment was instituted with a low-fat, low-cholesterol diet with special emphasis on weight loss. After 12 weeks the patient had lost 15 pounds, and the repeat lipoprotein profile showed

Total cholesterol	*=*	*192 mg/dL*
Triglyceride	*=*	*165 mg/dL*
HDL-C	*=*	*41 mg/dL*
LDL-C (measured)	*=*	*64 mg/dL*
IDL-C (measured)	*=*	*54 mg/dL*

As in the above case, the need for screening of family members cannot be overemphasized.

****

Case 7: An Elderly Patient

A 72-year-old women was found to have elevated cholesterol levels on routine screening. She has hypertension that is well controlled by diet and captopril and is otherwise in good health. She does not use any other medication on a chronic basis, quit smoking one year earlier (after 50 years of smoking one pack per day), does not drink alcohol, and does not exercise. Family history revealed that both her parents suffered from CHD after reaching age 80, and there was no other history of dyslipidemia or cardiovascular disease in the family.

Physical examination, chemistry profile, and thyroid function tests were within normal limits (weight 126 lbs, height 5'4", blood pressure 122/68). Lipoprotein analysis revealed:

Total cholesterol	*=*	*292 mg/dL*
Triglyceride	*=*	*292 mg/dL*
HDL-C	*=*	*42 mg/dL*
LDL-C	*=*	*192 mg/dL*

The patient was evaluated by a dietitian who suggested minor modifications in the P/S ratio of her dietary fat intake. Recommendations for increased exercise (walking 30 minutes every other day) also were given. The patient made the recommended modifications and met the dietitian twice during the next four months; however, her lipoprotein profile remained virtually the same after six months.

Comment: This healthy elderly patient has mixed dyslipidemia with an elevated LDL-C (> 190 mg/dL, 4.91 mmol/L), elevated triglyceride, and a relatively low HDL-C. Her other risk factors (smoking, hypertension) have been modified and both her parents had been blessed with longevity. This kind of situation is a common cause of controversy, and some physicians are reluctant to prescribe drug therapy in such patients. The NCEP guidelines do not consider age as a limiting factor for treatment, however, and several studies suggest that such treatment is beneficial in the elderly. We also believe that elderly people should be treated for dyslipidemia if the overall cardiovascular risk is substantial.

The physician decided to use a modified approach and recommended treatment with regular niacin at a low dose of 1.0 grams per day in two divided doses. The patient started treatment with 250 mg per day and increased the daily dose by 250 mg every week. She took the medication with meals and avoided concomitant hot drinks. Although she still have an occasional facial flush, she was reluctant to use aspirin because of its potential gastrointestinal effects. By four weeks she was taking 500 mg twice daily. Her new lipoprotein profile was

Total cholesterol   = 271 mg/dL
Triglyceride   = 104 mg/dL
HDL-C   = 56 mg/dL
LDL-C (calculated)= 194 mg/dL

Despite the improvement in the HDL-C and triglyceride levels, the patient decided to discontinue the niacin because of the associated flushing. No alternative treatment was prescribed. Eight months later the patient suffered an uncomplicated anterior wall MI. On discharge, the physician decided to begin treatment with lovastatin. After four months of treatment with

40 mg lovastatin per day, the patient's lipoprotein profile showed

Total cholesterol   = 227 mg/dL
Triglyceride   = 104 mg/dL
HDL-C   = 57 mg/dL
LDL-C (calculated)= 149 mg/dL

Comment: This case demonstrates several issues associated with the treatment of dyslipidemia among the elderly:

1. The decision to initiate treatment in primary prevention is still problematic. Although the physician decided to treat in the first place, he did not follow it through as soon as minor side effects developed (compare the approach in Case 3). Once this became a case of "secondary prevention," the decision to treat seemed easier. Although we are not implying that treatment would have prevented the myocardial infarction in this particular instance (some studies have shown that benefit from treatment occurs only after 1 to 2 years of treatment), this may be true in other instances.
2. Elderly patients tend to have more side effects to some medications, especially when there are other associated illnesses and multiple medications. Although the long-term safety with reductase inhibitors has not yet been demonstrated, this issue is less worrisome among the elderly than in younger persons. Lovastatin, and more recently pravastatin and simvastatin have been shown to be safe and easy to use in moderate-term studies (4 to 5 years) and thus is convenient for use in the elderly.

****

Case 8: A Hypertensive Patient

A 36-year-old man was found to have cholesterol levels of 230 mg/dL (5.95 mmol/L) and 236 mg/dL (6.10 mmol/L) on two nonfasting screening tests. Five years earlier he was found to have hypertension and had been treated since with hydrochlorothiazide, 50 mg per day and propranolol 80 mg per day. Two years ago he had an episode of acute arthritis in his right metatarsopharyngeal joint and was treated with probenecid for several months. There was no history of cardiovascular disease, diabetes mellitus, liver disease, kidney disease, or thyroid disease. He did not smoke cigarettes, drank 5 to 6 beers every week, and did not participate in any form of exercise. He has been on a low-salt diet but did not pay much attention to the caloric or fat content of his food.

Family history revealed that his mother had suffered from hypertension and diabetes mellitus and died from coronary heart disease at age 69. Two of his mother's sisters also suffered from hypertension and diabetes mellitus. His father died from cancer at age 72. He has two brothers, one of whom has hypertension; he has no children of his own.

Physical examination was unremarkable, except for excessive weight of 236 pounds, (height 5'6"). A chemistry profile revealed a fasting glucose of 142 mg/dL and uric acid, 8.6 mg/dL. Thyroid function tests were within normal limits. Fasting lipoprotein profile (average of two sets) showed

Total cholesterol   = 238 mg/dL
Triglyceride   = 324 mg/dL
HDL-C   = 34 mg/dL
LDL-C (calculated)= 139 mg/dL

Comment: This patient has a mixed dyslipidemia with elevated triglyceride levels, borderline LDL-C levels, and low HDL-C levels. His family history suggests the existence of insulin resistance (see Chapter 8). There are several secondary contributing factors to his lipoprotein abnormality: excessive weight, alcohol, and antihypertensive medications (a diuretic and a beta-adrenergic antagonist). He has mild fasting hyperglycemia and hyperuricemia, which also can be aggravated by the above factors.

The first course of action in this case should be to modify other risk factors (i.e., weight loss and increased exercise activity) and to remove as many secondary causes of dyslipidemia as possible.

The patient was evaluated by a dietitian and received instructions for a low-fat, low-cholesterol, 1600-calorie diet. He was advised to reduce his alcohol intake markedly and to begin an exercise program. The diuretics and propranolol were discontinued gradually; however, his blood pressure began to rise and a calcium antagonist was instituted.

Three months later the patient had lost 20 pounds, reduced his beer intake to two drinks per week, and was walking regularly 30 minutes every other day. His blood pressure was well controlled and he was feeling more energetic than previously. His fasting blood tests revealed a glucose level of 96 mg/dL and uric acid 7.2 mg/dL. Lipoprotein analysis showed

Total cholesterol   = 204 mg/dL
Triglyceride   = 180 mg/dL
HDL-C   = 42 mg/dL
LDL-C (calculated)= 126 mg/dL

Comment: This patient's cardiovascular risk profile improved markedly by a combined approach to diet, exercise, and removal of several secondary aggravating factors. Making such life-style changes is not an easy task and requires support from the medical staff as well as patient determination. The patient's ideal weight is around 150 pounds and he still has a long way to go; but, as in the case with blood pressure control, even mild reductions in weight often may achieve a substantial improvement in the lipoprotein profile. To ensure long-term compliance and success, continuous medical supervision and dietary support are imperative.

****

Case 9: A Diabetic Patient

A 45-year-old premenopausal woman is known to have noninsulin diabetes mellitus and is treated with an oral hypoglycemic agent (glypizide 5mg once daily). She has a history of effort-related angina pectoris and had recently began to complain of intermittent claudication after walking three to four blocks. She is not known to suffer from hypertension, and her liver and kidney function tests are within normal limits. Other medications include a long-acting nitrate and low-dose aspirin. She does not smoke and does not drink alcohol, avoids simple sugars, but does not pay much attention to salt or fat.

Family history revealed longevity and good health on her mother's side. There is a strong history of diabetes mellitus, hypertension, and premature CHD in her father's family. Her father died suddenly at the age of 49. The patient does not have any brothers or sisters, and her two children are healthy and nonobese.

On physical examination she was markedly overweight (weight of 262 lbs, height 5'7") and had decreased pulses in her lower extremities. Blood pressure was 160/85. No bruits were heard. A chemistry profile revealed a fasting glucose of 220 mg/dL, and her glycosylated hemoglobin was 11.5%. Thyroid function tests were within normal limits. The fasting lipoprotein profile (average of two sets) revealed

Total cholesterol   = 238 mg/dL
Triglyceride   = 324 mg/dL
HDL-C   = 38 mg/dL
LDL-C (calculated)= 143 mg/dL

Comment: This patient has overt noninsulin-dependent diabetes mellitus with presumptive evidence of atherosclerosis. She has systolic hypertension (although this requires confirmation on repeat examinations), is obese, and has a family history of premature CHD. Her diabetes is not adequately controlled.

As in Case 8, vigorous modifications of diet, weight, and exercise habits are required. Although these things might improve her diabetes control, an increase in the dose of her hypoglycemic medication may be warranted. Similarly, although her dyslipidemia is likely to improve with adequate dietary and life-style modifications, drug therapy may be required to achieve lipoprotein levels that are associated with regression of atherosclerosis. Niacin is the drug of choice for mixed hyperlipidemia, but it can increase insulin resistance and cause deterioration of diabetes. Both gemfibrozil and lovastatin are suitable alternatives, although

there are some differences in their effects on LDL-C and VLDL-C in this situation. (See discussion above on treatment of combined dyslipidemia.)

The patient was evaluated by a dietitian and received instructions for dietary and exercise modifications. Her physician decided to begin drug therapy for her dyslipidemia and prescribed gemfibrozil 600 mg BID. Three months later the patient had lost 8 lbs and had enrolled in a supervised exercise rehabilitation program. Her blood pressure was 140/80 and her fasting blood glucose was 156 mg/dL. Lipoprotein analysis showed:

Total cholesterol   = 236 mg/dL
Triglyceride   = 155 mg/dL
HDL-C   = 43 mg/dL
LDL-C (calculated)= 162 mg/dL

Comment: The intensive nonpharmacologic and pharmacologic interventions produced a desirable change in this patient's triglyceride and HDL-C levels, as well as in other risk factors (glucose intolerance, weight, and blood pressure). The LDL-C levels had increased, however, as occasionally happens in hypertriglyceridemic persons treated with gemfibrozil. Although the overall effect on CHD risk may have improved (as illustrated in the reduced CV mortality observed in such patients in the Helsinki Study), a better lipoprotein profile may be achievable by the addition of a resin (now that the triglyceride levels are reduced) or by switching treatment to an HMG CoA reductase inhibitor.

SECTION FIVE

EDUCATIONAL RESOURCES

CHAPTER 13

Educational Resources

Recent surveys indicate a continued lack of knowledge about the risk of heart disease from lipids and the need to treat lipid disorders. Yet positive changes regarding knowledge, attitudes, and behaviors regarding cholesterol are taking place. A 1986 survey revealed 64% of physicians and 72% of consumers believed that lowering cholesterol would have a significant impact on heart disease. Almost half of the persons questioned had had their blood cholesterol checked and nearly one-fourth were making dietary changes. Yet, knowledge and action varied considerably by geographic area (Fig. 13.1). Recent studies indicate that cholesterol screening and awareness were slightly higher among women and lower among younger persons, blacks, and those with less education. (Fig. 13.2). Persons who had diabetes, hypertension, or obesity were most likely to have their cholesterol level checked and know their cholesterol values. Persons who smoke or who are sedentary tend to be least likely to have their cholesterol levels checked, and if the cholesterol level is checked, to be least likely to know their cholesterol levels (Fig. 13.3).

Overall, screening and treatment practices still differ appreciably from national recommendations. The major barrier in translating guidelines into clinical practice appears to be the difficulties in integrating preventive medicine into medical practice. Common issues include the remaining skepticism about lowering lipids, insufficient information, excessive time requirements to communicate information to the patient, and insufficient skills in patient education and use of hypolipidemic agents.

Many of the lessons learned from the use of antihypertensives over the past 15 years can be used to gain wider patient acceptance about necessary drugs and life-style changes. Patient education directed toward reducing risk of cardiovascular disease and adherence to treatment regimens are especially important in dealing with lipid problems. Nutrition, a key element, requires a critical amount of basic information and

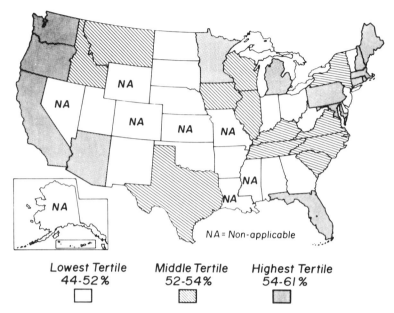

Figure 13.1. Cholesterol screening and awareness of cholesterol levels by state—behavioral risk factor surveillance system, 1989. Wide variations occur among adults who report having their cholesterol levels checked. The overall percentage ranged from 44% in the District of Columbia to 61% in Connecticut. These data are standardized for age, sex, race, and educational attainment using 1980 U.S. census data. (Adapted from: Leads from the Morbidity and Mortality Weekly Report. Factors related to cholesterol screening, cholesterol level awareness - United States, 1989. JAMA 1990;264:2985–2986.)

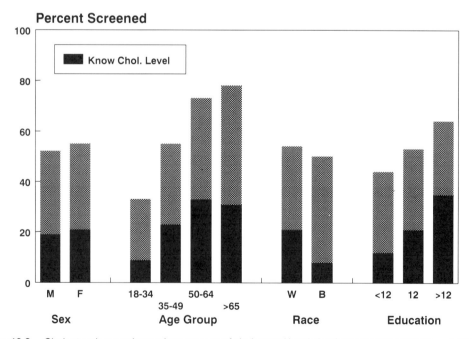

Figure 13.2. Cholesterol screening and awareness of cholesterol levels by demographic category—behavioral risk factors surveillance system, 1989. Among the various groups, females, the elderly, younger persons, whites, and the higher educated tend to be screened more often for cholesterol. In addition, there are noticeable differences in awareness of actual cholesterol levels among blacks and those in the youngest age group. (From: MMWR 1990;39:633–637.)

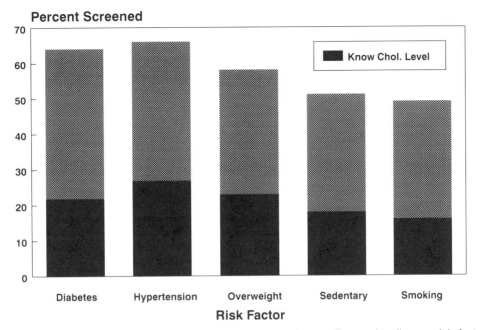

Figure 13.3. Awareness of cholesterol levels in relation to other cardiovascular disease risk factors—behavioral risk factors surveillance system, 1989. Those patients who have hypertension and diabetes appear to be screened most often for cholesterol, and persons who were sedentary or smoked were less likely to do so or to be aware of their cholesterol levels. The proportion of those who know their cholesterol levels varies from 16%, among the smokers, to 27% among those with hypertension. (From: MMWR 1990;39:633–637.)

attention to the changing composition of common foods and dietary practices. If treatment recommendations are to be carried out, then detailed instructions, often written, are necessary to ensure adherence to complex regimens.

Prepared handouts on specific hypolipidemic drugs can apprise the patients about the types of drugs, the dosage, possible side effects, and potential adverse reactions. An awareness of drug-induced side effects can reduce potential discomfort to patients and make them more tolerant of anticipated problems. The section on pharmacologic therapy (*see* Chapter 11) contains various suggestions on reducing side effects for specific agents. Repetition and reinforcement of this information will increase the likelihood of continued successful treatment. The need for long-term treatment invites continuous interaction between the patient and physician. Furthermore, if a patient is left alone for prolonged periods, compliance to the treatment regimen becomes increasingly more difficult.

Toward this end, the physician should help the patient obtain a level of support from family and friends alike and reinforce therapeutic efforts with contacts from his office, such as telephone calls from the office nurse or nutritionist. It is important that patients be given the opportunity to solve problems as they arise in conjunction with the physician or the staff. An uninterrupted habitual routine will do much to foster long-term adherence. Disruption of the patient's daily routine of diet and medication for any reason invariably will cause problems with maintaining treatment effects.

Excellent publications from various sources contain information that the physician may distribute or recommend to patients for a better understanding of lipid disorders and their treatment. Many of the currently available materials can be provided at little or no cost. Some physicians may want to maintain a library of educational materials for lending to patients, thus reducing the burden on them and their staff as patients become more involved in their own care. Furthermore, the technology for detecting CHD and the potential for treating lipid disorders is constantly evolving and changing. Topical lay educational materials should be an immense help to practitioners in learning what their patients are reading and in providing practical suggestions for management of dyslipidemias.

The following section contains a listing of selected publications and videotapes currently available to better understand and treat lipid disorders. In addition to professional publications, materials that can be recommended to patients are listed. Instructional sheets for patients who use hypolipidemic drugs contain the key features of each agent, precautionary measures, and specific directions for taking the drug. Because of changing indications and practice relating to adverse effects, the package inserts should be consulted for complete information.

Resources for the Practitioner

General

The American Heart Association (contact local affiliate office) is the source for the following information. Please consult

1. *Dietary Guidelines for Healthy American Adults: A Statement for Physicians and Health Professionals*—Circulation 1988;77: 721A–724A; and
2. *Dietary Treatment of Hypercholesterolemia—A Handbook for Counselors* (1990). This detailed booklet on dietary management of lipid disorders deals with basic and special eating plans. The appendix contains information on implementation of diets and reference data on food composition.
3. *The Physicians' Cholesterol Education Handbook: Recommendations for the Detection, Classification, and Treatment of High Blood Cholesterol* (Oct. 1987). This 28-page handbook discusses the detection and classification of dyslipidemias with brief guidelines on diet and drug therapy. The appendix summarizes the major cat-

egories of familial dyslipidemias. Other appendices deal with treatment of special groups of patients, such as those who have established CHD and the elderly.

4. *Heart Rx* (1987). A package of educational materials and guides for use in developing office-based, cardiovascular disease risk factor reduction and patient education programs. This booklet contains materials on diet modification and nutrition specific to heart disease; smoking and helping the smoker to quit; high blood pressure control; and early warning signs of heart attack and stroke. The nutrition section contains selected AHA professional publications listed in the patient section of this resource guide. In addition, special materials such as reproducible masters for a diet history sheet and behavior change contracts are included.

The American Medical Association (Campaign Against Cholesterol, 3575 Cahuenga Boulevard West, #440 Los Angeles, CA 90068) publishes the following information:

1. *Physician's Guide: Detection, Evaluation and Treatment of High Blood Cholesterol in Adults* (1988). This is a simple, well illustrated summary based on the Report of the Expert Panel on Detection, Evaluation and Treatment of High Blood Cholesterol in Adults.

Another good source of information is published by George F. Stickley Company (210 West Washington Square, Philadelphia, Pennsylvania 19106):

1. *Nutrition for Family and Primary Care Practitioners* by Anita B. Lasswell, Daphne A. Roe, and Louis Hochheiser (1986)—A listing of selected nutrition education texts and reference manuals suited for office practice.

The Johns Hopkins University Press (John Hopkins Preventive Cardiology Program; Harvey 402; Johns Hopkins Hospital, 600 North Wolf Street, Baltimore, Maryland 21205) publishes the following:

1. *The Johns Hopkins Physician's Lipid Education Program: A Practical Approach to the Patient With a Lipid Disorder* (1988), Second Edition. This monograph consists of six sections designed to provide practical guidelines for identifying and managing patients who have lipid disorders. The information presented is consistent with the 1987 report of the Expert Panel on Detection, Evaluation, and Treatment of High Blood Cholesterol of the National Cholesterol Education Program (NCEP) and the 1983 National Institutes of Health (NIH) recommendations for the treatment of hypertriglyceridemia.

The National Institutes of Health (National Cholesterol Education Program, NHLBI Education Information Center, (4733 Bethesda Avenue, Suite 530, Bethesda, Maryland 20814) has published the following:

1. *Cholesterol Counts: Steps for Lowering Your Patient's Blood Cholesterol.* This publication provides assistance in the area of cholesterol and coronary heart disease. It should not be regarded as a formal guide for treating elevated blood cholesterol; it reflects the experience of some noted authorities in the field and is designed to make it easier for the individual practitioner to draw from that expertise. (NIH Publication No. 85-2699 October 1985.)
2. *Cholesterol: Current Concepts for Clinicians.* This continuing medical education (CME) monograph is based on the National Cholesterol Education Program (NCEP) 1987 report. The monograph comes with a CME quiz, evaluation form, and a list of additional cholesterol resources. The CME is available to physicians only. (NIH Publication No. NN201 Oct. 1988.)
3. *Community Guide to Cholesterol Resources.* This is a detailed booklet to provide the physician with adequate resources for initiating treatment. The guide is divided into two sections: the first part is for physicians and the second part is for patients.

In the physician's section, a number of resources are given that include continued medical education opportunities, professional materials available, and publications given by title and cost. The patient's section has a number of practical tips for diet—primarily fat and cholesterol contents of foods—how to cook and choose foods, and how to locate a dietitian for counseling. The latter section may be photocopied and given to patients for their use. (NIH Publication No. 88-2927 Feb. 1988.)

4. *Current Status of Blood Cholesterol Measurement in Clinical Laboratories in the United States: A Report From the Laboratory Standardization Panel of the National Cholesterol Education Program.* This publication discusses the current status of blood cholesterol measurement and provides recommendations for ways to improve laboratory performance. It helps to provide the uniformly precise and accurate measurements required to diagnose and monitor treatment of persons who have high blood cholesterol. (NIH Publication No. 2928 1987.)
5. *Executive Summary: Current Status of Blood Cholesterol Measurement in Clinical Laboratories in the United States (1988).* This report summarizes the laboratory standards for cholesterol measurement and interprets the recommendations as they apply to selecting a laboratory. (NIH Publication No. NN185.)
6. *Highlights of the 1987 Report of The Expert Panel on Detection, Evaluation, and Treatment of High Blood Cholesterol.* This is a summary of the guidelines of the Adult Treatment Panel of the National Cholesterol Education Program that provides physicians and other health professionals with state-of-the-art cholesterol management recommendations. (NIH Publication No. 2926.)
7. *Kit '90 The Right Moves, Manual A, Information for the Planners.* This contains several elements that can assist physicians interested in community programs. It provides ideas, data and graphics to use

in developing materials for distribution, and a comprehensive listing of print and audiovisual resources.

8. *The Laboratory Standardization Panel Report Recommendations for Improving Cholesterol Measurement.* This comprehensive report provides detailed guidelines and recommendations to help implement the standardization of laboratory measurement and thus improve the reliability of cholesterol measurement in the United States. It addresses three specific groups: manufacturers of instruments and reagents, clinical laboratories, and users of desktop analyzers. The report outlines the technical and organizational elements needed to ensure the overall reliability of cholesterol measurement. In addition, it gives details on preanalytic and analytic factors that affect cholesterol values and national resources for cholesterol standardization. This report is available in two formats: the full report (NIH Publication No. 90-2964 1990) and an executive summary (NIH Publication No. 90-2964a 1990).

9. *Recommendations Regarding Public Screening for Measuring Blood Cholesterol: Summary of a National Heart, Lung, and Blood Institute Workshop, November, 1988.* This report discusses methods that could make public screening more effective in detecting high blood cholesterol in persons who might otherwise not be identified in the health care system. The recommendations for public screening include using recruitment approaches that attract adult segments of the community and developing special approaches to reach target groups; ensuring precise and accurate cholesterol measurements; including education material as part of the screening; ensuring a well-trained, supervised staff; establishing liaison with community health care resources; providing screening at reasonable cost; providing active referral and follow-up; and recommending referrals on the basis of NCEP guidelines. (NIH Publication No. 3045 1989.)

The United States Pharmacopoeial Convention (12601 Twinbrook Parkway, Rockville, Maryland 20892) has published the following:

1. *USP DI: Volume I: Drug Information for the Health Care Provider* (1988 Edition). This book has detailed information on many drugs, including those that lower cholesterol. It contains reproducible sheets of patient-oriented information on many drugs, including types of cholesterol-lowering drugs; it is updated with bimonthly supplements.

Related Risk Factors

The American Heart Association also makes the following publications available:

1. *Statement on Exercise: A Position Statement for Health Professionals by the Committee on Exercise and Cardiac Rehabilitation of the Council on Clinical Cardiology, American Heart Association.* (Circulation 1990; 81:396–398.) This article describes the known benefits of regular exercise, makes recommendations for implementation of exercise programs, and suggests areas in which additional research might enhance understanding of the effects of exercise on the cardiovascular system.

2. *The Exercise Standards and Guidelines.* This publication provides information on exercise testing and training both for those persons who show no evidence of CVD and for those who have CVD. The text is directed to physicians and allied health personnel alike, coinciding with a previous statement on exercise. The guidelines describe in detail the anticipated cardiovascular response to exercise, protocols for exercise testing, and evaluation of the exercise tests. Instructions for beginning an exercise program are classified by whether or not the persons are apparently healthy and concern general principles of the exercise prescription and monitoring of an exercise program. This is a timely and valuable publication. (Circulation 82;1990:2286–2322.)

The National Institutes of Health published the following aids for risk factor intervention:

1. *1988 Report of The Joint National Committee on Detection, Evaluation, and Treatment of High Blood Pressure.* This is a review of the national standards for the identification and management of high blood pressure. Physicians should familiarize themselves with this document, because it represents current guidelines based on the recommendations of an expert panel. (NIH Publication No. 1088.)
2. *Clinical Opportunities for Smoking Cessation: A Guide for the Busy Physician.* The guide highlights the clinical opportunities available to the physician and to other personnel in the physician's office to reduce smoking-related cardiovascular and lung disease. (NIH Publication No. 2178 1986.)
3. *How You Can Help Patients Stop Smoking: Opportunities for Respiratory Care Practitioners.* Respiratory care practitioners are largely hospital-based and provide care to patients at high risk of chronic obstructive pulmonary disease (COPD) and CHD. Accordingly, they are in a position to intervene with large numbers of smokers at high risk of disease. (NIH Publication No. 2961 1989.)
4. *How To Help Your Patients Stop Smoking: A National Cancer Institute Manual for Physicians.* This manual provides physicians, nurses, other health workers, and their associates with the information necessary to institute smoking cessation techniques in their practices. (NIH Publication No. 89-3064 1989.)
5. *Making a Difference: Opportunities for Cessation Counseling.* This is a training tool that demonstrates counseling techniques for smoking cessation through a series of vignettes that show physicians asking smokers about their smoking, strongly advising them to quit and addressing their concerns, teaching simple cessation techniques, and providing follow-up. The segment format enables an instructor to stop the tape at convenient times to involve the audience in a discussion on how to use the illustrated technique with patients. When combined with the booklet, *Clinical Opportunities for Smoking Intervention: A Guide for the Busy Physician,* this videotape can be used as a training program for practicing physicians. (Clinical opportunities booklet. NIH Publication No. 89-2178 1989.)

Resources for the Patient

General

The American Dietetic Association (ADA) (Suite 1100, 208 South LaSalle Street, Chicago, Illinois 60604-1003) has published the following:

1. *Children, Cholesterol, and Diet: Answers to Questions Parents Ask Most* (1989). This introductory nutrition education tool discusses how dietary patterns developed during childhood can contribute positively or negatively to the lipid profile as well as to normal growth and development. A registered dietitian can provide nutritional guidance to children and families that optimizes nutritional status and modifies food intake to prevent the development of high blood cholesterol with age.
2. *Cholesterol Countdown* (1989)—Facts on Cholesterol, Its Role in the Body, and Ways to Control It in Your Diet, (Revised Edition).
3. *Food 3—Eating the Moderate Fat and Cholesterol Way.* This is a dependable guide for patients who wish to reduce the levels of total fat, saturated fat, and cholesterol in their diets. The book includes tips on selecting and preparing foods, comparative charts, and 20 taste-tested recipes. (Published by ADA in cooperation with the USDA, 1982.)
4. *Lowfat Living* (1988). This illustrated brochure unfolds to reveal a large, convenient fact sheet that shows food choices to promote a low-fat life-style.

The American Heart Association has published the following:

1. *The American Heart Association Diet: An Eating Plan for Healthy Americans* (1985). This pamphlet includes goals of the American Heart Association diet, information on choices among the various food groups, and commonly asked questions with answers.
2. *Cholesterol and Your Heart* 1989. This pamphlet explains cholesterol as a risk factor for coronary heart disease.
3. *Coronary Risk Factor Statement for the American Public* (1987). The purpose of this booklet is to make the American public aware of the importance of coronary risk factors and how they can be controlled.
4. *Dining Out—A Guide to Restaurant Dining* (1984). This booklet has excellent guides for selecting proper foods from different cuisines, including many helpful tips on eating right foods when dining out.
5. *Dietary Treatment of High Blood Pressure and High Blood Cholesterol* (1990). This manual gives information about how to lower your blood cholesterol with food choices and eating patterns low in saturated fatty acids and cholesterol.
6. *Fact Sheet on Heart Attack, Stroke, and Risk Factors* (1990)—Consumer Guide to Fat and Cholesterol Control Food Products. This pamphlet lists products by brand names that are acceptable on a low-fat, low-cholesterol diet.
7. *A Guide to Losing Weight* (1986). This pamphlet presents calorie control diet plans (1200, 1500, 1800) based on food groups and controlled serving sizes and includes valuable tips on weight-loss dieting.
8. *Nutrition Counseling for Cardiovascular Health:* A Consumer Guide (1984). This includes chapters on Choosing and Using a Dietitian, Adopting a Healthier Lifestyle, and Special Dietary Needs.
9. *Nutrition for the Fitness Challenge* (1983). This booklet incorporates information on benefits of exercise and nutrition. It includes guidelines to help choose nutritious foods and help reduce your risk of developing coronary heart disease. This pamphlet is for persons who exercise regularly. It reviews the myths and truths about the relation between exercise and diet and gives some details of exercise physiology.
10. *Nutrition Labeling: Food Selection Hints for Fat Controlled Meals* (1978). This explains how to read labels and derive information on dietary fat from them.
11. *Nutritious Nibbles—A Guide to Healthy Snacking* (1984).
12. *Recipes for Low-Fat Low Cholesterol Meals* (1987). This includes recipe modification techniques and various new recipes that are heart healthy.
13. *Taking It Off—A Self-Help Guide to Weight Loss* (1987). This includes two pamphlets, *Winning for Life* and *How to Win at Losing,* that address behavioral issues related to weight management.
14. *When Your Doctor Says, Eat Less Fat and Fewer High-Cholesterol Foods* (1987). This fact sheet provides information to prevent and treat elevated blood lipids by following the American Heart Association's low-fat, low-cholesterol diet plan.
15. *When Your Doctor Says, Lose Weight* (1987). This fact sheet gives suggestions designed to aid those persons who want to lose weight.

The American Medical Association (Campaign Against Cholesterol, 3575 Cahuenga Boulevard West, #440, Los Angeles, CA 90068) makes the following publications available:

1. *What You Should Know About High Blood Cholesterol* (1988). This brochure discusses briefly the risk for developing atherosclerosis and coronary heart disease. It also gives recommended dietary modifications to help lower blood cholesterol levels.
2. *Count Out Cholesterol* (1989). This slide chart gives the cholesterol, saturated fat, and fat content of commonly used foods.

It also answers questions about the risk of cholesterol and how to make dietary changes.

Bantam Books, a division of Bantam Doubleday (Dell Publishing Group, Inc., 666 Fifth Avenue, New York, New York 10103) has published a nutritional bestseller:

1. *Controlling Cholesterol* (1988) by Kenneth H. Cooper, M.D., M.P.H. Dr. Cooper, the founder of the Cooper Clinic and director of the Aerobics Center in Dallas, Texas, reviews all the literature to date that gives lay persons an easy to understand summary of where we now stand with current recommendations based on current research. The effects of all the blood lipids on heart disease are discussed as well as the effects of coffee, alcohol, exercise, and estrogen on blood cholesterol. The book provides a two-week menu plan either basic, moderate, or strict.

The Publications Department of the Center for Science in the Public Interest, a nonprofit consumer "watchdog" agency located at 1501 16th Street NW, Washington, DC 20036, sells booklets, "slide rule" guides and wall charts to determine fat and calories in foods.

Fisher Books (P.O. Box 38040, Tucson, Arizona 85740-8040) has published (in 1989) the following:

1. *Good Fat, Bad Fat: How to Lower Your Cholesterol and Beat the Odds of a Heart Attack*, by Glen Griffin, MD and William Castelli, MD. This authoritative, but practical text discusses the 1987 NCEP guidelines and presents a "Bad-Fat Tally" system. The main portion of the text deals with healthy recipes.

The Johns Hopkins University Press in support of the John Hopkins Preventive Cardiology Program (Harvey 402, Johns Hopkins Hospital, 600 North Wolf Street, Baltimore, Maryland 21205) has published

1. *Beyond Cholesterol* (1989) by Peter O. Kwiterovich, Jr, MD. This is a thorough discussion of cholesterol and other car-

diac risk factors, including five chapters that relate to diet. Much background material is made available to the reader.

The National Institutes of Health National Cholesterol Education Program, along with the National Heart, Lung, and Blood Institute (C-200 Bethesda, Maryland 20892) have collaborated to produce

1. *Cholesterol: Your Guide for a Healthy Heart* (1989). Prepared by the editors of Consumer Guide with assistance from NHLBI, this guide presents recent information about cholesterol. It discusses how blood cholesterol levels affect heart disease risk, explains cholesterol testing, provides an analysis of various methods for controlling cholesterol, and gives tools to help people adopt a heart-healthy diet, including recipes and a cholesterol and fat counter. (It is available in bookstores and other book outlets, or from Publications International, Limited, 7373 North Cicero Avenue, Lincolnwood, Illinois 60646.)

2. *Dietary Guidelines for Americans: Avoid Too Much Fat, Saturated Fat, and Cholesterol.* This presents tips for choosing and preparing foods that are low in fat, saturated fat, and cholesterol. It is reprinted from a U.S. Department of Agriculture Bulletin. (NIH Publication No. NN171 1986.)

3. *Eating to Lower Your High Blood Cholesterol.* This booklet gives dietary information on eating patterns to lower blood cholesterol. Charts list foods by type of fat and cholesterol content. A tear-out shopping list and menus for heart-healthy foods are provided. (NIH Publication No. 90-2972 September, 1990.)

4. *Facts About Blood Cholesterol.* This booklet provides NCEP recommendations and answers common questions about blood cholesterol and discusses ways to lower blood cholesterol levels. The booklet also contains a fat and cholesterol comparison chart. (NIH Publication No. 90-2696 1990.)

5. *Community Guide to Cholesterol Resources.* This guides the reader to information sources that offer practical tips and information on fat and cholesterol content of foods, guidelines for cooking and choosing foods, and how to locate a dietitian and receive counseling for blood cholesterol lowering efforts. (NIH Publication No. 88-2927 February, 1988.)

6. *Eating for Life.* This booklet contains information to improve chances for a long and healthy life by helping make the right food choices and reduce the risk of developing cardiovascular disease and cancer. (NIH Publication No. 3000 1988.)

7. *Heart Attacks.* The causes and symptoms of heart attacks are discussed as well as treatment, prevention, and bypass surgery. (NIH Publication No. 2700 1986.)

8. *Heart to Heart: A Manual on Nutrition Counseling for Reduction of Cardiovascular Disease Risk Factors.* (NIH Publication No. 85-1528 1985.)

9. *Kit '90 The Right Moves, Manual B, Information for Consumers.* This contains a variety of educational materials on hypercholesterolemia, hypertension, smoking cessation, and other CHD risk factors that can be used to supplement and expand health care programs. They are appropriate for year-round use or for specials events and health observances such as National High Blood Pressure Education Month (May) and National Cholesterol Education Month (September). Because NHLBI Kit '90 has no copyright, these materials may be copied as is, or any organization's logo, name, address, and telephone number can be added.

10. *So You Have High Blood Cholesterol—An Explanation For Patients.* The brochure answers many questions that people have regarding blood cholesterol levels. It contains a glossary with terms used in the brochure and other information about cholesterol. (NIH Publication No. 89-2922 June, 1989.)

11. *The Healthy Heart Handbook for Women.* This answers many questions about women and cardiovascular disease and explains that cigarette smoking, high blood pressure, and high blood cholesterol are the three major modifiable risk factors for cardiovascular disease. Also the booklet addresses other risk factors, such as diabetes, obesity, stress, oral contraceptives, and alcohol and includes self-help strategies for controlling modifiable risk factors as well as a list of resources for gaining additional education materials. (Written for women.) (NIH Publication No. 2720 1987.)

The United States Department of Agriculture, Human Nutrition Information Service (Public Information Office, Room 360, Federal Building, 6505 Belcrest Road, Hyattsville, Maryland 20782) has made useful publications including the following:

1. *Mini Bulletin: Avoid Too Much Fat, Saturated Fat, and Cholesterol* (1986).—Home and Garden Bulletin 232-3. This bulletin has tips on choosing foods low in saturated fat and cholesterol.

2. *Nutritive Value of Foods* (June 1991)—Home and Garden Bulletin 72. This includes amounts of total fat, saturated fat, mono- and polyunsaturated fat and cholesterol in 908 common foods.

Related Risk Factors

The American Cancer Society makes the following pamphlets available:

1. *Fresh Start: Participant's Guide.* This pamphlet is designed to help people stop smoking and stay off cigarettes.

2. *How To Quit Cigarettes* (1987). This pamphlet contains daily instructions to follow in the quest to stop smoking.

The American Heart Association publishes numerous health care pamphlets, including

1. *Calling It Quits* (1984). This packet contains two helpful pamphlets on how to quit smoking and a guide to becoming an ex-smoker.

2. *About Your Heart and Exercise* (1989). This pamphlet discusses the function of the heart and blood vessels as well as types of exercise, exercise programs, exercise and weight control, and exercise and pulse rate.
3. *E Is for Exercise* (1989). This pamphlet discusses the different kinds of exercise and how to exercise properly.
4. *Exercise and Your Heart* (1989). This pamphlet presents information on the effects of physical activity on your heart and guidelines for starting and sticking to your exercise program.
5. *Exercise Diary* (1989). This pamphlet will help you monitor your exercise program. Goals can be set according to your accomplishments recorded in this booklet each week.
6. *Running for a Healthy Heart* (1989). This pamphlet teaches you the benefits of regular exercise, what to do before you start, and how to monitor your progress. A sample jogging program is included.
7. *Walking for a Healthy Heart* (1989). This pamphlet includes information about exercise benefits, beginning a walking program, and monitoring your progress. It also includes a sample walking program for beginners.

Cookbooks

The following cookbooks all address vital nutritional and health-care issues:

1. *The American Heart Association Cookbook,* by the staff of the American Heart Association, 5th ed., 1991, Times Books/Random House. This is an authoritative book with many useful recipes.
2. *The American Cancer Society Cookbook,* Ann Lindsay, 1988, Hurst Books, New York. This is a treasure of useful, informative tips and recipes for diets to lower the risk of developing heart disease and cancer. It has several hundred recipes designed with calories and fat per serving. There are also features on ways to substitute to reduce fat intake.
3. *Cooking Light '91,* Oxmoor House, Birmingham 1991. These are recipes devel-

oped by a team of dietitians, home economists, health professionals, and cooks. The recipes are relatively light in calories, sodium, and sugar. There are many gourmet recipes. Each recipe lists not only the number of calories but also the amount of fat, cholesterol, and sodium per serving. General cooking tips are included.
4. *Delicious Ways to Lower Cholesterol,* Oxmoor House, 1989. This is a collection of sound recipes, cooking tips, and a guide for heart-healthy eating written by the Registered Dietetics Department of Nutrition Sciences, University of Alabama at Birmingham, Birmingham, Alabama.
5. *Eat Smart for A Healthy Heart Cookbook,* Denton A. Cooley, MD and Carolyn E. Moore, PhD, RD, 1987, Barron's Educational Services, Inc., New York. This book includes over 400 recipes developed at the Texas Heart Institute. It also includes body weight tables, an exchange system based on limiting fat to 30% of daily calories, and a short section on behavior modification.
6. *Eaters Choice: A Food Lover's Guide to Lower Cholesterol,* Ron Goor, PhD and Nancy Goor, 1989, Houghton Mifflin Company, Massachusetts. This useful book has great flexibility and includes 200 recipes and a sample two-week menu plan. The Goor's build a complete picture of blood cholesterol's relation to health, describing risk factors that relate to heart disease.
7. *Jane Brody's Good Food Book* and *Jane Brody's Nutrition Book,* Jane Brody, paperback, 1988, 1989. *Jane Brody's Good Food Gourmet,* Jane Brody, 1990. All three books provide recipes that are heart healthy, low in fat and low in sodium, in conjunction with cooking principles, shopping tips, and generic nutritional advice.
8. *Light Hearted Seafood,* Janis Harsila, RD, and Evie Hansen, National Seafood Educators, 1989. This book gives information about how to shop, store, and clean seafood. Microwave directions and gen-

eral tables of food composition also are provided.

9. *The Living Heart Diet*, Michael E. DeBakey, MD, Antonio M. Gotto, Jr, Lynne W. Scott, John P. Forety, Simon & Schuster, Inc., 1986. These authors, associated with the Baylor College of Medicine, present not only a good eating program but also explanations of its importance. There are over 500 medically tested recipes, along with nutrition and sodium tables for common foods.

10. *Low-Cholesterol Cuisine*, Anne Lindsay, 1989, Hearst Book. This book contains everyday and upscale recipes.

11. *Low-Cholesterol Gourmet*, Lynn Fischer and W. Virgil Brown, MD, 1989, Acropolis Books. This book provides practical food preparation information, tips on dining out, and on making holiday meals and a list of 500 foods with their cholesterol and saturated fat contents.

12. *Low-Fat, Low-Cholesterol Cookbook, An Essential Guide for Those Concerned About Their Cholesterol Levels*, American Heart Association, Scott Grundy, MD, PhD and Mary Winston, EdD (eds.), Times Books, Random House, Inc., New York, 1989. The cookbook provides a step-by-step guide to lowering cholesterol effectively. In addition to 200 recipes specifically developed by the AHA, the book provides tips about grocery shopping, eating out, and adapting recipes. A discussion of how the body produces cholesterol and information about cholesterol and cholesterol goals also is included. An extensive appendix suggests diet plans, discusses alternate cholesterol-lowering therapies, and describes behavior modification techniques that can ease the transition to a new diet.

13. *Mediterranean Light: Delicious Recipes from the World's Healthiest Cuisine*, Martha Rose Shulman, 1989, Bantam Books. This book modifies traditional Mediterranean recipes by reducing most of the oil.

14. *More Lean and Luscious*, Bobbie Hinman and Millie Snyder, 1988, Prima Publishing and Communications. Although primarily a cookbook, there are many "hint" sections that provide such things as tips for adding fiber to your meals and tips for entertaining.

15. *The Low-Cholesterol Oat Plan*, Barbara Earnest and Sarah Schlesinger, 1989, Hearst Books. This book contains 300 recipes, all of which contain oat products.

16. *The New American Diet*, Sonja L. Conner and William E. Conner, 1986, Simon and Schuster. This is a handbook for the person attempting to develop a diet conducive to lipid reduction. The first part of the book discusses the evidence that links the current American diet to various disorders. The second part spells out the approach to dietary changes and contains a lot of practical advice for patients; physicians will find this section helpful. The last part of the book contains a number of recipes and nutrient analyses for people who want to expand their menus. Finally, the book has a number of unusual items for people who wish to maintain a low-fat diet.

Videotapes

Hall-Fousheé Productions (Suite 214B, 1313 5th Street SE, Minneapolis, Minnesota 55414) has developed the following set of nutritional tapes:

1. The HeartCare Program (1989). This is a composite of four videos under two topics: Introduction to Low-Fat Eating; A Close-Up Look at Low-Fat Foods that taste good and Dealing with The Barriers to Change and Secrets to Low-Fat Eating.

Lowfat Lifeline (Department 302, 52 Condolea Court, Lake Oswego, Oregon 97035) has produced the following:

1. A Fare for the Heart (1989).—A 60-minute VHS video that demonstrates healthy gourmet cooking. This video is considered a unique video/cookbook combination.

2. Eat to Your Heart's Content (1989).—A lively 30-minute VHS video that shows

how to cook quick meals for a healthy heart. Twenty recipe booklets are included with each video.

The Supermarket Savvy Tour Video (1987), produced by Leni Reed, MPH, RD is 52-minutes video available from NASCO, (Supermarket Savvy, P.O. Box 7069, Reston, Virginia 22091). Six people representing various dietary needs make an aisle-by-aisle supermarket nutrition tour led by a nutritionist. The shoppers learn how to read labels, how to calculate the percent of calories from fat, and other valuable nutrition information for healthier food selection and cooking.

Schwartz & Ecker Productions (6121 Hollis Street, Emeryville, CA 94608) has produced the following tape:

1. Cholesterol Control: An Eater's Guide (1989).—A 30-minute VHS color videotape involving a step-by-step approach. Suggestions on selecting foods in the supermarkets, cooking demonstrations, meal planning, and a discussion of cholesterol are included.

Patient Instructions

Bile Acid Resin (Cholestyramine [Questran®, Questran Light®, and Cholybar®]; Colestipol [Colestid®])

The medications are available in packets (Cholestyramine [4 grams] or Colestipol [5 grams]), as a flavored powder (sugar or sugarless) in a can with scoop (Cholestyramine [378 grams]), as a flavorless powder in a bottle with scoop (Colestid [500 grams]), or as a flavored bar (Cholestyramine [4 grams]). The medicines have approximately the same effect, namely lowering the total cholesterol level as much as 20%. The drugs usually are given in two to four divided doses up to 16 to 20 grams per day with meals (Table 13.1).

The bile acid resins are useful drugs for treating high cholesterol with relatively normal triglyceride values. These drugs are considered to be the first line of treatment because of their safety and proven efficacy. This medicine is an insoluble powder that stays in the intestinal tract and is not ab-

Table 13.1. Dosage Schedule[a] for Bile Acid Resin: Start with 1 Pack or Scoop and Take the Indicated Dose with Meals or Just before Meals

Week	Breakfast	Lunch	Dinner
1	—	—	1 pk/scoop
2	—	—	2 pks/scoops
3[a]	—	1 pk/scoop	2 pks/scoops
	—	2 pks/scoops	2 pks/scoops
4[a]	—	2 pks/scoops	2 pks/scoops

[a] Dose may be given at breakfast rather than at lunch or divided doses at preferred meals.

sorbed by the body. When taken, bile acids needed for cholesterol production adhere to the drug in the intestine and are excreted by the body. This leads to a lower plasma cholesterol level. At least one major study has shown that these drugs help reduce the risk of coronary heart disease among men who have high blood cholesterol.

These medicines should never be taken without liquids. The best liquids to mix with the bile acid resins depend on individual taste preference. Some patients prefer to put the drug in cold juice, others add it to water or even to soups or cereals. The mixture can be prepared ahead of time and stored in the refrigerator overnight to reduce any gritty taste, but not more than a two-to-three-day supply should be prepared ahead of time.

Because the drugs are present in the intestine, they may cause some gastrointestinal side effects such as abdominal fullness, indigestion, or constipation. Side effects can be avoided by starting on small doses of and gradually building up to a maintenance dose. In addition, the use of water-soluble fiber, such as oat bran or psyllium, with ample liquids helps to ease any possible constipation problems.

Bile acid resins can block certain other medications from being absorbed. This problem can be eliminated by taking these other medications at least 1 hour before or 4 hours after taking the bile acid resins, to reduce possible interference with absorption.

Nicotinic Acid (Niacin)

Niacin is available in 100-mg, 250-mg, or 500-mg tablets. The dose ranges generally from 1.0 to 3.0 grams up to 4 times per day, but the total dose usually does not exceed 6.0 grams (Tables 13.2 and 13.3).

Niacin is a vitamin but its effect on lipids is unrelated to that function. Niacin, an effective drug for treatment of high blood cholesterol, is especially useful because it also lowers triglycerides and raises HDL-cholesterol. Niacin has been shown to reduce cardiac events in persons who already have had a heart attack.

Niacin sometimes is difficult to take because of flushing, burning, and itching of the skin. Sustained-release or slow-release forms reduce the flushing effect but more readily affect the liver and also may have a lesser effect on lipid reduction.

Table 13.2. Dosage Schedule for Niacin: Start with 100-mg Tablet and Take the Indicated Dose with Meals

Week	Breakfast	Lunch	Dinner
1	100 mg	100 mg	100 mg
2	200 mg	200 mg	200 mg
3[a]	300 mg	300 mg	300 mg
4	400 mg	400 mg	400 mg
5	500 mg	500 mg	500 mg
6	600 mg	600 mg	600 mg
7	700 mg	700 mg	700 mg
8	800 mg	800 mg	800 mg
9	900 mg	900 mg	900 mg
10	1000 mg	1000 mg	1000 mg

[a] Once dosage at each meal exceeds 200 mg, then a 250-mg or 500-mg tablet can be substituted for ease of administration.

Table 13.3. Alternative Dosage Schedule for Niacin: Start with 250-mg Tablet and Take the Indicated Dose with Meals

Week	Breakfast	Lunch	Dinner	Bedtime Snack
1	—	—	250 mg	—
2	250 mg	—	250 mg	—
3	250 mg	250 mg	250 mg	—
4	250 mg	250 mg	250 mg	250 mg
5[a]	500 mg	250 mg	250 mg	250 mg
6	500 mg	250 mg	500 mg	250 mg
7	500 mg	500 mg	500 mg	250 mg
8	500 mg	500 mg	500 mg	500 mg
9	1000 mg	500 mg	1000 mg	500 mg

[a] Once dosage at each meal exceeds 250 mg a 500-mg tablet can be substituted for ease of administration.

For most patients, the side effect of flushing is no more than a nuisance that generally subsides with continued therapy and is easily tolerated once a constant dosage is reached. Side effects can be minimized by starting niacin at a low dose, increasing the dose gradually, and taking it with meals. If needed, taking aspirin (one-half to one tablet) about one-half hour before the dose of niacin will diminish flushing. Also, nonsteroidal antiinflammatory drugs may be used. Hot drinks (e.g., coffee) will increase flushing. Cold drinks will decrease flushing.

People who have gastrointestinal disorders may be excluded from taking the drug. Symptoms of abdominal discomfort, palpitations, or appearance of jaundice (yellow appearance in skin or eyes) should be reported immediately to the physician. Another problem includes lowering of blood pressure, especially when standing, which may require that sudden changes in posture be avoided.

Gradually start niacin according to the given schedule to reduce flushing and other symptoms. **DO NOT ACCEPT NIACIN-AMIDE AS A SUBSTITUTE FOR NIA-CIN, BECAUSE THIS DRUG DOES NOT LOWER CHOLESTEROL. DO NOT SUBSTITUTE SUSTAINED- OR SLOW-RELEASE TABLETS FOR REGULAR NIACIN UNLESS RECOMMENDED BY YOUR PHYSICIAN.**

HMG CoA Reductase Inhibitors— Lovastatin (Mevacor, Pravachol®, Zocor®)

These drugs are available in 10-mg, 20-mg and 40-mg tablets. The usual dose of lovastatin is 20 to 40 mg once daily at dinner. The dose may be increased up to 80 mg per day as a single evening dose or as a divided dose with meals (Table 13.4).

Lovastatin is the first of a group of new drugs that slows the production of cholesterol in the liver and increases the rate of removal of cholesterol from the blood. Although the long-term effects of this drug on preventing cardiovascular disease have not been established, decreases up to 40% in LDL-cholesterol have been noted with treatment.

Table 13.4. Dosage Schedule for Lovastatin[a]: Start with 20-mg Tablet and Take at Bedtime

Week	Breakfast	Lunch	Dinner
1	—	—	1 tab (20 mg)
2	—	—	1 tab (20 mg)
3	—	—	1 tab (20 mg)
4	—	—	1 tab (20 mg)
5	—	—	1 tab (40 mg)
6	—	—	1 tab (40 mg)
7	—	—	1 tab (40 mg)
8	—	—	1 tab (40 mg)
9	—	1 tab (20 mg)	1 tab (40 mg)
10	—	1 tab (20 mg)	1 tab (40 mg)
11	—	1 tab (20 mg)	1 tab (40 mg)
12	—	1 tab (40 mg)	1 tab (40 mg)

[a]Pravastatin (Pravachol®) and simvastatin (Zocor®), second-generation HMG CoA reductase inhibitors, are now available and can be administered in similar fashion at approximately one-half of the dose of lovastatin.

The primary concerns about lovastatin are its possible effects on liver function and episodes associated with tender, painful, or weak muscles. Liver function tests should be performed regularly every 4 to 6 weeks during the first 15 months of therapy, and periodically thereafter. The drug should be used with caution in patients who have had liver disease. Lovastatin in combination with gemfibrozil can result in severe muscle damage and should be used only in special circumstances.

Gemfibrozil (Lopid®)

This drug is available as a 600-mg tablet. The recommended adult dose is almost always 1.2 grams per day given in two divided doses 30 minutes before meals. Higher or lower doses are used infrequently.

Gemfibrozil is a drug in the fibric acid series that lowers blood lipids primarily by decreasing blood triglycerides with variable effects on total blood cholesterol. In most instances it also raises HDL levels. Gemfibrozil causes a decrease of at least 20% to 30% in triglyceride levels and an increase of 10% to 20% in HDL-C levels. Its effect on LDL-C levels is less marked and is therefore not primarily used to lower LDL levels. In the recently completed Helsinki Heart Study, the use of gemfibrozil was shown to be successful in preventing coronary heart disease in some patients.

The side effects of this medication generally are few but may include gastrointestinal upset, disturbances in liver function, rash, and muscle problems. Persons who use fibric acid drugs over a long time may be more predisposed to gallstones. Mild decreases in hematocrit or white blood cell counts have been observed but tend to stabilize with continued use.

The drug should not be used by those who have liver disease or severe kidney problems. Also preexisting gallbladder disease may limit use of the drug. Gemfibrozil in combination with lovastatin can result in severe muscle damage and the combination should be used only with great caution.

Probucol (Lorelco®)

This drug is available in 250-mg or 500-mg tablets. The dose is 1000 mg daily given in two divided doses before the morning and evening meals.

Probucol is a drug used primarily to lower LDL cholesterol. It does so by taking LDL-C out of the circulation. Probucol is infrequently used because it can cause the HDL cholesterol to decrease. Also, the LDL cholesterol response is not always consistent. It has been shown to be effective in reducing cholesterol deposits in the body and may turn out to be a very important drug.

For most patients, adverse effects are not a problem with probucol. Although usually well tolerated, patients may experience diarrhea, increased urination at night, and tearing of the eyes. Headaches and rash also are possible with the drug. Heart rhythm problems found experimentally have not been reported in humans, but it is desirable to maintain careful follow-up for dysrhythmias.

Suggested Readings

Meichenbaum D, Turk DC. Facilitating treatment adherence. New York: Plenum Press, 1987.

United States Department of Health and Human Services, Public Health Service, National Institutes of Health. Community Guide to Cholesterol Resources. NIH Publication No. 88-2927, 1988.

Wenger NK, ed. The education of the patient with cardiac disease in the twenty-first century. New York: Le Jacq Publishing, Inc., 1986.

GLOSSARY

Abbreviations (Frequently Used)
ACAT- Acyl-CoA: cholesterol acyltransferase
AHA- American Heart Association
Apo- Apolipoprotein (used for specific entity, as apo C-II)
CDC- Centers for Disease Control
CDP- Coronary Drug Project
CE- Cholesteryl (cholesterol) ester
CETP- Cholesteryl ester transfer protein
CHD- Coronary heart disease
CHO- Carbohydrate
CLAS- Cholesterol Lowering Atherosclerosis Study
CPPT- Coronary Primary Prevention Trial
CV- Coefficient of variation
CVD- Cardiovascular disease
DHA- Docosahexaenoic acid
EPA- Eicosapentaenoic acid
FATS- Familial Atherosclerosis Treatment Studies
FCHL- Familial combined hyperlipidemia
FCR- Fractional catabolic rate
FFA- Free fatty acid
FH- Familial hypercholesterolemia
FHTG- Familial hypertriglyceridemia
HDL- High density lipoprotein
HLP- Hyperlipoproteinemia
HMG CoA Reductase- 3-hydroxy-3-methylglutaryl-coenzyme A reductase
HTGL- Hepatic triglyceride lipase
HypoHDL- Hypoalphalipoproteinemia
IDL- Intermediate density lipoprotein
LCAT- Lecithin:cholesterol acyltransferase
LDL- Low density lipoprotein
Lp(a)- Lipoprotein a
LPL- Lipoprotein lipase
LpX- Lipoprotein X
LRC- Lipid Research Clinic
MI- Myocardial infarction
MRFIT- Multiple Risk Factor Intervention Trial
MUFA- Monounsaturated fatty acid
NCEP- National Cholesterol Education Program
NIDDM- Noninsulin-dependent diabetes mellitus
PUFA- Polyunsaturated fatty acid
PVD- Peripheral vascular disease
RI- HMG CoA Reductase Inhibitor
Sf- Svedberg unit
SI- Systéme International
TC- Total cholesterol
TG- Triglyceride
VLDL- Very low density lipoprotein
VLDL-TG- Very low density lipoprotein triglyceride
Accuracy The degree of agreement of a measurement, X, (or an average of measurements of the same thing) with an accepted reference or true value, T, usually expressed as the difference between the two values, $X - T$, or the difference as a percentage of the reference or true value, $100 (X - T)/T$.

Antioxidants A class of substances that prevents oxidation of molecules including lipids. Examples include probucol, and possibly vitamins C and E.

Apolipoproteins (Apoproteins, apo) Certain proteins in the blood that bind with lipids such as cholesterol, triglycerides and phospholipids to form a complex called lipoproteins. Apoproteins act as the structural elements in lipoproteins, regulate the enzymes of lipid metabolism, serve in the recognition and uptake of lipoprotein particles by specific receptors, and facilitate the transfer of cholesteryl esters from HDL to VLDL, LDL and chylomicrons.

Apoprotein A The major apoprotein of HDL. There are two major subgroups, both of which are found in HDL. A-I serves as a cofactor for LCAT and A-II as a cofactor for HTGL.

Apoprotein B There are two major subgroups, B-100 and B-48, both of which are involved in the structural role for lipoproteins. Apoprotein B-100 also is responsible for the recognition of LDL by the LDL receptor. The blood level of apo B is associated with the risk of CHD.

Apoprotein C There are three major subgroups: C-I serves as a cofactor for LCAT, C-II serves as a cofactor for LPL, and C-III serves as an inhibitor for LPL.

Apoprotein D Appears to regulate the transfer of cholesteryl esters between various lipoproteins and specifically from HDL to VLDL, IDL to LDL; it is present in a lipoprotein complex containing apo A-I, CETP and LCAT.

Apoprotein E A major apoprotein of VLDL and chylomicrons but also found on IDL and HDL. It serves as a binding protein with the cell receptor. There are 3 alleles: E^2, E^3, and E^4, which bind with varying efficiency to the LDL receptor. The inheritance of certain alleles of apo E can lead to disorders in lipoprotein metabolism, such as Type II dyslipidemia.

Atherosclerosis The process by which lipid-rich plaques (atheromas) develop in medium and large-sized arteries and impede blood flow. With progression of the disease through adulthood, three evolutionary lesions characterize the disease—the fatty streak, the fibrous plaque, and the complicated lesion.

Attributable risk The additional risk that follows exposure to a certain factor beyond that experienced by persons who are not exposed. Attributable risk is calculated as the incidence of disease in exposed persons minus the incidence in nonexposed persons, presumably from other causes.

Bias A quantitative measure of inaccuracy or departure from accuracy. A consistent signed (− or +) difference between two values. In general, the difference between the true, accepted, or expected value and the observed value, expressed in the units of the measurement or as a percentage.

Bile salts Derivatives of cholesterol with detergent properties that are produced by the liver and aid in the solubilization of lipid molecules in the digestive tract. Removal of these salts by bile acid sequestrants lowers the blood cholesterol level.

Bile acid sequestrant (Bile acid binding resin) A class of drugs that binds bile acids in the intestines, blocking reabsorption of cholesterol and lowering the blood cholesterol level.

Bran The skin or husk of grains from wheat, rye, or oats that can be separated from the flour by sifting.

Cholesterol Structurally different from other major lipids, cholesterol contains a steroid-ring nucleus. A sterol found only in animals, it is an essential component of cell membranes, supports hormone synthesis in the adrenal gland and gonads, and is a precursor of bile acids.

Cholesterol acyltransferase (ACAT:Acyl-CoA) ACAT is an intracellular enzyme that catalyzes the formation of cholesteryl esters. It acts in conjunction with HMG CoA reductase and the LDL-receptor system to minimize fluctuations in the concentration of free cholesterol in intracellular membranes.

Cholesteryl ester The form of cholesterol in which a fatty acid is attached to its third carbon through a chemical bond called an ester bond. This is the form required for storage of cholesterol in cells. Because it is hydrophobic, it is situated in the core of the lipoprotein molecules.

Cholesteryl ester transfer protein (CETP) A protein that transfers cholesteryl esters between lipoprotein particles and may play a key role in reverse cholesterol transport. A deficiency results in high HDL and low LDL.

Cholestyramine A cholesterol-lowering agent of the bile acid resin class that primarily reduces blood LDL-C by binding bile acids in the intestine.

Chylomicron The largest of the lipoproteins, it contains 98% lipid by weight, mainly triglyceride. Chylomicrons transport dietary triglyceride and cholesterol from the intestine into the bloodstream to sites of storage and utilization. Chylomicrons normally are not present in the blood after an overnight fast.

Chylomicron remnant The breakdown product of a chylomicron particle that results when its triglyceride is broken down by lipoprotein lipase. This particle is relatively enriched in cholesterol and is removed rapidly from the blood by the chylomicron remnant receptor in the liver.

Chylomicron remnant receptor A protein on the surface of liver cells that recognizes apoprotein E on the chylomicron remnant that leads to the removal of the remnant from blood.

Cis fatty acids In the cis configuration of unsaturated fatty acids, the two hydrogen atoms attached to the double bond lie on the same side.

Clofibrate An early fibric acid derivative used to reduce blood cholesterol and triglyceride. This agent has been found to produce a high incidence of adverse effects and has been replaced largely by gemfibrozil in the United States.

Coefficient of variation (CV) A measure of precision calculated as the standard deviation (SD) of a set of values divided by the average (X̄). It is multiplied by 100 to be expressed as a percentage:

$$CV = \frac{SD}{\bar{X}} \times 100$$

Colestipol A cholesterol-lowering agent of the bile acid resin class that primarily reduces LDL-C by binding bile acids in the intestine.

Corneal arcus A crescent-shaped deposit of lipid in the cornea with a clear interval between the deposit and the limbus. This lesion, especially in younger persons, is frequently associated with elevated blood lipids.

DHA (Docosahexaenoic acid) An omega-3 fatty acid found in fish oil.

Downregulation An intrinsic autoregulatory mechanism in which the concentration of a particular substance (such as a hormone) inhibits the synthesis of another substance (such as receptors) by a feedback mechanism.

Dyslipidemia (Dyslipoproteinemia) The general term for lipid disorders. An abnormal quantity or condition of total blood lipid levels or lipoprotein subgroups often characterized by aberrant lipoprotein biochemical properties.

EPA (Eicosapentaenoic acid) An omega-3 fatty acid found in fish oil.

Familial combined hyperlipidemia (FCH) An inherited disorder associated with premature coronary heart disease. Elevated cholesterol, triglyceride, or both may be present. The lipoprotein phenotype may vary over time as well as among family members.

Familial hypercholesterolemia (FH) A disorder characterized by an inherited defect in the LDL receptor that prevents the removal of LDL from the blood at a normal rate, resulting in high levels of blood cholesterol and LDL cholesterol but normal triglyceride levels. In homozygotes there are few if any LDL receptors present, resulting in LDL cholesterol levels > 500 mg/dL and CHD usually before puberty, whereas, in heterozygotes, there are effective LDL receptors present but not in sufficient numbers, resulting in LDL cholesterol levels usually > 250 mg/dL and CHD in middle age.

Familial hyperalphalipoproteinemia A condition characterized by HDL levels greater than 70 mg/dL; inherited as a dominant trait and apparently associated with longevity.

Familial hypertriglyceridemia (FHTG) An inherited

disorder involving high VLDL and triglyceride levels with low HDL levels; FHTG often is associated with obesity, diabetes, and gout.

Familial hypobetalipoproteinemia A genetic condition associated with low LDL cholesterol and apo B levels and a decreased risk of CHD.

Fatty acid A long-chain hydrocarbon that contains a carboxyl group at one end. Saturated fatty acids have completely saturated hydrocarbon chains and no double bonds. Unsaturated fatty acids have one or more carbon-to-carbon double bonds in their hydrocarbon chains.

Fatty streak A lesion found in arteries and thought to be the precursor of atherosclerotic plaques in the intima of coronary arteries.

Fiber Vegetable material that is resistant to digestion. Fiber may be water-insoluble, such as wheat bran, or water soluble, such as oat bran, guar gum from legumes, and pectin from fruits. The properties of fibers differ depending on water solubility.

Fibric acid derivatives A class of drugs such as gemfibrozil or clofibrate. These drugs are thought to act by increasing LPL activity and decreasing VLDL synthesis. Their main use is to decrease triglyceride levels in forms of dyslipidemia associated with hypertriglyceridemia.

Foam cell A macrophage characterized by accumulation of modified (such as oxidized or acetylated) LDL or beta VLDL. Its major lipid component is cholesteryl ester. It is thought to play a major role in the formation of fatty streaks and atherogenesis.

Formulae

1. Conversion of plasma cholesterol (EDTA) to serum cholesterol:

   Serum cholesterol (mg/dL) = plasma cholesterol (mg/dL) $\times$ 1.03

2. Calculation of LDL cholesterol

   Conventional units (mg/dL)

   $$LDL\text{-}C = \text{total cholesterol} - \left[HDL\text{-}C + \frac{\text{triglyceride}}{5}\right]$$

   SI units (mmol/L)

   $$LDL\text{-}C = \text{total cholesterol} - \left[HDL\text{-}C + \frac{\text{triglyceride}}{2.18}\right]$$

   VLDL cholesterol is estimated from the blood triglyceride level. This equation does not apply if the triglyceride values exceed 400 mg/dL or 4.52 mmol/L.

3. Estimated calories from food

   Calories from alcohol = 7 $\times$ grams of alcohol.
   Calories from carbohydrate = 4 $\times$ grams of carbohydrate.
   Calories from fat = 9 $\times$ grams of fat.
   Calories from protein = 4 $\times$ grams of protein.

4. Conversion of lipid data to Système International (SI) Units (mmol/L)

Cholesterol Conversion (multiply by 0.02586)

mg/dL	mmol/L	mg/dL	mmol/L
190	4.91	260	6.72
200	5.17	270	6.98
210	5.43	280	7.24
220	5.69	290	7.50
230	5.95	300	7.76
240	6.20	310	8.02
250	6.47	320	8.28

Triglyceride Conversion (multiply by 0.01129)

mg/dL	mmol/L	mg/dL	mmol/L
200	2.26	700	7.90
300	3.39	800	9.03
400	4.52	900	10.16
500	5.65	1000	11.29
600	6.77	1500	16.94

HDL Cholesterol Conversion (multiply by 0.02586)

mg/dL	mmol/L	mg/dL	mmol/L
25	0.65	55	1.42
30	0.78	60	1.55
35	0.90	65	1.68
40	1.03	70	1.81
45	1.16	75	1.94
50	1.29	80	2.07

5. Conversion of Système International Units (mmol/L) to Conventional Laboratory Units (mg/dL)

   Cholesterol conversion: multiply by 38.7
   Triglyceride conversion: multiply by 88.6

6. Approximation of Desired Body Weight

   Women: Desirable weight (lbs) = 100 lbs + 5 lbs for every inch above 5 feet
   Men: Desirable weight (lbs) = 110 lbs + 6 lbs for every inch above 5 feet

   Conversion to metric measures for weight and height
   1 lb. = 0.45 kg   1 inch = 2.54 cm

Fractional catabolic rate (FCR) The catabolic rate of a lipoprotein expressed as a fraction of the total pool of the lipoprotein.

Free fatty acids Fatty acids are transported in the blood largely in the form of firmly bound but rapidly reversible complexes with plasma albumin. Most of the albumin-bound free fatty acids in the blood are derived form lipolysis of triglyceride in adipose tissue during fasting. In addition they are released postprandially during hydrolysis of chylomicron and VLDL triglyceride by lipoprotein lipase.

Gemfibrozil A fibric acid derivative used to lower triglyceride and cholesterol levels. It also acts to raise HDL-C levels. The mechanism appears to occur through increasing LPL activity and inhibiting VLDL synthesis.

Hepatic triglyceride lipase, hepatic lipase (HTGL) An enzyme on the surface of liver cells that breaks down the triglyceride in the core of triglyceride-rich lipoproteins, such as intermediate density lipoprotein, to produce a modified particle. It facilitates catabolism of HDL.

High density lipoprotein (HDL) A blood lipoprotein that carries 20% to 25% of the cholesterol in the blood. There are at least two major separate subclasses of HDL, HDL_2 and HDL_3. About 50% of HDL is protein, the major apoprotein is A-1. HDL is thought to be involved in the reverse cholesterol transport system but may have other qualities that protect against atherosclerosis as well. The blood level of HDL cholesterol is inversely related to CHD risk.

HMG CoA reductase inhibitor (RI) A class of drugs lovastatin, pravastatin, or simvastatin that inhibits the activity of the enzyme HMG CoA leading to reduced cholesterol synthesis.

Homozygote A term used to indicate that a person carries faulty genes from both the father and mother for a specific trait.

Hydroxymethoxyglutaryl CoA reductase (HMG CoA reductase) This enzyme governs the rate-limiting step in the synthesis of cholesterol. This enzyme is regulated by intracellular cholesterol levels.

Hyperapobetalipoproteinemia (Hyperapo B) A disorder of apoproteins associated with elevated blood levels of apo B and premature CHD, despite normal or only moderately elevated levels of LDL cholesterol. Hyperapo B leads to overproduction of VLDL in the liver, which results in increased synthesis of its breakdown product, LDL.

Hyperlipidemia A term loosely used to indicate an elevation of any blood lipid but more strictly defined as an elevation of the blood triglyceride resulting from increased VLDL, chylomicrons, or both.

Hypoalphalipoproteinemia (Hypo HDL) A dyslipidemia associated with premature CHD in which the blood levels of HDL-C are very low. The LDL-C, LDL apo B, and triglyceride levels may be normal.

Intermediate density lipoprotein (IDL) An intermediate product in the conversion of VLDL to LDL; the lipoprotein can bind to the LDL receptor but accumulates in plasma of patients who have type III hyperlipidemia (dysbetalipoproteinemia).

Isozymes Multiple forms of an enzyme that differ from one another in one or more properties.

LCAT (Lecithin-cholesterol acetyl transferase) An enzyme that mediates the esterification of free cholesterol, primarily in HDL particles, by removing a fatty acid from lecithin and transferring it to cholesterol.

Lecithin The major phospholipid found in plasma.

Ligand An organic molecule that reacts to form a complex with another molecule.

Lipid Any of the various substances that are insoluble in water but will dissolve in organic solvents, such as chloroform or ether. Examples of lipids are cholesterol, cholesteryl esters, triglyceride, phospholipids, and fatty acids.

Lipid cascade The sequential metabolic reduction in the size of chylomicrons and VLDL that occurs with lipoprotein lipase-mediated hydrolytic cleavage of the triglyceride transported by the particles. Resultant free fatty acids are released to tissues to satisfy metabolic requirements or for storage. Chylomicron remnants are removed from the circulation by the liver, and VLDL remnants are either removed from the circulation by the liver or converted to LDL.

Lipoprotein The lipoproteins are particles made up of a hydrophobic core of cholesteryl esters and triglyceride surrounded by a single shell composed of apoproteins, unesterified cholesterol, and phospholipids. This structure enables lipoprotein particles to transport water-insoluble lipids through the circulatory system from the sites of absorption in the intestines or synthesis in the liver to storage areas. There are five major lipoprotein subgroups—chylomicrons, very low density lipoproteins (VLDL), intermediate density lipoproteins (IDL), low density lipoproteins (LDL), and high density lipoproteins (HDL).

Lipoprotein lipase (LPL) An enzyme found on the surface of blood vessels and responsible for the breakdown of triglyceride in either chylomicrons or VLDL to form free fatty acids and glycerol. It facilitates HDL production.

Lipoprotein phenotypes (Fredrickson classification) An older classification system that is useful as a shorthand in describing the lipid patterns. It does not deal with HDL-C.

TYPE I: Excessive fasting chylomicronemia.

TYPE IIA: Hypercholesterolemia (elevated LDL) without hypertriglyceridemia.

TYPE IIB: Elevated cholesterol and triglyceride (elevated VLDL + LDL).

TYPE III: Broad betalipoproteinemia (dysbetalipoproteinemia). A disorder caused by the accumulation of VLDL and chylomicron remnants, and characterized by equally high levels of blood cholesterol and triglyceride. The circulating VLDL is cholesterol-rich and electrophoretically resembles LDL.

TYPE IV: High triglyceride (elevated VLDL) levels only.

TYPE V: Fasting chylomicronemia and high triglyceride (elevated VLDL).

Lipoprotein X (LpX) An abnormal lipoprotein that accumulates in the plasma of patients who have obstructive liver disease. Free cholesterol cannot be excreted properly into the bile and is directed into the plasma, where it binds to albumin phospholipid to form an abnormal lipoprotein.

Lovastatin A hypolipidemic agent that is an HMG CoA reductase inhibitor. It acts early in the pathway of the hepatic synthesis of cholesterol.

Low density lipoprotein (LDL) The major lipoprotein carrier of cholesterol in the blood. About 50% of LDL is cholesterol and cholesteryl ester; 25% is protein, mostly apo B; LDL is thought to be the most athero-genic of the lipoproteins.

Lp(a) A minor lipoprotein in the density range between LDL and HDL formed by the attachment of apo (a) to the LDL molecule. Structurally, apo (a) resembles plasminogen. Lp(a) is thought to play a role in prema-ture CHD and stroke.

Monounsaturated fatty acid (MUFA) A clear, oily sub-stance that is liquid at room temperature. Monoun-saturated fatty acids have one unsaturated double bond. Olive oil, peanut oil, and canola oil contain pre-dominantly monounsaturates.

Nicotinic acid (niacin) A naturally occurring vitamin (B_3) that reduces triglyceride and cholesterol levels at pharmacologic doses. It is thought to act by inhibiting the synthesis of VLDL, lowering LDL-C, and trigly-ceride levels while raising HDL-C levels by interfering with HDL catabolism. The amide derivative nicotina-mide (niacinamide) has no lipid effects.

Oleic acid A monounsaturated fatty acid that has a chain of 18 carbons and contains one area of unsatur-ation between carbons 9 and 10.

Omega-3 fatty acids (n-3 fatty acids) A group of poly-unsaturated fatty acids in which the last area of un-saturation occurs at the third position from the "tail" end, or omega carbon. Omega-3 fatty acids (EPA, DHA) are the main constituents of fish oils.

Omega-6 fatty acids (n-6 fatty acids) A group of poly-unsaturated fatty acids in which the last area of un-saturation occurs at the sixth position from the omega end (e.g., linoleic acid).

P/S ratio A term used to describe the ratio of polyunsat-urated fat (P) to saturated fat (S) in the overall diet. The P/S ratio ignores monounsaturated fat. Gener-ally, a diet with a high P/S ratio will lower blood cho-lesterol levels.

Palmitic acid A saturated fatty acid that has a chain of 16 carbons and no areas of unsaturation. Foods high in palmitic acid increase blood levels of cholesterol.

Phospholipids A class of lipids that consists of glycerol, two fatty acids, and a third component in which a chemical compound is attached to a glycerol through its phosphorous component.

Polar group A hydrophilic (water-loving) group.

Polygenic hypercholesterolemia The most common dyslipidemia associated with an elevated blood cho-lesterol. This disorder, associated with CHD, arises from a combination of unknown genetic and environ-mental factors.

Polyunsaturated fat A clear, oily substance that is liquid at room temperature and whose main component is polyunsaturated fatty acids (PUFA).

Pravastatin A second-generation HMG CoA reductase inhibitor.

Probucol A drug that lowers both LDL-C and HDL-C levels in the plasma in varying amounts and also may prevent oxidation of lipoproteins.

Prostaglandin An oxygenated eicosanoid (20-carbon fatty acid) that has a hormonal function. Prosta-glandins are unusual hormones in that they generally have effects only in that region of the organism in which they are synthesized.

Receptor, LDL A protein on the surface of cells that binds LDL apo B and allows the LDL to be taken into the cell and processed. It also binds apo E on VLDL remnants and IDL and facilitates their uptake from blood into the liver.

Relative risk (Risk ratio) The ratio or risk for a particu-lar clinical event among the exposed persons relative to those who are nonexposed. It represents the strength of the association between a factor and the likelihood of disease.

Reliability of measurements This involves two com-ponents: Precision is the within-day or day-to-day re-producibility; accuracy is the true value traceable to a reference or definitive method.

Risk factors Observations or measurements that indi-cate a heightened risk for future cardiovascular events independent of other known associations. Risk is usu-ally expressed as relative risk or attributable risk.

Saturated fat A white, oily substance that is solid at room temperature and whose main component is sat-urated fatty acids. A saturated fatty acid is one in which all positions in the carbon chain are saturated with hydrogen atoms.

Simvastatin A second-generation HMG CoA reductase inhibitor.

Stearic acid A saturated fatty acid that has a chain of 18 carbons. It may be an exception to the rule that satu-rated fatty acids raise blood cholesterol, possibly be-cause stearic acid is converted in the body into oleic acid, a monounsaturated fatty acid.

Step-1 Diet A diet in which less than 30% of the calo-ries are from total fat, less than 10% from saturated fat, no more than 10% from polyunsaturated fat, and between 10% to 15% from monounsaturated fat. Cholesterol is limited to less than 300 mg per day. In this diet, 50% to 60% of the calories are from carbo-hydrates and 10% to 20% are from protein.

Step-2 Diet A more restricted diet than a Step-1 diet. Total fat is less than 30% of the calories, with less than 7% from saturated fat, no more than 10% from poly-unsaturated fat, and 10% to 15% from monoun-saturated fat. Cholesterol is restricted to less than 200 mg per day. Like the Step-1 diet, 50% to 60% of the calories are from carbohydrates and 10% to 20% are from protein

Sterol A solid complex cyclic alcohol with lipid-like sol-ubility. Examples are cholesterol and ergosterol.

Svedberg unit (Sf) The unit is used in ultracentri-fugation methods to express the sedimentation con-stant S: 1 S = 10^{-13} sec. The sedimentation constant S is proportional to the rate of sedimentation of a mole-cule in a given centrifugal field and is related to the size and shape of the molecule.

Tangier disease An HDL deficiency syndrome associated with atherosclerosis and CHD despite low levels of circulating LDL cholesterol.

Thromboxane A$_2$ A product derived from arachidonic acid that causes blood vessels to become narrower and promotes platelet aggregation. **Thromboxane A$_3$** is derived from eicosapentanoic acid and is biologically inactive.

Trans fatty acids Isomers of the "cis" polyunsaturated fatty acids in which the hydrogen atoms on carbon atoms linked by double bonds are on opposite sides. Trans fatty acids are found abundantly in hydrogenated oils and margarine.

Transport protein A protein whose primary function is to transport a substance from one part of the cell to another, from one cell to another, or from one tissue to another.

Triglyceride (triacylglycerols or neutral fats) A true fat that consists of one molecule of glycerol and three fatty acid molecules. One of the principal blood lipids, triglyceride transported by either chylomicrons or VLDL provides a major source of energy for the body. Triglyceride may relate to the risk of CHD through several possible mechanisms.

Upregulation An intrinsic autoregulatory mechanism in which the concentration of a particular substance such as hormones stimulates the synthesis of a substance by a positive feedback mechanism.

Very low density lipoprotein (VLDL) The major carrier of triglyceride in the blood of a fasting patient, VLDL is made primarily in the liver and contains about 65% of its weight as triglyceride. It also contains cholesterol and phospholipid. The major apolipoproteins of VLDL are apo B-100, apo C, and apo E.

VLDL remnant The breakdown product of VLDL with VLDL remnants containing triglyceride and relatively more cholesterol than VLDL. Some VLDL remnants may be taken up by the liver; the rest are converted into intermediate density lipoprotein.

Xanthelasma A yellowish/orange, flat deposit of lipids on the eyelids or under the eyes. This lesion may, but not necessarily, be a marker for lipid problems.

Xanthoma Yellow or orange dermal macules or papules, subcutaneous plaques, or tendinous nodules associated with lipid disorders. Tendon xanthomas occur in the presence of hypercholesterolemia and are characteristic of familial hypercholesterolemia. Eruptive xanthomas occur on the abdomen and thighs as punctate yellowish papules surrounded by a red halo and occur primarily with disorders of triglyceride metabolism.

APPENDICES

APPENDIX A

Foods—Calorie, Fat, and Cholesterol Content

Dairy and Egg Products

	Calories (Approximate)	Total Fat (gm)	Saturated Fat (gm)	Monounsat. Fat (gm)	Polyunsat. Fat (gm)	Cholesterol (mg)
CHEESE AND MILK:						
Cheese, American, 1 ounce	106	8.9	5.6	2.5	0.3	27
Cheese, blue, 1 ounce	100	8.2	5.3	2.2	0.2	21
Cheese, camembert, 1 ounce	85	6.9	4.3	2.0	0.2	21
Cheese, cheddar, 1 ounce	114	9.4	6.0	2.7	0.3	30
Cheese, cottage, creamed, 4% fat,						
large curd, 1 cup	232	10.1	6.4	2.9	0.3	34
small curd, 1 cup	217	9.5	6.0	2.7	0.3	31
Cheese, cottage, lowfat, 1% fat, 1 cup	164	2.3	1.5	0.7	0.1	10
Cheese, cottage, uncreamed, dry						
curd, less than 1/2% fat, 1 cup	123	0.6	0.4	0.2	Tr[b]	10
Cheese, cream, 1 ounce (2 tbsp)	99	9.9	6.2	2.8	0.4	31
Cheese, mozarella, made with part-						
skim milk, 1 ounce	72	4.5	2.9	1.3	0.1	16
Cheese, Muenster, 1 ounce	104	8.5	5.4	2.5	0.2	27
Cheese, parmesan, grated, 1 tbsp	23	1.5	1.0	0.4	Tr	4
Cheese, ricotta, part skim milk,						
1/2 cup	171	9.8	6.1	2.8	0.3	25
Cheese, Swiss, 1 ounce	107	7.8	5.0	2.1	0.3	26
Milk, whole, 3.3% fat, 1 cup	150	8.2	5.1	2.4	0.3	33
Milk, lowfat, 2% fat, 1 cup	125	4.7	2.9	1.4	0.2	18
Milk, lowfat, 1% fat, 1 cup	104	2.4	1.5	0.7	0.1	10
Milk, nonfat, skim, 1 cup	90	0.6	0.4	0.2	Tr	5
Milk, buttermilk, cultured, 1 cup	99	2.2	1.3	0.6	0.1	9
Milk beverages, eggnog, 1 cup	342	19.0	11.3	5.7	0.9	149
Milk beverages, shakes, thick,						
vanilla, 1 cup	275	14.7	9.2	4.2	0.6	61
CREAM AND COFFEE CREAMERS:						
Cream, half-and-half, 1 tbsp	20	1.7	1.1	0.5	0.1	6
Cream, light, coffee, or table, 1 tbsp	29	2.9	1.8	0.8	0.1	10
Cream, heavy, fluid, unwhipped,						
1 tbsp	52	5.6	3.5	1.6	0.2	21
Cream, sour, cultured, 1 tbsp	26	2.5	1.6	0.7	0.1	5
Coffee whitener, liquid, frozen (con-						
tains coconut or palm kernel),						
1 tbsp	20	1.5	1.4	Tr	Tr	0
Coffee whitener, powdered (contains						
coconut or palm kernel oil), 1 tsp	11	0.7	0.7	Tr	Tr	0
Dessert toppings (nondairy), pow-						
dered, made with whole milk,						
1 tbsp	8	0.5	0.4	Tr	Tr	Tr

	Calories (Approximate)	Total Fat (gm)	Saturated Fat (gm)	Monounsat. Fat (gm)	Polyunsat. Fat (gm)	Cholesterol (mg)
Imitation cream (made with vegetable fat),	11	0.9	0.8	0.1	Tr	0
DESSERTS:						
Ice cream, regular (about 10% fat), 1 cup	269	14.3	8.9	4.1	0.5	59
Ice cream, rich (about 16% fat), 1 cup	349	23.7	14.7	6.8	0.9	88
Ice milk, hardened (about 4.3% fat), 1 cup	184	5.6	3.5	1.6	0.2	18
Ice milk, soft serve (about 2.6% fat), 1 cup	223	4.6	2.9	1.3	0.2	13
Ice milk, sherbet (about 2% fat), 1 cup	270	3.8	2.4	1.1	0.1	14
Yogurt, made with lowfat milk, 8 ounces	194	2.8	1.8	0.8	0.1	11
Yogurt, made with nonfat milk, 8 ounces	127	0.4	0.3	0.1	Tr	4
Yogurt, made with whole milk, 8 ounces	139	7.4	4.8	2.0	0.2	29
Egg white, 1	16	Tr	0	0	0	0
Egg, whole, 1 large	79	5.6	1.7	3.2	0.7	213
Egg yolk, 1 large	63	5.6	1.7	3.2	0.7	213

Fats, Oils and Related Products

	Calories (Approximate)	Total Fat (gm)	Saturated Fat (gm)	Monounsat. Fat (gm)	Polyunsat. Fat (gm)	Cholesterol (mg)
FATS (solid at room temperature):						
Butter, 1 tbsp	102	11.4	7.1	3.3	0.4	31
Lard, 1 tbsp	116	12.8	5.0	5.8	1.4	12
Shortening (animal and vegetable fat), 1 tbsp	115	12.8	5.2	5.7	1.4	7
Shortening (vegetable), 1 tbsp	113	12.8	3.2	5.7	3.3	0
Tallow, edible, 1 tbsp	116	12.8	6.4	5.3	0.5	14
Margarine, regular (at least 80% fat), stick, corn oil, 1 tbsp	102	11.4	2.0	5.5	3.4	0
stick, soybean oil, 1 tbsp	102	11.4	2.4	5.4	3.0	0
tub, corn oil, 1 tbsp	102	11.4	2.0	4.5	4.4	0
tub, soybean oil, 1 tbsp	102	11.4	1.8	5.1	3.9	0
Margarine, diet (about 40% fat), tub, 1 tbsp	50	5.7	1.2	2.4	2.1	0
Oils (liquid at room temperature), coconut, 1 tbsp	117	13.6	11.8	0.8	0.2	0
corn, 1 tbsp	120	13.6	1.7	3.3	8.0	0
olive, 1 tbsp	119	13.5	1.8	9.9	1.1	0
palm, 1 tbsp	120	13.6	6.7	5.0	1.3	0
palm kernel, 1 tbsp	117	13.6	11.1	1.5	0.2	0
peanut, 1 tbsp	119	13.5	2.3	6.2	4.3	0
safflower	120	13.6	1.2	1.6	10.1	0

	Calories (Approximate)	Total Fat (gm)	Saturated Fat (gm)	Monounsat. Fat (gm)	Polyunsat. Fat (gm)	Cholesterol (mg)
soybean oil (partially hydrogenated), 1 tbsp	120	13.6	2.0	5.9	5.1	0
rapeseed oil, 1 tbsp	120	13.6	1.0	8.0	4.0	0
sunflower, 1 tbsp	120	13.6	1.4	2.7	8.9	0
Mayonnaise, 1 tbsp	99	11.0	1.6	3.1	5.7	8
Light mayonnaise, 1 tbsp	44	4.3	0.7	1.8	1.6	4
Peanut butter, 1 tbsp	95	8.3	1.7	3.8	2.4	0
SALAD DRESSINGS:						
Russian, 1 tbsp	75	7.8	1.1	1.8	4.5	0
French, 1 tbsp	65	6.4	1.5	1.2	3.4	2
Ranch, 1 tbsp	50	5.0	0.9	1.3	2.9	4
Low-calorie Italian, 1 tbsp	16	1.5	0.2	0.3	0.9	1
Italian, 1 tbsp	70	7	1.0	1.7	4.1	0
Bleu cheese, 1 tbsp	80	8	1.5	1.9	4.3	9
Mayonnaise-type, 1 tbsp	60	5	0.7	1.3	2.6	4
Thousand Island, 1 tbsp	60	6	0.9	1.3	3.1	5

Fish, Shellfish, Meat, Poultry, and Related Products

	Calories (Approximate)	Total Fat (gm)	Saturated Fat (gm)	Monounsat. Fat (gm)	Polyunsat. Fat (gm)	Cholesterol (mg)
FISH:						
Fish, cooked, flounder or sole (a lean fish), baked, 3 ounces	82	1.0	0.3	0.2	0.4	59
Fish, cooked, salmon, red (a fatty fish), baked, 3 ounces	140	5.4	1.2	2.4	1.4	60
Fish, canned, salmon, pink, water pack, 3 ounces	120	5.0	0.9	1.5	2.1	34
Fish, canned, sardines, atlantic, oil pack, 3 ounces	173	9.4	2.1	3.7	2.9	85
Fish, canned, tuna, chunk light, oil pack, 3 ounces	167	7.0	1.4	1.9	3.1	55
Fish, canned, tuna, chunk light, water pack, 3 ounces	95	1.0	N/A	N/A	N/A	90
Shellfish, raw, 3 ounces clams, unspecified	65	1.4	0.3	0.3	0.3	42
oysters, Eastern	56	1.5	0.5	0.2	0.5	42
shrimp	84	0.9	0.3	0.2	0.4	166
MEATS:						
Meat, beef, eye of round, lean only, roasted, 3 ounces	156	5.9	2.4	2.7	0.2	56
Meat, beef, rib roast, lean and fat, roasted, 3 ounces	330	28.4	11.7	13.6	1.0	70
Meat, beef, ground beef, cooked, well done, 3 ounces	224	15.6	7.6	8.5	0.7	88

	Calories (Approximate)	Total Fat (gm)	Saturated Fat (gm)	Monounsat. Fat (gm)	Polyunsat. Fat (gm)	Cholesterol (mg)
Meat, pork, ham, roasted, 3 ounces	187	9.4	3.2	4.2	1.1	80
Meat, pork, bacon, fried crisp, 2 slices	73	6.2	2.2	3.0	0.7	11
Meat, lamb, loin chop, lean only, 3 ounces	183	8.5	3.5	3.2	0.5	80
Meat, lamb, loin chop, lean and fat, 3 ounces	250	17.0	7.7	6.8	1.0	82
Meat, veal cutlet (1 cutlet), 3 ounces	185	9.4	4.0	4.0	0.4	86
Poultry, chicken, dark meat, baked without skin, 3 ounces	174	8.3	2.3	3.0	1.9	79
Poultry, chicken, light meat, baked without skin, 3 ounces	147	3.8	1.1	1.3	0.8	72
Poultry, chicken, dark meat, fried with skin, 3 ounces	242	14.4	3.9	5.7	3.3	78
Poultry, chicken, light meat, fried with skin, 3 ounces	209	10.3	2.8	4.1	2.3	74
Beef liver, fried, 3 ounces	195	9.0	2.5	3.6	1.3	372
Frankfurters (beef), 1 frank	184	16.8	6.8	8.2	0.7	27
Bologna (beef, pork), 1 ounce	89	8.0	3.0	3.8	0.7	16
Salami, cooked (beef, pork), 1 ounce	71	5.7	2.3	2.6	0.6	18
Braunschweiger, 1 ounce	102	9.1	3.1	4.2	1.1	44

Miscellaneous Items (With Ingredients of

Animal Origin as Sources of Cholesterol)

	Calories (Approximate)	Total Fat (gm)	Saturated Fat (gm)	Monounsat. Fat (gm)	Polyunsat. Fat (gm)	Cholesterol (mg)
Beef pot pie, 1	515	30.5	7.9	12.9	7.4	42
Beef stew, 1 cup	220	10.5	4.4	4.5	0.5	72
Chicken pot pie, 1 piece	545	31.3	10.3	15.5	6.6	56
Chicken a la king, 1 cup	470	35.5	12.9	13.4	6.2	220
Chili with beef, 1 cup	340	15.6	5.8	7.2	1.0	28
MISCELLANEOUS:						
Cakes, pound, 1 slice	160	10.0	5.9	3.0	0.6	68
Cakes, white, 2 layer with chocolate icing, 1 piece	250	7.7	3.0	2.9	1.3	3
Cakes, yellow, 2 layer with chocolate icing, 1 piece	235	7.9	3.0	3.0	1.4	36
Cookies, brownies, with chocolate icing, 1 brownie	105	5.3	2.0	2.3	0.7	13
Cookies, chocolate chip, 4 cookies	205	12.0	3.5	4.6	3.2	21
Cookies, vanilla wafers, 10 cookies	185	6.7	1.7	2.8	1.7	25
Crackers, graham, 2 crackers	55	1.3	0.3	0.5	0.4	0
Crackers, saltines, 4 crackers	50	1.0	0.4	0.4	0.2	3
Cupcakes, with chocolate icing, 1 cupcake	130	4.6	2.0	1.7	0.7	15
Doughnuts, cake type, 1 doughnut	100	4.7	1.2	1.2	2.0	10

	Calories (Approximate)	Total Fat (gm)	Saturated Fat (gm)	Monounsat. Fat (gm)	Polyunsat. Fat (gm)	Cholesterol (mg)
Doughnuts, yeast-leavened, 1 doughnut	176	11.3	4.9	3.4	0.8	14
Milk, chocolate (20% milk solids), 1 ounce	145	9.0	5.4	3.0	0.3	5
Pizza with cheese, ⅛ of 12″	145	4.0	2.1	1.2	0.5	13
Potatoes, french fried (fried in edible tallow), 10 strips	158	8.3	3.4	4.0	0.5	6

[a] Provisional Table on the Fatty Acid and Cholesterol Content of Selected Foods, United States Department of Agriculture, Human Nutrition Information Service, Washington, D.C.: United States Government Printing Office, June 1984.
[b] Tr = trace; N/A = data not available.

APPENDIX B

Dietary Habits Worksheet

This worksheet allows tabulation of basic dietary information for guiding dietary recommendations. (Adapted from Heart to Heart, a Manual on Nutrition Counseling for the Reduction of Cardiovascular Disease Risk Factors. USDHHS, NIH, 1983.)

Date _____

Patient's name _____

Nutrition counselor _____

1. Height: _____
   Weight: Current _____ Ideal _____ Difference _____
   % Ideal Weight _____

2. How long has patient been at present weight? (± 5 lb) _____ yr

3. Has patient tried losing or gaining weight before _____ yes _____ no
   By diet (specify) _____ Weight change _____
   How long did it last? _____
   By drugs (specify) _____ Weight change _____
   How long did it last? _____

4. Has the patient every tried to change his/her diet in any other way?
   _____ yes _____ no
   If yes—what changes did he/she try?
   Salt-restricted _____ Low-fat, low-cholesterol _____ Diabetic _____
   Other _____
   Comments: _____

5. Typical weekly eating pattern:
   (Check how often and where meals are eaten:)

	Home (times/wk)	Meal from Home (times/wk)	Restaurant Cafeteria, etc. (times/wk)	Other Places (times/wk)	Never Eat This Meal (times/wk)
Morning meal	_____	_____	_____	_____	_____
Midday meal	_____	_____	_____	_____	_____
Evening meal	_____	_____	_____	_____	_____
Snacks	_____	_____	_____	_____	_____

6. Is any other member of the household on a special diet? _____ yes _____ no
   Explain: _____

7. Who usually prepares the food eaten at home? _____

8. Who usually goes grocery shopping for the household? _____

9. Does the patient drink alcoholic beverages (e.g., beer, wine, cocktails)?
   _____ yes _____ no
   If yes: How many beers? _____ (No. 12-oz cans/wk)
   How much wine? _____ (No. 4-oz glasses/wk)
   How many cocktails? _____ (No. /wk)
   At home? _____ Out? _____

10. Does the patient take vitamin, mineral, or food supplements?
   _____ yes _____ no   If yes: specify kind and amount per day

   _____

APPENDIX C

Food Frequency Checklist

This checklist provides a means for assessing usual dietary preferences and intake. (From Heart to Heart, a Manual on Nutrition Counseling for the Reduction of Cardiovascular Disease Risk Factors. USDHHS, NIH, 1983.)

Check the frequency the following foods are consumed	Never or less than 1 time/wk	1-2 times/wk	3-7 times/wk	More than once/d
Regular ground beef (serving)				
Lean ground beef (serving)				
Sausage, bacon, luncheon meat (piece)				
Lean meats (round, flank, etc) (serving)				
High fat meats (prime rib, steak) (serving)				
Poultry (piece)				
Fish (serving)				
Shellfish (serving)				
Organ meats (liver, heart, brains, etc) (serving)				
Beans (navy beans, black-eyed peas, soybeans, etc) (1 cup)				
Peanut butter (tbsp)				
Pizza (piece)				
Whole milk (cup)				
Cream (tbsp)				
Nonfat milk (cup)				
Hard cheese, cheese spread, regular cottage cheese (oz)				
Cheese, low fat (oz)				
Ice cream (1/2 cup)				
Eggs (1)				
Oils (in salad dressing, cooking, etc) kind of oil (tbsp)				
Butter (tsp or pat)				
Margarine (tsp or pat)				
Vegetables (serving)				
Fruits (serving)				
Fruit juice (cup)				
Breads, white or whole grain (slice or roll)				
Cereals (cup)				
Pasta, noodles, rice, etc (cup)				
Potatoes (1)				
Commercial baked goods (cookies, pies, cakes, etc) (serving)				
Homemade baked goods (cookies, pies, cakes, etc) (serving)				

	Never or less than 1 time/wk	1-2 times/wk	3-7 times/wk	More than once/d
Soft drinks (nondiet) (serving)				
Snack crackers (serving)				
Nut and seeds (1/4 cup)				
Potato chips or corn chips (cup)				
Sherbets and ices (1/2 cup)				
Candy (1 to 1 1/2 oz)				
Other foods, such as ethnic foods (Chinese, African-American, Mexican, etc) List _____				

APPENDIX D

Sample Menus

Average American Diet
(37% fat)

Breakfast
1 fried egg[b]
2 slices white toast
   with 1 teaspoon butter
1 cup orange juice
black coffee or tea

Snack
1 doughnut

Lunch
1 grilled cheese (2 ounces)
   sandwich on white bread
2 oatmeal cookies
black coffee or tea

Snack
20 cheese cracker squares

Dinner
3 ounces fried hamburger with
   ketchup
1 baked potato with sour cream
3/4 cup steamed broccoli with
   1 teaspoon butter
1 cup whole milk
1 piece frosted marble cake

A New Low-Fat Diet
(30% fat)

Breakfast
1 cup corn flakes with blueberries
1 cup 1% milk
1 slice rye toast
   with 1 teaspoon margarine
1 cup orange juice
black coffee or tea

Snack
1 toasted pumpernickel bagel
   with 1 teaspoon margarine

Lunch
1 tuna salad (3 ounces) sandwich
   on whole wheat bread with
   lettuce and tomato
1 graham cracker
tea with lemon

Snack
1 crisp apple

Dinner
3 ounces broiled lean ground
   beef with ketchup
1 baked potato with low-fat
   plain yogurt and chives
3/4 cup steamed broccoli with
   1 teaspoon margarine
tossed garden salad with 1 tablespoon
   oil and vinegar dressing
1 cup 1% milk
1 small piece homemade gingerbread[a]
   with a maraschino cherry

Nutrient Analysis		**Nutrient Analysis**	
Calories	2,000	Calories	2,000
Total fat (% of calories)	37	Total fat (% of calories)	30
Saturated fat (% of calories)	19	Saturated fat (% of calories)	10
Cholesterol	505 mg	Cholesterol	186 mg

[a]Homemade desserts should be made with unsaturated fats instead of saturated fats.
[b]Two egg whites may be substituted for one egg yolk.
(From National Cholesterol Education Program, National Heart, Lung, and Blood Institute: Eating to Lower Your High Blood Cholesterol. NIH Publication No. 89-2920, June, 1989.)

APPENDIX E

I. Foods to Decrease Saturated Fatty Acid and Total Fat Content of School Breakfasts

Offer 1% low-fat milk and low-fat yogurt[a]
Provide ready-to-eat cereal with 2 grams or less of fat per ounce
Offer a variety of breads, low-fat muffins, and whole grain products
Provide margarine low in saturated fatty acids
Provide a variety of fruit and fruit juices
Offer only low-fat meats

[a]The panel recognizes that low-fat dairy products currently are not considered foods in the National School Lunch Program.
(From National Cholesterol Education Program Report on the Expert Panel on Blood Cholesterol Levels in Children and Adolescents. Pediatrics 1992)

II. Foods to Decrease Saturated Fatty Acid and Total Fat Content of School Lunches

Meat, Poultry, and Fish
Use lean cuts of beef and pork that are well trimmed before cooking
Reduce fat content of ground beef served as burgers
Cook ground beef and discard fat before incorporation in recipes
Remove skin from chicken before cooking

Dairy
Offer 1% low-fat milk, low-fat yogurt, and low-fat cheese
Use ice milk in place of ice cream

Fruits and Vegetables
Provide a variety of raw and cooked vegetables and decrease use of fat in preparation
Increase use of fruit for dessert
Offer a salad bar to include fresh vegetables and fruits, pasta, and vegetable salads, with reduced-fat mayonnaise or salad dressing

Breads and Cereals; Dried Beans and Peas
Provide a variety of low-fat breads and cereals, especially whole grain products
Include dried beans and peas in entrees and as a salad bar offering

Food Preparation

Decrease amount of fats in desserts, such as cookies, cakes, and cobblers, and use fats low in saturated fatty acids

Substitute margarine for butter in food preparation and as a spread

Decrease frequency of frying foods and use unsaturated oil or margarine when sauteeing or frying

---

III. Snack Bar Foods Consistent with Recommended Eating Pattern

---

1% low-fat or skim milk, low-fat cheese, low-fat, or nonfat yogurt (plain or with fruit)

Fresh fruit

Pretzels, popcorn popped in unsaturated oil, bagels, and bagel chips (no fat added)

Chef salads prepared with lean meat or water-packed tuna and low-fat cheese served with low-fat or fat-free salad dressing

Sandwiches made with sliced turkey, lean roast beef, lean ham, low-fat cold cuts, and tuna salad prepared with water-packed tuna and reduced-fat mayonnaise or salad dressing

Peanut butter* and jelly sandwiches

Hamburgers or sloppy joes made with lean, well-drained ground beef or ground turkey

Tacos made with lean, well-drained ground beef and soft corn tortillas with low-fat cheese or a small amount of regular cheese

Beef, chicken, or bean chalupa with toasted (not fried) corn tortilla and low-fat cheese or small amount of regular cheese

Pizza made with lean, well-drained ground beef and low-fat cheese or small amount of regular cheese

Nachos with toasted (not fried) corn tortilla chips and con queso made with low-fat cheese

Cookies, cupcakes, and muffins prepared with unsaturated oil of margarine

Frozen yogurt, ice milk, frozen fruit bars, sherbet, fruit sorbets, low-fat pudding pops

---

*High in total fat; low in saturated fatty acid

INDEX

Page numbers in *italic* denote figures; those followed by "t" denote tables.

Apoprotein A-II—*continued*
glycosylation, in diabetes mellitus, 159
in high density lipoprotein particles, 101
Apoprotein B, 148
absence of, 152
amino acid sequence, abnormalities in, 145
glycosylation, in diabetes mellitus, 159
levels
coffee effects on, 76
HMG CoA reductase inhibitor effects on, 240t
in hyperbetalipoproteinemia, 148
in hypertriglyceridemia, 225
in hypobetalipoproteinemia, 152
race factors in, 118, 119
modification of, 7
in palpebral xanthelasma, 176
synthesis
in familial hypertriglyceridemia, 142
increased, 140
thyroid hormone effects on, 156
Apoprotein B-48, 92
levels, in thyroid disorders, 156
Apoprotein B-100, 6, 56, 57, 90, 92, 95
levels, in thyroid disorders, 156
of low density lipoprotein, 96
production of, obesity and, 74
Apoprotein C, 93, 95, 99
Apoprotein C-I, glycosylation, in diabetes mellitus, 159
Apoprotein C-II, 90, 92
deficiency, 90
chylomicronemia with, 139–140
familial, 151
Apoprotein C-III, 90
deficiency, 48
Apoprotein D, *See* Cholesteryl ester transfer protein
Apoprotein E, 53, 90, 92, 93, 95, 99
alleles, 148–149
in dysbetalipoproteinemia, 148
glycosylation, in diabetes mellitus, 159
levels, in diabetes, 158–159
in palpebral xanthelasma, 176
phenotype frequencies, 148
Apoprotein E$_2$, 148, 149
Apoprotein E$_4$, 148, 149
Arachidonic acid, 194, 195
omega-3 fatty acids effects on, 196
Arterial regression, nonlipid factors affecting, 21–23, 25, 249–250
Arterial wall
cholesterol accumulation in, 3
lipids entering of, 6–7
Arteries
thickening and hardening of, 4, *See also* Atherosclerosis
dilation of, 10
Arthritides, inflammatory, dyslipidemias associated with, 164
Arthritis, 176
Aspergillus, 237
Atherogenesis
triglyceride levels and, 53
very low density lipoprotein and, 53
Atheroma, 3
defined, 5
pathogenesis of, 5
progression of, 10–11

Atherosclerosis
cholesterol levels and, 32, 33
cholesterol lowering clinical trials, 17–27
current concepts of, 5–6
complicated lesions, 9–11
fatty streaks, 9
fibrous plaques, 6, 9
pathogenesis, 6–9
early works on, 4–5
dietary studies, 14–17
glucocorticoids and, 168
intervention studies on, 14
lipid hypothesis and, 3–4
population studies on, 11–14
premature, 48
progression of, 106, 227
risk factors for, 59–60, *See also* Risk factors
secondary prevention of, 172
spontaneous and accelerated, comparisons of, 11, 11t
symptoms and signs of, 172
Atherosclerotic lesions
growth of, 23
morphologic stages of, 5–6
Atromid-S®, *See* Clofibrate
Attitudes about high cholesterol levels, 43
Attributable risk, 37, 38
information provided by, 38
Attributable risk percent (ARP), 38
Autoimmune mechanisms, dyslipidemia caused by, 154, 156
Automated procedures, for cholesterol measurement, 128
Avocado, 210
Azathioprine, 168

Basal energy requirements, 212
Beckman Astra 4 & 8™, 126
Bed rested patient, exercise for, 219
Behavior pattern, Type A, coronary heart disease risk and, 62
Behavioral strategies, for diet changes, 215–217
Beta blockers, 58, 148, *See also* specific agent
dyslipidemia associated with, 165
Beta-quantification, 130
Beta-VLDL receptor, 7
Bezafibrate, 240
Bias in cholesterol testing, 126
Bile, 203
Bile acid, 147, 197
cholesterol conversion to, 99
Bile acid resins (sequestrants), 26, 239, *See also* Cholestyramine, Colestipol
clinical uses of, 227, 228t, 228–229, 231–233, 252
for combined dyslipidemia, 226
in diabetes mellitus, 257
for hypercholesterolemia, 224, 225, 254
combination therapy, 237, 243, 244
cost of, 223
efficacy of, 224, 224t
mechanism of action, 229–230
patient instructions for, 283
Bilirubin, 128
Binding protein, apoproteins as, 90
lipoprotein level variations, 123–125
Birth control pills, *See* Oral contraceptives
Black population, *See* Race factors